SPREAD SPECTRUM SYSTEMS

SPREAD SPECTRUM SYSTEMS

Second Edition

Robert C. Dixon

A Wiley-Interscience Publication

JOHN WILEY & SONS

New York · Chichester · Brisbane · Toronto · Singapore

Library of Congress Cataloging in Publication Data:

Dixon, Robert C. (Robert Clyde), 1932–
Spread spectrum systems.

"A Wiley-Interscience publication."
Bibliography: p.
Includes index.

1. Spread spectrum communications. I. Title.
TK5102.5.D55 1984 621.38′043 83-26080
ISBN 0-471-88309-3

Printed in the United States of America

10 9 8 7 6 5 4

PREFACE TO THE FIRST EDITION

Spread spectrum systems encompass communications, data transmission, message privacy, signal hiding, and position location within their repertoire. These systems are a unique blend of analog (usually RF) and digital disciplines. The coding methods used are the key to providing a spread spectrum system's capabilities.

This book represents an effort to unify at least the terminology used by those working in the spread spectrum field. It is intended to introduce working engineers and students to the philosophy and some of the details of spread spectrum technology. As far as is practical, this book is self-contained. Those who wish to dig deeper, pursue historical development, or investigate rigorous proofs will find in Appendix 3 a comprehensive listing of references in spread spectrum and related areas.

Most of the concepts of spread spectrum systems have been understood for many years, but the components and techniques for building systems capable of reliable operation have been available for a much shorter time. J. P. Costas concluded in 1959 that "for congested-band operation, broadband systems appear to offer a more orderly approach to the problem and a potentially higher average traffic volume than narrowband systems." At that time, however, transistors and other components were not available to build a reliable, reasonably sized spread spectrum system.

Today components have advanced to the point at which large parts of a spread spectrum system can be contained in a single integrated circuit. A code generator, for instance, which even in 1967 would have required at least 100 discrete transistors, can easily be incorporated in a single small package only slightly larger than one of the transistors. In the future a complete subsystem may well be reduced to one similarly small package. The point is that the use of spread spectrum techniques is no longer constrained by *constituent electronic* components, within limits.

Spread spectrum applications started with the first communicator who set up a scheduled time to send and receive messages. This scheduling may have come about through a desire to avoid heavy traffic (consider, for instance, 10 Indian smoke signalers talking at once) or a desire to avoid interception by surprising the would-be interceptor. The same technique of timing was adapted by radio operators, but they added a new dimension—frequency. The radio operator not only could schedule his transmissions for a time unknown to an interceptor but could transmit at one of many frequencies, which forced the interceptor to "find" his transmission in addition to guessing his schedule. Encoding of messages for error correction and improved time and frequency selection naturally followed.

Modern spread spectrum communications circuits and systems have evolved from just these simple concepts. Both of the most important communications and data systems (frequency hopping and direct sequence) use code-controlled frequency-time keying to avoid interception and minimize interference. These systems have grown in capability and in application since the late 1940s and have been applied today in many areas. Though almost all of the applications to date have been military, commercial equipment is available, and the major test equipment manufacturers have introduced instruments specific to the spread spectrum area.

The material in this book has been gathered from many sources—technical articles, internal memos, military contract reports, private conversations, and my personal experience. References are listed whenever they were available.

The first chapter is an introduction to spread spectrum systems and the reasons for their being. Chapter 2 describes the different types of systems in some detail. Subsystems are described in Chapters 3 through 8, with the emphasis placed on describing advantages and disadvantages of various implementations. Also, successful designs typical of real systems are included. Application of spread spectrum techniques is the subject of Chapter 9. I hope that some ideas for new applications will be generated by this chapter.

The gestation period for this book has spanned a number of years. The idea for it came about in 1963. Since then many people have contributed to its growth. A list of those who have influenced it would fill another book. Edward Guyer of the Northrop Corporation, P. M. Hooten and C. F. White of the Naval Research Laboratories, R. D. Matson and H. J. Schmidt of the Air Force Avionics Laboratories, D. R. Bitzer of D. O. D., and Dr. I. J. Gabelman of Rome Air Development Center all are due special thanks for their help in bringing the book to reality.

To those who insist on strict rigor and abundance of mathematical precedence no apology is given. There is enough of this material already existing to define completely all that is said here many times over. That is precisely the problem I have tried to overcome—hoping to separate enough of the trees from the forest that the usefulness and practical aspects of spread spectrum systems can be seen—simply and in their own context—without the myriad qualifying statements and assumptions that would otherwise be required. There are, listed within the bibliography of this book, enough of the works of those whose world is rigor and precedence to satisfy the needs of their fellows.

Those who, like me, are afflicted with the need for practical applications to secure their understanding have, I hope, found a book useful to them.

ROBERT C. DIXON

Cypress, California
June 1975

PREFACE

Since the first edition of this book in 1976, spread spectrum communication techniques have been recognized as a viable method to gain an advantage in interference environments. Many new military-oriented systems have been initiated and some civil systems have been attempted, but few are in regular operation. Some of the new systems are mentioned in Chapter 9.

Some of the recent technical literature regarding spread spectrum systems and techniques are listed in Appendix 2. There has been so much interesting literature generated since 1976 that it is difficult to provide a comprehensive listing. However, it is hoped that enough of the titles in the areas of interest have been included to help the reader narrow the field of search in particular subjects.

I am indebted to many people for their kindness and encouragement, and hope that this somewhat expanded version of *Spread Spectrum Systems* is not a disappointment to them. I am also indebted to my daughter, Theresa Vanderpool, to Linda Nellany, and to Carol Minikus for their diligent and painstaking typing. Any mistakes, misstatements, and misunderstandings included here are strictly my own.

The organization of this edition is the same as the earlier book, with expansion for further interpretation and inclusion of updated material, and a few deletions in the case of superseded information.

The purpose here, as always, is to aid system designers and others to avoid the mistakes I have made in spread spectrum system design and development; and to speed their efforts. If this much is accomplished, then this book can be counted as a success.

ROBERT C. DIXON

Cypress, California
March 1984

To

Nancy
David and Suzi
Theresa and Jeff
Robbie
Matthew and Sean

CONTENTS

SPREAD SPECTRUM SYSTEMS

1

THE WHAT AND WHYS OF SPREAD SPECTRUM SYSTEMS

Spread spectrum techniques, applied in recent years, have produced results in communications, navigation, and test systems that are not possible with standard signal formats. In many applications the advent of high-speed transistors and/or integrated circuits was the key to practical-sized-and-powered equipment based on spread spectrum modulation. But what is a spread spectrum system? What are the advantages of spread spectrum modulation? To be sure, there are also disadvantages, but what are they?

These questions and others are posed in this chapter and in those that follow. It is hoped that the reader will find the answers he or she needs in a useful form.

1.1 WHAT IS A SPREAD SPECTRUM SYSTEM?

Before we attempt to define a spread spectrum system, let us be sure that we understand what is meant by a spectrum. Every transmitting or modulating system has a characteristic signature that includes not only the frequency at which the signal is centered, but also the bandwidth of the signal when modulated by the intended signaling waveform. A spectrum, as we speak of it here, is the frequency-domain[1703] representation of the signal and, for our purposes, especially the modulated signal. We most often see signals presented in the time domain (that is, as functions of time). Any signal, however, can also be presented in the frequency domain, and transforms (mathematical operators) are available for converting frequency- or time-domain functions from one domain to the other and back again. The most basic of these operators is the Fourier transform, for which the relationship between the time and frequency domains is defined by the pair of integrals

$$F(f) = \int_{-\infty}^{\infty} f(t) \exp - j\omega t \, dt$$

which transforms a known function of time to a function of frequency, and

$$f(t) = \int_{-\infty}^{\infty} F(f) \exp j\omega t \, df$$

which performs the inverse operation.

Fourier transforms do not exist for some functions because they require the existence of the integral

$$\int_{-\infty}^{\infty} f(t) \, dt$$

Therefore discontinuous signals must often be transformed by use of the Laplace integral

$$L(s) = \int_{0}^{\infty} f(t) \exp - st \, dt$$

Tables of Fourier and Laplace transform pairs for many functions have been generated and well documented (e.g., see 1714 and 1716).

Figure 1-1 illustrates some of the Fourier transforms that are most important in our considerations of spread spectrum systems. For example, we will be interested in the spectra of carriers modulated by pseudorandom binary data streams. Also of some interest will be the spectra of frequency hopped carriers—especially where those carriers are to be used in multiple-access applications, and it is necessary to restrict any interference between mutual users of the same band of frequencies. Note that the frequency spectrum produced by modulation with a square pulse is a $(\sin x)/x$ function, while modulation having a $(\sin x)/x$ envelope produces a square spectrum. Other spectra that will be of special interest to us are those that are produced by triangular or trapezoidal envelopes and the Gaussian-shaped pulse. Still others will be introduced in the following chapters.

Even as an oscilloscope is a window in the time domain for observing signal waveforms, so is a spectrum analyzer a window in the frequency domain. The many spectrum presentations in this book are almost all spectrum analyzer representations, generated by sweeping a filter across

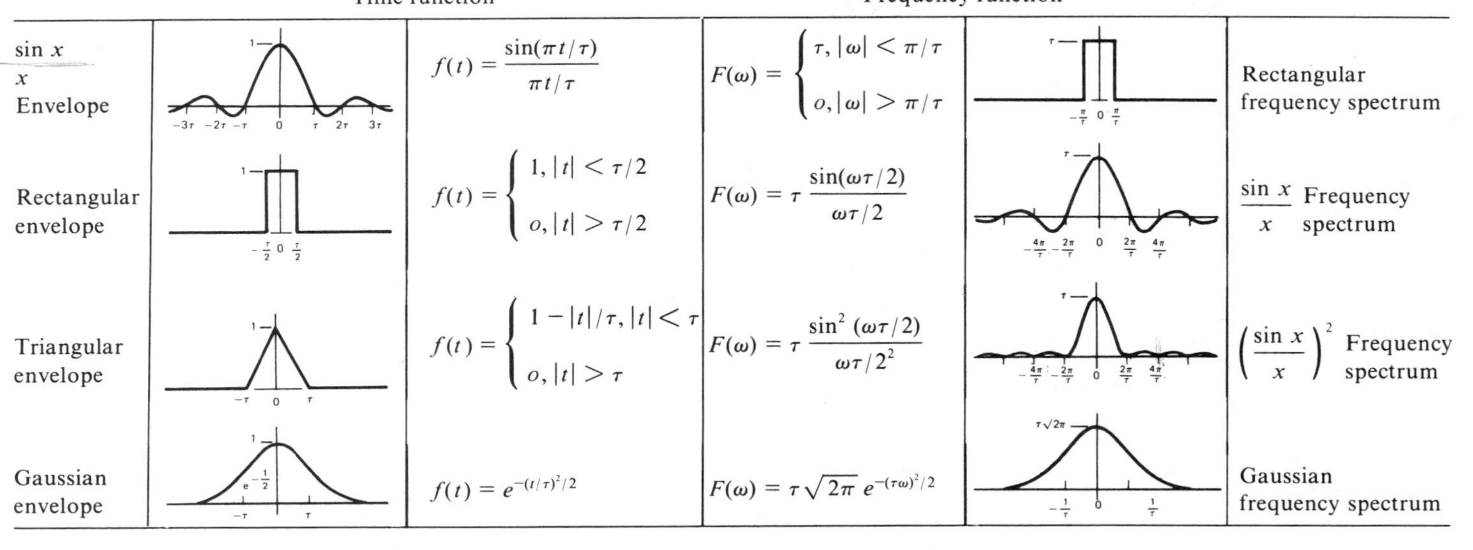

	Time function	Frequency function	
$\frac{\sin x}{x}$ Envelope	$f(t) = \frac{\sin(\pi t/\tau)}{\pi t/\tau}$	$F(\omega) = \begin{cases} \tau, \lvert\omega\rvert < \pi/\tau \\ o, \lvert\omega\rvert > \pi/\tau \end{cases}$	Rectangular frequency spectrum
Rectangular envelope	$f(t) = \begin{cases} 1, \lvert t\rvert < \tau/2 \\ o, \lvert t\rvert > \tau/2 \end{cases}$	$F(\omega) = \tau \frac{\sin(\omega\tau/2)}{\omega\tau/2}$	$\frac{\sin x}{x}$ Frequency spectrum
Triangular envelope	$f(t) = \begin{cases} 1 - \lvert t\rvert/\tau, \lvert t\rvert < \tau \\ o, \lvert t\rvert > \tau \end{cases}$	$F(\omega) = \tau \frac{\sin^2(\omega\tau/2)}{\omega\tau/2^2}$	$\left(\frac{\sin x}{x}\right)^2$ Frequency spectrum
Gaussian envelope	$f(t) = e^{-(t/\tau)^2/2}$	$F(\omega) = \tau\sqrt{2\pi}\, e^{-(\tau\omega)^2/2}$	Gaussian frequency spectrum

Figure 1.1 Four important Fourier transforms and their corresponding time and frequency functions.

the band of interest and detecting the power falling within the filter as it is swept. This power level is then plotted on an oscilloscope. All spectra referred to are power spectra.

Literally, a spread spectrum system is one in which the transmitted signal is spread over a wide frequency band, much wider, in fact, than the minimum bandwidth required to transmit the information being sent. A voice signal, for example, can be sent with amplitude modulation in a bandwidth only twice that of the information itself. Other forms of modulation, such as low deviation FM or single sideband AM, also permit information to be transmitted in a bandwidth comparable to the bandwidth of the information itself. A spread spectrum system, on the other hand, often takes a baseband signal (say a voice channel) with a bandwidth of only a few kilohertz, and distributes it over a band that may be many megahertz wide. This is accomplished by modulating with the information to be sent and with a wideband encoding signal.

The most familiar example of spectrum spreading is seen in conventional frequency modulation in which deviation ratios greater than one are used. Bandwidth required by an FM signal is a function not only of the information bandwidth but of the amount of modulation. As in all other spectrum spreading systems, a signal-to-noise advantage is gained by the modulation and demodulation process. For FM signals this gain advantage (referred to as "process gain") is

$$3\beta^2\left(\frac{S}{N}\right)_{\text{info}}$$

where β = deviation ratio, or $\Delta f_{\text{carrier}}/f_{\text{modulation}}$,
$(S/N)_{\text{info}}$ = signal-to-noise ratio in the baseband or information bandwidth.

Wideband FM could be classified as a spread spectrum technique from the standpoint that the RF spectrum produced is much wider than the transmitted information. In the context of this book, however, only those techniques are of interest in which some signal or operation other than the information being sent is used for broadbanding (or spreading) the transmitted signal.

Three general types of techniques will be accepted here as examples of spread spectrum signaling methods:

1. Modulation of a carrier by a digital code sequence whose bit rate is much higher than the information signal bandwidth. Such systems are called "direct sequence" modulated systems.
2. Carrier frequency shifting in discrete increments in a pattern dictated by a code sequence. These are called "frequency hoppers." The

transmitter jumps from frequency to frequency within some predetermined set; the order of frequency usage is determined by a code sequence.

3. Pulsed-FM or "chirp" modulation in which a carrier is swept over a wide band during a given pulse interval.

Closely akin to the frequency hoppers are "time hopping" and "time-frequency hopping" systems whose chief distinguishing feature is that their time of transmission (usually of low duty cycle and short duration) is governed by a code sequence. In time-frequency hoppers it follows that the code sequence determines both the transmitted frequency and the time of transmission.

Figure 1.2 shows frequency spectra typical of direct sequence and frequency-hopping systems. These spectra are often tens to hundreds of megahertz wide, where conventional signals are usually confined to bandwidths in the range of 10–100 kiloHertz.

The basis of spread spectrum technology is expressed by C. E. Shannon in the form of channel capacity:

$$C = W \log_2 \left(1 + \frac{S}{N}\right)$$

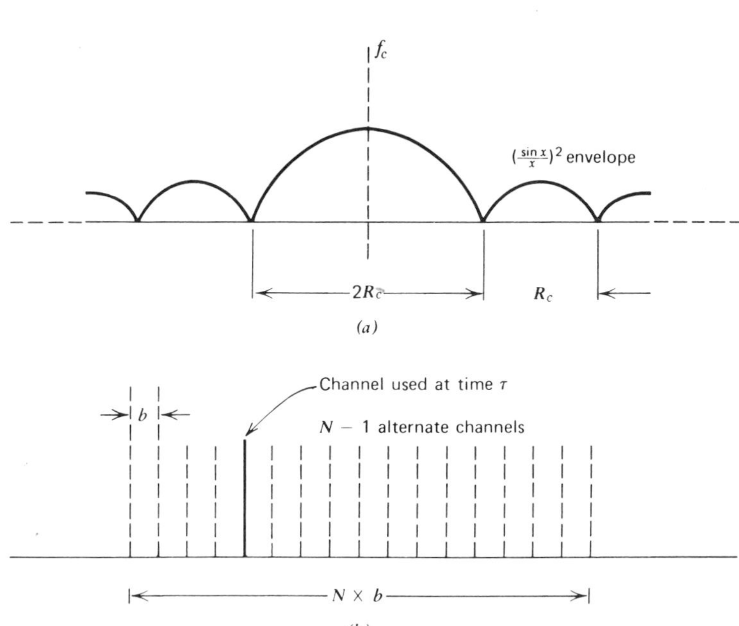

Figure 1.2 Ideal spectra of typical spread spectrum signals: (*a*) direct sequence signal spectrum; (*b*) frequency-hopping signal spectrum.

where C = capacity in bits per second,
W = bandwidth in hertz,
N = noise power,
S = signal power.

This equation shows the relationship between the ability of a channel to transfer error-free information, compared with the signal-to-noise ratio existing in the channel, and the bandwidth used to transmit the information.

Letting C be the desired system information rate and changing bases, we find

$$\frac{C}{W} = 1.44 \log_e \left(1 + \frac{S}{N}\right)$$

and for S/N small, say $\leqslant 0.1$ (as one would wish it to be in an antijam system),

$$\frac{C}{W} = 1.44 \frac{S}{N}$$

since

$$\log_e \left(1 + \frac{S}{N}\right) = \frac{S}{N} - \frac{1}{2}\left(\frac{S}{N}\right)^2 + \frac{1}{3}\left(\frac{S}{N}\right)^3 - \frac{1}{4}\left(\frac{S}{N}\right)^4 \cdots \left(-1 < \frac{S}{N} < 1\right)$$

by the logarithmic expansion.

From this equation we find

$$\frac{N}{S} = \frac{1.44W}{C} \approx \frac{W}{C}$$

and

$$W = \frac{NC}{S}$$

thus we see that for any given noise-to-signal ratio we can have a low information-error rate by increasing the bandwidth used to transfer the information; for example, if we want a system to operate in a link in which the interfering noise is 100 times greater than the signal and our information rate 3 kilobits per second (kbps) then our 3-kbps information must be transmitted with a bandwidth of

$$W = \frac{1 \times 3 \times 10^5}{1.44} = 2.08 \times 10^5 \text{ Hz}$$

Incidentally, the information itself may be embedded in the spread spectrum signal by several methods. The most common is that of adding the information to the spectrum-spreading code before its use for spreading modulation. This technique is applicable to any spread spectrum system that uses a code sequence to determine its RF bandwidth (either direct sequence or frequency hopping systems are good candidates). Of course, the information to be sent must be in some digital form because addition to a code sequence involves modulo-2 addition to a binary code. Alternately, information may be used to modulate a carrier before spreading it. This is usually done by some form of angle modulation, for the need in spread spectrum systems is often to output a constant-power RF envelope.

A spread spectrum system, then, must meet two criteria: (1) the transmitted bandwidth is much greater than the bandwidth or rate of the information being sent and, (2) some function other than the information being sent is employed to determine the resulting modulated RF bandwidth.

This is the essence of spread spectrum communications—the art of expanding the bandwidth of a signal, transmitting that expanded signal, and recovering the desired signal by remapping the received spread spectrum into the original information bandwidth. Furthermore, in the process of carrying out this series of bandwidth trades the purpose is to allow the system to deliver error-free information in a noisy signal environment.

1.2 WHY BOTHER?

The first reaction common among those encountering spread spectrum techniques is "why bother?" The answers are varied and seldom the same. In a world beset by too little RF spectrum to satisfy the ever-growing demands of military, commercial, and private users the question "why spread spectrum" must certainly be considered valid, for spread spectrum systems have almost as many reasons for being as they have users.

Some of the properties that may be cited are the following:

1. Selective addressing capability.
2. Code division multiplexing is possible for multiple access.
3. Low-density power spectra for signal hiding.
4. Message screening from eavesdroppers.

5. High-resolution ranging.
6. Interference rejection.

These properties come about as a result of the coded signal format and the wide signal bandwidth that results. A single receiver or group of receivers may be addressed by assigning a given reference code to them, whereas others are given a different code. Selective addressing can then be as simple as transmitting the proper code sequence as modulation.

Not all of these characteristics, however, are necessarily available from the same system at the same time. It is somewhat anomalous, for instance,

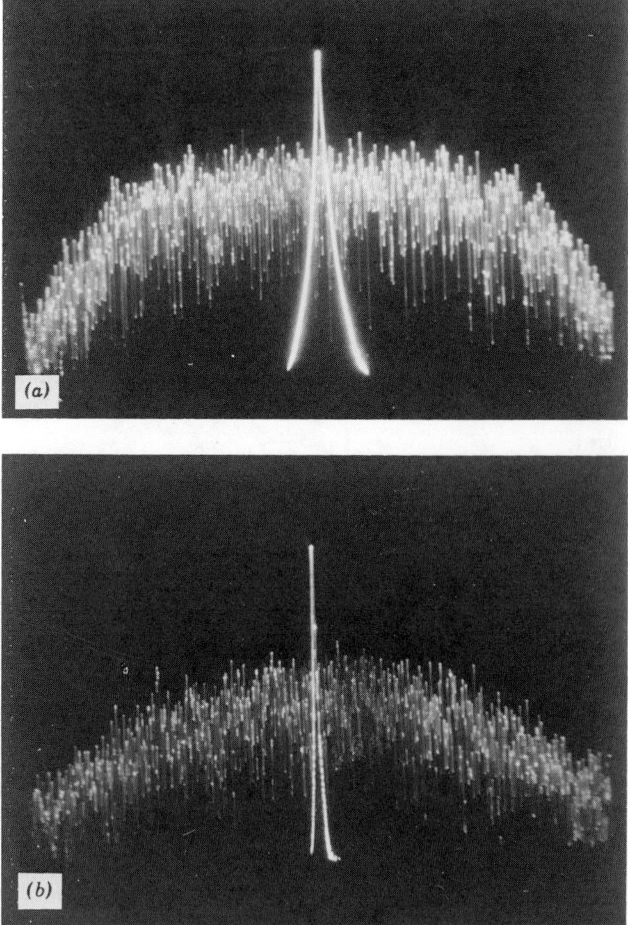

Figure 1.3 Illustration of power density showing equal power spread spectrum and CW signals, for spread spectrum bandwidth of (*a*) 2 MHz (0.88 MHz, 3 dB) and (*b*) 10 MHz (4.4 MHz, 3 dB).

to expect, at the same time, a signal that is easily hidden but can also be received in the face of a large amount of interference. Signal-hiding requirements and inferference rejection are often at odds, but the same system might be used for both by using low-power transmissions when low detectability is desired and high-power transmissions when maximum interference rejection is needed.

When codes are properly chosen for low cross correlation, minimum interference occurs between users, and receivers set to use different codes are reached only by transmitters sending that code. Thus more than one signal can be unambiguously transmitted at the same frequency and at the same time; selective addressing and code-division multiplexing are implemented by the coded modulation format.

Because of the wideband signal spectra generated by code modulation, the power transmitted is low in any narrow region. At any rate, the density of a spread spectrum signal is far less than that of more conventional signals in which all the transmitted power is sent in a band of frequencies commensurate with the baseband information bandwidth. Again, because of the coded signals employed, an eavesdropper cannot casually listen to messages being sent. Though the systems may not be "secure," some conscious effort must be made to decode the messages.

Figure 1.3 illustrates the difference in the power density of continuous-wave and typical spread spectrum signals.

Resolution in ranging is afforded in accordance with the code rates used, and the sequence length determines maximum unambiguous range. Ranging has been the most prominent and certainly the best known use of spread spectrum systems.

1.3 PROCESS GAIN AND JAMMING MARGIN

The most commonly used quantity in spread spectrum systems is that of "process gain," although it must be pointed out that what is usually intended is not process gain but "jamming margin." Process gain is a readily available quantity, if the bandwidth employed in a system is known and the information rate is available. In Section 1.1, we considered Shannon's information-rate theorem, which shows that one can send information without error, if some method can be devised that employs a wide enough bandwidth to transmit the information. The process gain that we consider here is just an embodiment of that theorem, in which the signal is spread, and the process gain produced by the spreading and despreading process is equal to the bandwidth ratio between the information and the RF bandwidth used to send it.

A spread spectrum system develops its process gain in a sequential signal bandwidth spreading and despreading operation. The *transmit* part of the process may be accomplished with any one of the bandspreading modulation methods. Despreading is accomplished by correlating the received spread spectrum signal with a similar local reference signal. When the two signals are matched, the desired signal collapses to its original bandwidth before spreading, whereas any unmatched input is spread by the local reference to its bandwidth or more. A filter then rejects all but the desired narrowband signal; that is, given a desired signal and its interference (atmospheric noise, receiver noise, or jamming), a spread spectrum receiver enhances the signal while suppressing the effects of all other inputs. This process is described in detail in Chapter 2.

The difference in output and input signal-to-noise ratios in any processor is its "process gain." A given processor, for instance, with an input signal-to-noise ratio of 10 dB and an output signal-to-noise ratio of 16 dB would have a process gain of 6 dB.

In spread spectrum processors the process gain available may be estimated by the rule of thumb:

$$\text{process gain} = G_p = \frac{\mathrm{BW}_{\mathrm{RF}}}{R_{\mathrm{info}}}$$

where the RF bandwidth ($\mathrm{BW}_{\mathrm{RF}}$) is the bandwidth of the transmitted spread spectrum signal and the information rate (R_{info}) is the data rate in the information baseband channel. This does not mean, however, that a processor can perform when faced with an interfering signal having a power level larger than the desired signal by the amount of the available process gain. Another term, "jamming margin," which expresses the capability of a system to perform in such hostile environments, must be introduced.

Jamming margin is that quantity that is usually intended in the specification of spread spectrum systems, but it is less readily predicted from bandwidth and information-rate information. One can be sure, however, that jamming margin in any given system is always less than the process gain available from that system.

Jamming margin takes into account the requirement for a useful system output signal-to-noise ratio and allows for internal losses; that is,

$$\text{jamming margin} = G_p - \left[L_{\mathrm{sys}} + \left(\frac{S}{N}\right)_{\mathrm{out}}\right] = M_j$$

where L_{sys} = system implementation losses,
$(S/N)_{\mathrm{out}}$ = signal-to-noise ratio at the information output;

for example, a system with 30-dB process gain, minimum $(S/N)_{\mathrm{out}}$ of 10 dB, and L_{sys} of 2 dB would have an 18-dB jamming margin (M_j). It could not be expected to operate with interference more than 18 dB above the desired signal.

The "jamming threshold" of a particular system is of interest in determining how well that system will operate in the presence of interference. Consider the following processor model with signal and noise inputs (the noise input includes interference):

$$\left.\begin{matrix}\text{signal in}\\ \text{noise in}\end{matrix}\right\} \left(\frac{S}{N}\right)_{\text{in}} \bullet\!\!-\!\!\boxed{G_p}\!\!\rightarrow \left(\frac{S}{N}\right)_{\text{out}} = \left(\frac{S}{N}\right)_{\text{in}} \times G_p$$

For the region of interest $S \ll N$. Converting to decibels, we have

$$\left(\frac{S}{N}\right)_{\text{out}} (\text{dB}) = \left(\frac{S}{N}\right)_{\text{in}} (\text{dB}) + G_p\,(\text{dB})$$

but

$$\left(\frac{S}{N}\right)_{\text{in}} (\text{dB}) = -\frac{J}{S}\,(\text{dB})$$

Therefore

$$\left(\frac{S}{N}\right)_{\text{out}} (\text{dB}) = G_p(\text{dB}) - \frac{J}{S}\,(\text{dB})$$

for the region above the jamming threshold, where the threshold for a particular system is that level for which (see Figure 1.4)

$$\left[G_p(\text{dB}) - \frac{J}{S}\,(\text{dB})\right]_{\text{Ideal}} - \left[G_p(\text{dB}) - \frac{J}{S}\,(\text{dB})\right]_{\text{meas'd}} = 1 \text{ dB}$$

The cause of thresholding lies in such things as tracking loss, nonlinearities, and thresholding of the postcorrelation detector. No system can be designed without a jamming threshold, but with care the threshold point can be placed beyond the normal operating region. All possible thresholding sources would tend to degrade at the same point, for when one part of the system operates at interference levels beyond the others the overall system performance does not improve.

Table 1.1 compares the process gain from spread spectrum processing techniques with that provided by other methods, such as antenna pointing or electronic interference cancellation.

Antenna-cancellation techniques employ the directivity of an antenna to reject a would-be jammer. Such techniques have been used but are not always practical because of positioning restrictions. When geometry is fixed, antenna rejection of interference is limited only by the particular antenna's directivity.

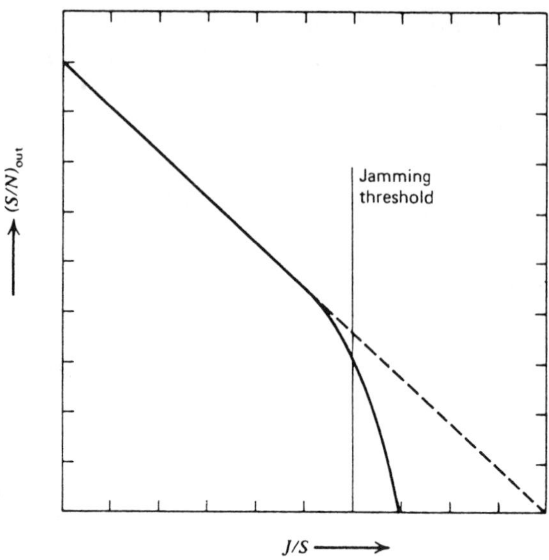

Figure 1.4 Typical process gain curve.

For effective electronic interference cancellation a duplicate of the signal to be rejected must be generated. This replica is then subtracted from the combined interfering and desired signals. The result is a residue consisting of the desired signal only. This technique has proved to be useful but is limited to those situations in which few interfering signals occur (a separate signal simulator and subtractor is required for each signal) and in which relatively noncomplex interfering signals are encountered.

Table 1.1 Comparison of Process Gains Available for Various Techniques, Including Signal Rejection and Cancellation

System	Process Gain
Direct sequence	$\frac{\text{BW}_{\text{RF}}}{R_{\text{info}}} = TW$
Frequency hopping	$\frac{\text{BW}_{\text{RF}}}{R_{\text{info}}} = TW =$ number of frequency choices
Time hopping	$\frac{1}{\text{transmit duty cycle}}$
Chirp	Compression ratio $= \tau\, dF = TW$
Antenna rejection	Antenna gain
Electronic cancellation	Depends on accuracy of replica; sometimes 40 dB
Selective rejection	High but useful only against narrowband interference.

Selective rejectors such as notch filters are useful for removing interference signals but are limited to narrowband signals such as CW. Rejection of a wideband signal with a filter could also reject much of the desired signal, which would degrade system operation and net no gain in performance, even though the interference were rejected.

In general, spread spectrum processing offers the most flexible means of providing unwanted signal rejection because it is not necessary to design for rejection of any particular kind of interference and geometric considerations are not normally important. Spread spectrum processing does not, however, offer the highest process gain (or undesired signal rejection) for every situation. The alternatives mentioned here may be combined with spread spectrum techniques to produce a compatible system with the advantages of spread spectrum and other signal-to-noise improving methods.

Conventions in This Book

The following conventions will be employed:

Direct Sequence Bandwidth. All direct sequence signal bandwidths are assumed to be equal to the 3-dB bandwidth of the spectrum. Thus, the bandwidth will vary, depending on the particular type of carrier modulation employed such as biphase, quadriphase, or minimum shift key.

Frequency Hopper Bandwidth. Frequency hoppers are assumed to have BW_{RF} equal to m times the channel bandwidth, where m is the number of frequency channels available.

Information Rate. Information rate is a function of the type of information transferred over the links discussed and is the minimum bit rate necessary for satisfactory reproduction of the information at the receiver. For voice transmission 3 kHz is the analog bandwidth. Data are consistent with the minimum rate sufficient to convey analog signals,* or digital data.

Chip Rate. Chip rate is equivalent to the code generator clock rate, or in frequency-hopping systems, the hop rate.

Postcorrelation Bandwidth. The signal bandwidth following the receiver correlation process is the two-sided information bandwidth produced by the baseband modulating signal or by a frequency hopped carrier.

All other bandwidths are 3-dB bandwidths.

*Data rate representing an analog signal is at least the Nyquist rate, or 2 bits per Hertz. The actual rate depends on the A/D technique used.

PROBLEMS

1. Given a requirement for operating in an environment in which an interfering signal is 250 times the desired, how much jamming margin is required?
2. If in the system described above a 10-dB output signal-to-noise ratio is required, what is the minimum process gain allowable?
3. Information rate is 6 kbps. What is the minimum RF bandwidth needed?
4. Assume that a direct sequence approach has been employed. What code rate should be used?
5. How many channels would be required by a frequency hopper for a similar performance?
6. Given that chirp modulation is required and the information rate is 10 kbps, what RF sweep bandwidth is the minimum that would produce a 20-dB process gain?
7. A direct sequence system has a code rate of 1 Mcps (cps = chips per second) and an information rate of 100 kbps. Is it worth it?
8. A frequency hopper has a 5-MHz RF bandwidth available and an information rate of 100 kbps. Comment.
9. How does a chirp system meet the criterion that the RF transmission bandwidth employed is not a function of the information sent?
10. How does a direct sequence system meet the same criterion?
11. Same question for frequency hopping.
12. What is the prime limiting factor in the processing gain of a time-hopping system?

2

SPREAD SPECTRUM TECHNIQUES

Several basic spread spectrum techniques available to the communications system designer, are listed in Chapter 1 and described in a general way. This chapter gives detailed descriptions of the various techniques and the signals generated. In addition to the two most important (or at least most prevalent) forms of spread spectrum modulation, other useful techniques such as chirping, time hopping, and various hybrid combinations of modulation forms are described. Each is important in that it has useful application. Curiously enough, there has been little competition between the different techniques. The historical tendency has been to confine each form to a particular field. Chirp modulation, for instance, has been used almost exclusively in radar. Here all are presented together to give the reader a reasonably complete picture of contemporary spread spectrum technology.

2.1 DIRECT SEQUENCE SYSTEMS

Direct sequence (or, to be more exact, directly carrier-modulated, code sequence modulation) systems are the best known and most widely used spread spectrum systems. This is because of their relative simplicity from the standpoint that they do not require a high-speed, fast-settling frequency synthesizer. The JPL ranging technique described in Golomb et al. *Digital Communications with Space Applications*[219] and used in various space applications is a direct sequence approach. Today, direct sequence modulation is being used for communication systems and test systems, and even laboratory test equipment capable of producing a choice of a number of code sequences or operating modes is available. It is reasonable to expect that direct sequence modulation will become a familiar form of modulation in many areas in the foreseeable future. Even now, commercial applications of such systems are being contemplated.

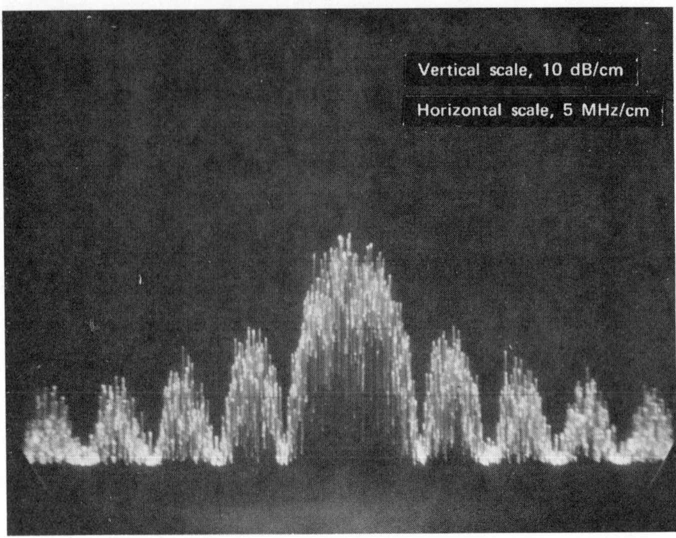

Figure 2.1 Direct sequence spread spectrum signal 180° biphase-modulated by a 5 Mcps code.

Characteristics of Direct Sequence Signals

Direct sequence modulation, as mentioned in Chapter 1, is just exactly that—modulation of a carrier by a code sequence. In the general case, the format may be AM (pulse), FM, or any other amplitude- or angle-modulation form. Very common, however, is 180° biphase phase-shift keying, though the reasons for it may not be immediately obvious. The basic form of direct sequence signal is that produced by a simple, biphase-modulated (PSK) carrier. A spectrum typical of this signal format is shown in Figure 2.1. The main lobe bandwidth (null-to-null) of the signal shown is twice the clock rate of the code sequence used as a modulating signal. Each of the sidelobes has a bandwidth from null-to-null that is equal to the clock rate; that is, if the code sequence being used as a modulating waveform has a 5-Mcps (cps = chips per second) operating rate, the main lobe bandwidth will be 10 MHz and each sidelobe will be 5 MHz wide. This is exactly the case in Figure 2.1.

Typically, the direct sequence biphase modulator has the form shown in Figure 2.2. A balanced mixer whose inputs are a code sequence and an RF carrier operates as the biphase modulator. In the time domain the biphase-modulated carrier looks like the signal shown in Figure 2.3. There the carrier is transmitted with one phase when the code sequence is a "one" and a 180° phase shift when the code sequence is a "zero." In the usual biphase-modulated system, which phase is of little consequence. It is worthy of note that, although other modulation forms such as PAM

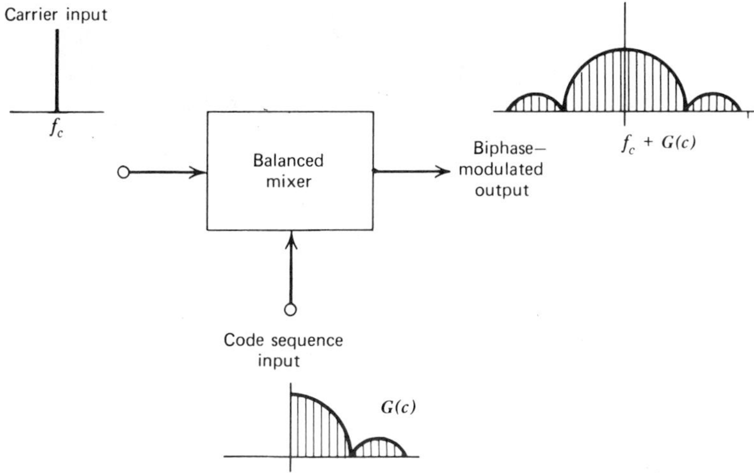

Figure 2.2 Direct sequence modulator (biphase).

(pulse-amplitude modulation) could be used, balanced* modulation is the most common in direct sequence systems. This is true for several reasons:

1. The suppressed carrier produced is difficult to detect without resorting to somewhat sophisticated methods. Figure 2.4 shows a typical direct sequence signal (as does Figure 2.1 in less detail) with carrier suppression. It is obvious that a conventional receiver would not be useful for detecting the carrier here because it is well below the level of the "noise" produced by the code modulation. For comparison, Figure 2.5 shows the same modulator but without carrier balance; therefore, the carrier is no longer suppressed.
2. More power is available for sending useful information because the transmitter power is used to send only the code-produced signal.
3. The signal has a constant envelope level so that transmitted power efficiency is maximized for the bandwidth used. PAM modulation in which a carrier would be pulse modulated by a code as in Figure 2.6 is acceptable; it produces a similar $[(\sin x)/x]^2$ power spectrum but lacks the same effective power at the receiver. Thus for the same range of operation higher peak power would be necessary.

*Biphase-balanced modulation, for example, is well adapted to code modulation in that one carrier phase may be transmitted for a one in the code and the opposite phase for a zero. This produces a signal $A \cos \omega_c t \cos \omega_m t = A \cos (\omega_c \pm \omega_m) t$ after filtering in which the carrier term is absent. Balanced modulation is discussed in detail in Chapter 4.

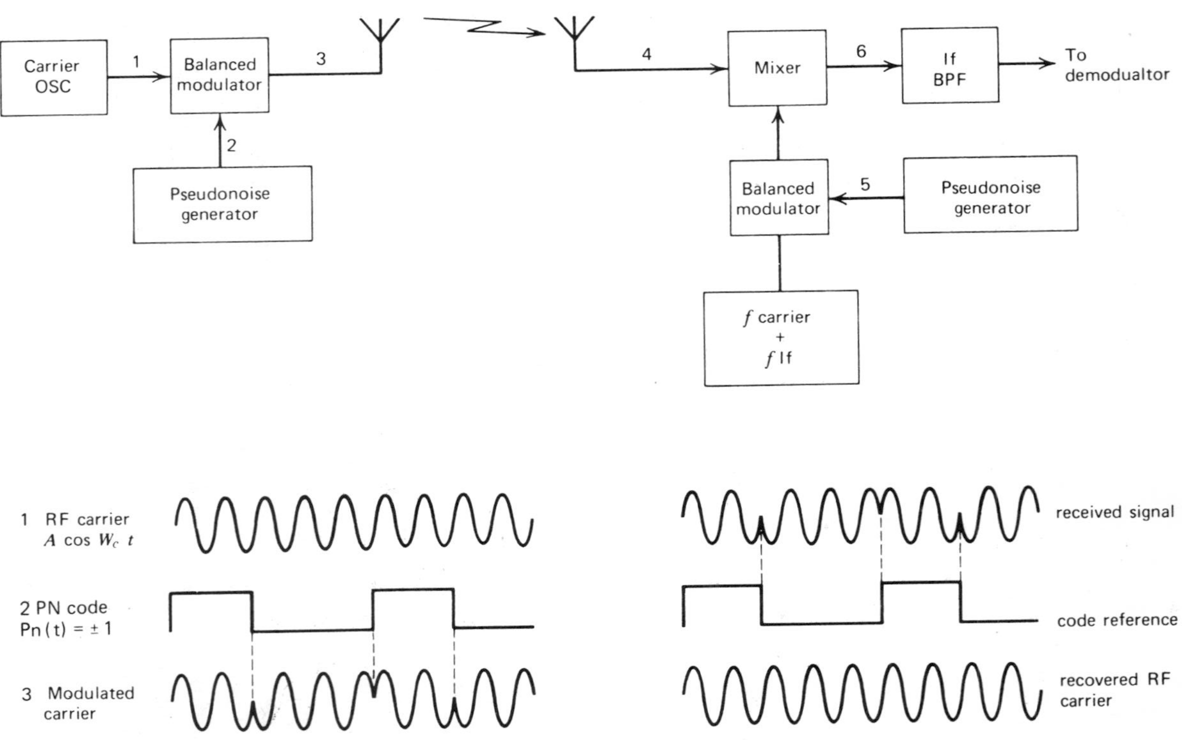

Figure 2.3 Overall direct sequence system showing waveforms.

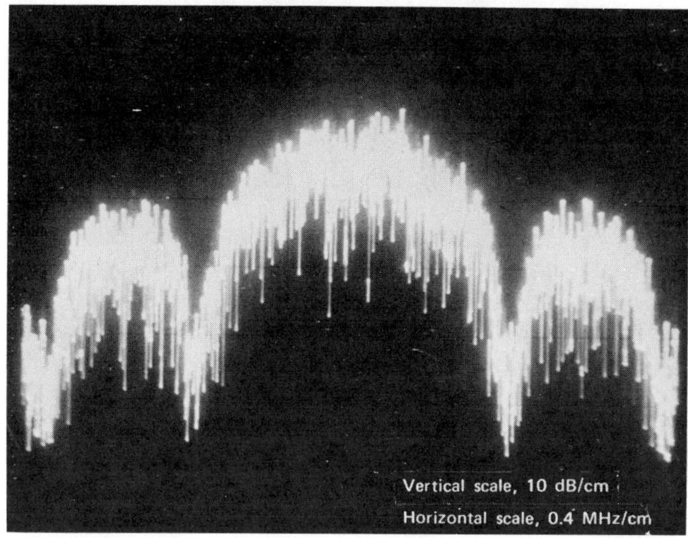

Figure 2.4 Direct sequence, suppressed carrier spectrum-biphase, code-modulated.

The simplified block diagram in Figure 2.3 illustrates a typical direct-sequence communications link. It shows that the direct sequence system is similar to (and, in fact, can be) a conventional AM or FM communications link with code modulation overlaid on the carrier. In actual practice the carrier is not usually modulated by baseband information. The baseband information is digitized and added to the code sequence. For

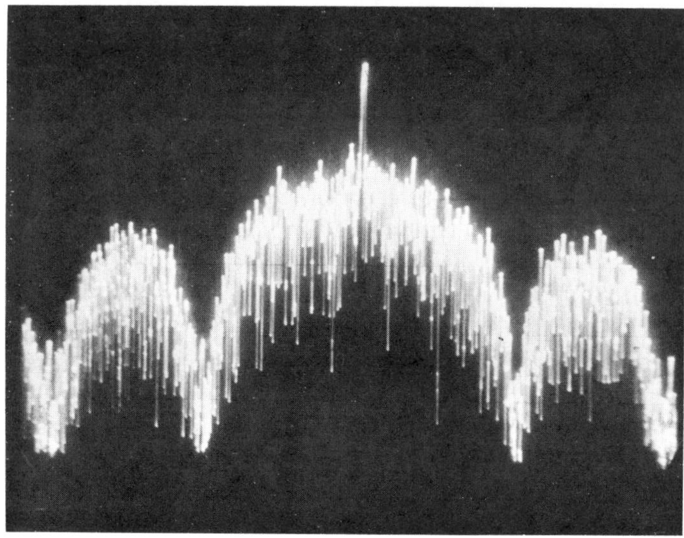

Figure 2.5 Direct sequence unsuppressed carrier, spectrum biphase and code modulated.

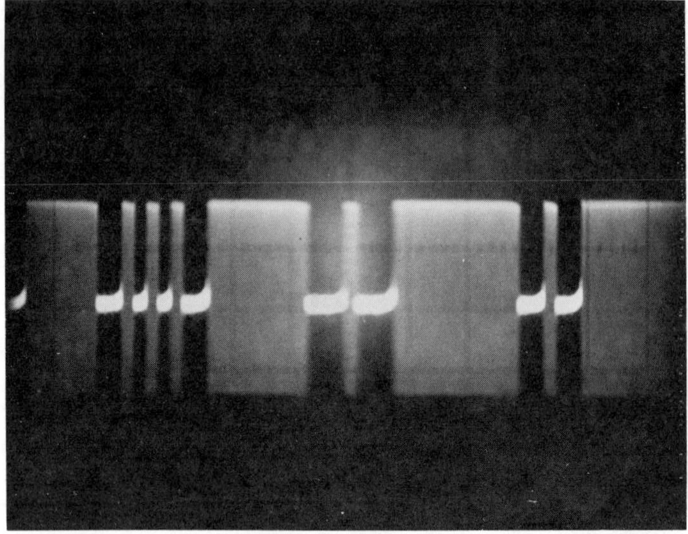

Figure 2.6 Pulse amplitude-modulated carrier, code modulated.

this discussion, however, we will assume that the RF carrier has been modulated before code modulation because this somewhat simplifies discussion of the modulation–demodulation process.

After being amplified, a received signal is multiplied by a reference with the same code and, assuming that the transmitter's code and receiver's code are synchronous, the carrier inversions transmitted are removed and the original carrier restored. This narrowband restored carrier can then flow through a bandpass filter designed to pass only the baseband-modulated carrier.

Undesired signals are also treated in the same process of multiplication by the receiver's reference that maps the received direct sequence signal into the original carrier bandwidth. Any incoming signal not synchronous with the receiver's coded reference (a wideband signal) is spread to a bandwidth equal to its own bandwidth plus the bandwidth of the reference; that is, the receiver's multiplier output signal variance is the covariance of its input signals.

Because an unsynchronized input signal is mapped into a bandwidth at least as wide as the receiver's reference, the bandpass filter can reject almost all the power of an undesired signal. This is the mechanism by which process gain is realized in a direct sequence system; that is, the receiver transforms synchronous input signals from the code-modulated bandwidth to the baseband-modulated bandwidth. At the same time non-synchronous input signals are spread at least over the code-modulated bandwidth.

System information bandwidth sets the postcorrelation filter bandwidth, and this filter in turn determines the amount of power from an unsynchronized signal which reaches the information demodulator. Therefore we see that, as already stated, the multiplication-and-filtering sequence provides the desired signal with an advantage or "process gain."

Radio-Frequency Bandwidth in Direct-Sequence Systems

Radio-frequency (RF) bandwidth in direct sequence systems, as discussed in the preceding sections, directly affects system capabilities. If a 20-MHz bandwidth is available, the process gain possible is limited by that 20 MHz.

Several approaches are useful in choosing the proper bandwidth in a signal-hiding application; the interest is in reducing the power transmitted per Hertz of bandwidth, and wide bandwidths are used. When maximum process gain for interference rejection is desired, bandwidth again should be large. When either frequency allocation or the propagation medium does not permit wide RF bandwidth, some restraint must be applied.

A prime consideration in spread spectrum systems (and especially direct sequence systems) is the bandwidth of the system with respect to the interference furnished to other systems (that may not necessarily be spread spectrum systems) operating in the same or adjacent channels. Referring to Figures 2.4 and 2.5 will show that high sidelobe energy is present in some direct sequence formats, even though the content of the signal is not enhanced by transmitting the sidelobes. In systems such as JTIDS (Joint Tactical Information Distribution System), for example, the band employed is shared with both IFF and TACAN systems. Therefore, a special form of direct-sequence modulation called MSK (minimum shift key) is used.

There are, in fact, a number of waveforms, all closely related, that can be employed in direct sequence systems to control the level of energy contained in the unneeded sidelobes. We shall discuss these in detail in Chapter 4, but for now we simply summarize the characteristics of the available waveforms in Table 2.1. That the simple biphase or quadriphase

Table 2.1 Comparison of Direct Sequence Waveforms

Waveform	Null-to-null Main Lobe BW	3-dB BW	First Sidelobe	Rolloff Rate
BPSK	2 × code clock	0.88 × code clock	−13 dB	6 dB/octave
PAM	2 × code clock	0.88 × code clock	−13 dB	6 dB/octave
QPSK	2 × code clock[a]	0.88 × code clock	−13 dB	6 dB/octave
QQPSK	2 × code clock[a]	0.88 × code clock	−13 dB	6 dB/octave
MSK (classic)	1.5 × code clock	0.66 × code clock	−23 dB	12 dB/octave

[a]Requires two codes at same rate as single BPSK code.

direct sequence signal has a $[(\sin x)/x]^2$ spectrum can be simply shown: Given a rectangular pulse whose duration is T and whose amplitude is A over $T \pm T/2$ and zero elsewhere:

$$C_n = \int_{-T/2}^{T/2} A \exp(-j\omega t)\,dt = \frac{A}{j\omega} \exp(-j\omega t) \int_{-T/2}^{T/2}$$

$$= \frac{A \exp(j\omega\, T/2) - A \exp(-j\omega\, T/2)}{j\omega}$$

$$= \frac{2A}{\omega} \sin \frac{\omega T}{2} = TA \frac{\sin \omega T/2}{\omega T/2}$$

This expression is obviously of the form $(\sin x)/x$: It is the voltage distribution for the signal whose power distribution is then $[(\sin x)/x]^2$. This shows what is readily apparent when using a spectrum analyzer—the spectrum of biphase and quadriphase signals has a $[(\sin x)/x]^2$ distribution. Pulse shaping, as we shall see, can provide significantly improved power distribution, and that is the approach used in the other forms of direct sequence modulation that we will discuss.

In any case, the direct sequence system has a main lobe bandwidth that is a function of the waveshape and the code rate used. For example, in Figure 2.1 the code rate is 5 Mcps, Figure 2.4 is generated with a code rate of 25 Mcps, and the chip rate in Figure 2.7 is a very high 100 Mcps. All of these signals are from biphase modulators, but the spectra are identical to

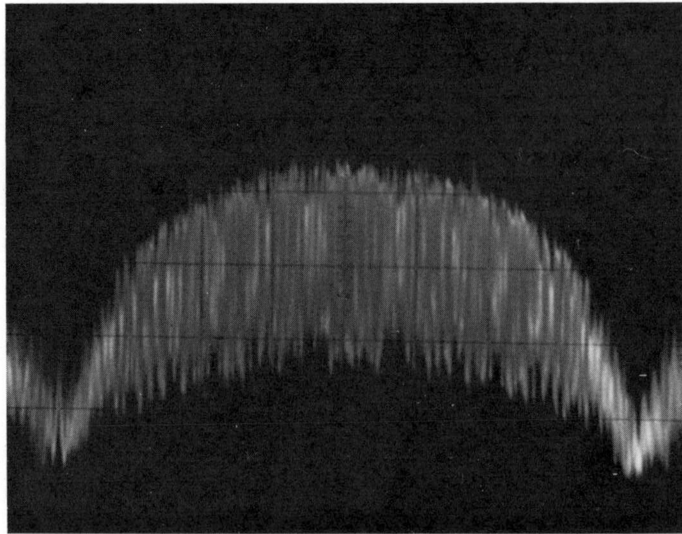

Figure 2.7 High-speed code-modulated direct sequence (BPSK) spectrum.

those that would be produced by the equivalent quadriphase or offset quadriphase modulator. Comparison of these spectra will show that the spectral shape is $[(\sin x)/x]^2$ and the null-to-null bandwidth is $2R_c$.

The signal lost as a result of accepting only the main lobe is small. Only 10% of the power in a BPSK or QPSK signal is contained in the sidelobes (see Figure 2.8). Signal-power loss is not the only effect of bandwidth restriction, however. The sidelobes contain much of the harmonic power of the modulation. Thus restriction to a narrow RF bandwidth is the equivalent of restricting the rise and fall times of the modulating code. Also, the sharply peaked triangular correlation of a coded signal is rounded by a bandwidth restriction. Figures 2.9 and 2.10 show the effect of band restriction on the RF envelope and correlation function of a direct sequence signal.

Quadriphase (QPSK) transmission (discussed in Chapter 4) is one method of restricting RF bandwidth when a given code rate is to be used. QPSK does halve the RF bandwidth required, but processing gain is then a function of the halved bandwidth. More than four phases can also be used to restrict bandwidth further, but the same rationale holds; for example, a system with a 22.75 Mcps code rate and 10 kbps data to be sent would require a 20 MHz (null-to-null) bandwidth using the standard form of BPSK modulation. Its process gain would be 20 MHz/10 kbps = 2×10^3. On the other hand, QPSK could reduce the bandwidth

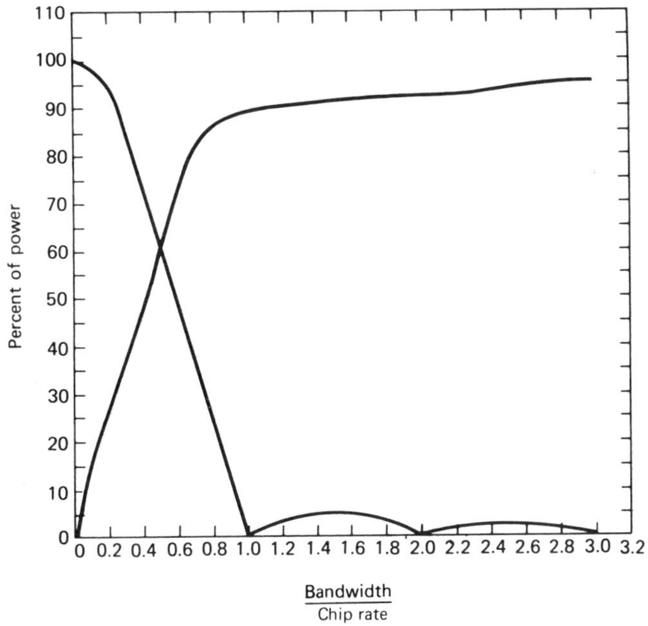

Figure 2.8 Power distribution in $[(\sin x)/x]^2$ spectrum.

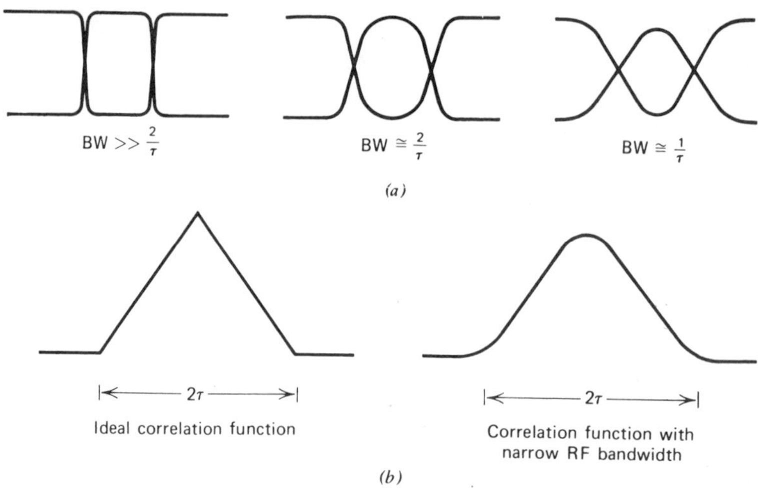

Figure 2.9 RF bandwidth restriction and its effect on typical direct sequence signals: (a) RF envelope of direct sequence signal for various RF bandwidths; (b) effect of bandwidth restriction on correlation function.

required to only 10 MHz but would also reduce process gain to 10 MHz ÷ 10 kbps = 1×10^3, or 3 dB less.

The choice of modulation and code rate is highly dependent on the system in which they are to be used. One must consider bandwidth availability, process gain required, and the basic data rates. RF bandwidth restriction is of more than usual importance in direct-sequence

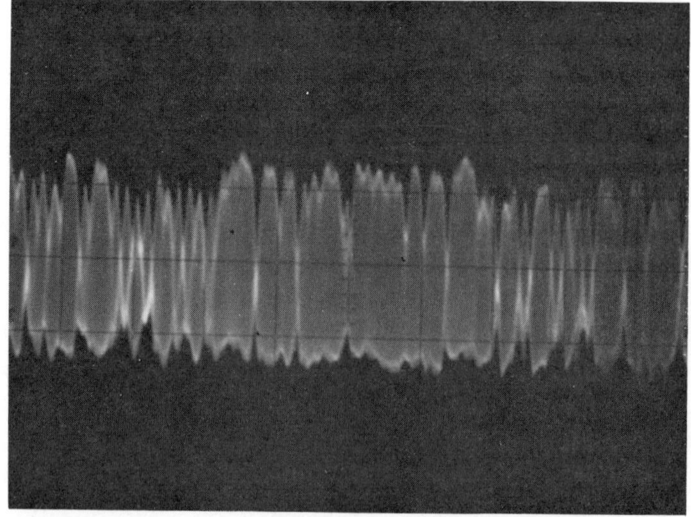

Figure 2.10 This photograph shows amplitude distortion of a bandlimited direct sequence signal.

ranging systems, for degradation of the correlation function, as shown in Figure 2.9, could result in a loss of ability to measure timing precisely. Bandwidth restriction, therefore, can reduce range resolution (i.e., increase the minimum measurement).

Distribution of the power in a direct sequence signal (whether biphase or quadriphase) is plotted in Figure 2.8 along with a plot that shows the relative amplitudes of the first two sidelobes over a region equal to three times the code rate. Figure 2.8 shows that 90% of the total power is contained in a bandwidth equal to twice the bit rate, 93% of the total power in four times the bit rate, and 95% of the total power is in six times the bit rate.

Further consideration of RF bandwidth in both direct sequence and frequency-hopping systems is discussed in Chapter 7 with respect to the RF transmission and reception link. Here it is sufficient to say that direct sequence-signal bandwidth is a direct function of code rate, modulation envelope, and the number of phase positions employed.

Direct Sequence Process Gain

Process gain in a direct sequence system is a function of the RF bandwidth of the signal transmitted compared with the bit rate of the information. The gain in question is exhibited as a signal-to-noise improvement resulting from the RF-to-information bandwidth tradeoff.

The usual assumption taken is that the RF bandwidth is that of the main lobe of the $[(\sin x)/x]^2$ direct sequence spectrum, which is 0.88 times the bandwidth-spreading code clock rate. Therefore, for a system having a 10 Mcps code clock rate and a 1 kbps information rate the process gain would be $(0.88 \times 10^7)/(1 \times 10^3) = 8.8 \times 10^3$ or 39 dB. (Such parameters are typical for practical systems.)

Let us consider the limitations that exist with respect to expanding the bandwidth ratio arbitrarily so that process gain may be increased indefinitely. (Unfortunately, physical limitations prevent this increase.) Only two parameters are available to adjust process gain. The first, RF bandwidth, depends on the code rate used. If we wish to have an RF (null-to-null) bandwidth 100-MHz wide, the code rate would be at least 50 Mcps. On this basis, how wide is it practical to make the system RF bandwidth? At present, integrated circuits are available that allow limited code generation at rates up to 200 or even 300 Mcps. Is there a profit in going higher or, for that matter, in using these bit rates? Consider higher bit rates: doubling the present state-of-the-art code rates would increase process gain by only 3 dB, which is at best a modest gain when compared to the effort required to double the operating speed of present circuits.

Another consideration is that as code rates go higher, operating errors must go lower in inverse proportion. To be operationally useful a code generator should be able to operate for hours or even days without error. 

If we consider what this means in error rate, we see that a 100-Mcps code sequence generator must operate error free for 3.6×10^{11} bits, just to allow a system to operate for 1 hr. On the other hand, high-speed logic circuits tend toward noise sensitivity and are more susceptible to error.

Finally, but not of least concern, is that high-speed digital circuits consume large amounts of current and power dissipation is high. High-speed code generators often consume as much as 1.0 W/stage.

The foregoing reasons, coupled with the problems of spectrum occupancy, equipment implementation (both RF and digital), and propagation constraints, tend to limit the code rates used for band spreading. For the near future, it appears that code rates of 50–100 Mcps are the highest that are practical for general use.

Reiterating an important point, we state that to improve process gain significantly beyond that obtainable with a 100-Mcps code, we would be forced to generate codes in the 1×10^9 bps region. This, at present, is not practical for the long, multifeedback sequence generators required for useful communications systems. Though codes generated at rates near 500 Mcps have been shown as laboratory curiosities, they have been extremely short and exhibited errors at rates in the 1×10^{-3} range. A 100-Mcps code generator with even a 1×10^{-8} error rate could be expected to make one mistake every second. This would prevent data transmission even if it permitted system synchronization. (It is unlikely that the system could be synchronized.)

Data rate reduction to improve process gain is limited in its extent by the willingness of a user to slow information transfer and by the overall stability of the transmission link. Once data rate is slowed to the tens of bits-per-second area or lower such things as local oscillator phase noise or instability in the propagation medium become significant and can cause errors. It is conceded that systems with 1 bps and lower data rates have been built. Here, however, 10 bps is recommended as a lower limit.

Now, if we consider systems with code rate limited at 100 Mcps and data rate limited at 10 bps, we see that for practical systems process gain is limited at about

$$\frac{0.88 \times 10^8}{10} = 0.88 \times 10^7, \quad \text{or } 69 \text{ dB}$$

It is realized that future systems will transmit data at gigabit rates and that even as this paragraph is being written techniques for building such systems are being worked out. Thus, in setting a seat-of-the pants practical limit of 100 Mcps on spread spectrum code generators for direct sequence communications, it is realized that this limit appears paradoxical. However, there is a difference in high-speed codes used for RF spectrum spreading and in high-speed data (the high-speed data would also result in a wide RF spectrum, of course), even though both are bit streams. The difference in the two lies in the action of errors on the system; that is, an error (or, for that matter, many errors up to some correctible limit) might be tolerated in a data-modulated system, whereas an error in a code generator would completely disable the code-modulated system until resynchronization occurs.

Direct Sequence Code and Spectrum Relationships

It has been repeatedly stated here, and emphasized in illustrations, that the spectral power envelope of a direct sequence signal is of a $[(\sin x)/x]^2$ form. This is an expected result since any good set of Fourier transform pairs[1714] will show that the frequency function corresponding to a square pulse is $(\sin x)/x$ (which is a voltage distribution function) and code modulation produces a series of pulses. Because the power envelope is a function of the voltage squared, the $[(\sin x)/x]^2$ power spectrum results. {As an aside, we point out that a spectrum-analyzer presentation of a direct sequence signal always appears to have a $[(\sin x)/x]^2$ distribution because spectrum analyzers do not display phase information, only amplitude.}

Any code sequence used to modulate a direct sequence system produces a similar power envelope, but what about the details of distribution within that envelope?

Observation of a pseudonoise code sequence will show that it is made up of a series of variable-period pulses, which could be viewed as half-period square waves. Figure 2.11 illustrates this concept. These square wave half-periods vary in duration from one code clock chip to n chips for a linear maximal sequence $2^n - 1$ bits in length. Each of these half-periods has a $(\sin x)/x$ spectrum associated with it. As a code sequence modulates a direct sequence transmitter, the output spectrum is actually

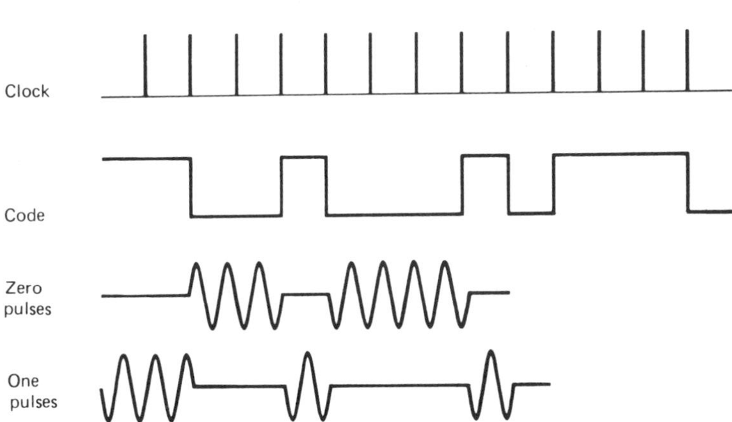

Figure 2.11 Direct sequence transmission as a pulse series.

a composite made up of a series or group of spectra produced by the various half-wave components of the code.

Because the pulse of shortest duration in any code sequence is equal in length to the code sequence clock period, it follows that the frequency spectrum for the composite modulation containing that code sequence must have a main-lobe bandwidth such that its first nulls fall at the code clock rate. Reinforcing this expectation is the fact that the distribution of single ones and zeros in a code sequence is such that they outnumber all other pulse lengths. Chapter 3 includes a detailed discussion of the distribution of the half-periods (or runs) in a pseudonoise code sequence. For now, it is enough that one-fourth of the runs in a maximal code are one chip long and that the number of runs or lengths longer than one decreases by half for each chip of increase in run length.

The number of frequency sets available is a function of the length of the code (and code generator being used). For an n-chip sequence generator, there are $n + 1$ frequency sets, and the spacing of individual frequency components is as narrow as $R_c/(2^n - 1)$, where the code length is $2^n - 1$ (maximal), n is the sequence generator length, and R_c is the code chip or clock rate.

The Minimum Shift Key Technique

Minimum shift keying (MSK) is a direct sequence modulation technique that offers exceptionally good suppression of signal sidelobes and fast side lobe rolloff. MSK has a main lobe whose power distribution is approximately the same as BPSK and QPSK, but the null-to-null bandwidth is $1.5R_c$.* Although somewhat more difficult to implement than BPSK or QPSK, MSK does offer some advantages that other forms of direct sequence modulation cannot match. We hasten to state that there are many forms of MSK modulation, each with different spectrum and characteristics. Offset QPSK is, in fact, one form of MSK modulation. We will further discuss these forms and compare their characteristics in detail in Chapter 4.

2.2 FREQUENCY HOPPING†

"Frequency hopping" modulation is more accurately termed "multiple-frequency, code-selected, frequency shift keying." It is nothing more than FSK (frequency shift keying) except that the set of frequency choices is greatly expanded. Simple FSK most often uses only two frequencies; for example f_1 is sent to signify a "mark," f_2 to signify a "space." Frequency

*Classic MSK.

†See the Bibliography, Sections 1, 7, and 8.

"hoppers," on the other hand, often have thousands of frequencies available. One real system[111] has 2^{20} discrete frequency choices, randomly chosen, each selected on the basis of a code in combination with the information transmitted. The number of frequency choices and the rate of hopping from frequency to frequency in any frequency hopper is governed by the requirements placed on it for a particular use.

Characteristics of Frequency Hopping Signals

A frequency hopping system or "frequency hopper" consists basically of a code generator and a frequency synthesizer capable of responding to the coded output from the code generator. A great deal of effort has been expended in developing rapid-response frequency synthesizers for spread spectrum systems.

Ideally, the instantaneous frequency hopper output is a single frequency. Practically, however, the system user must be satisfied with an output spectrum which is a composite of the desired frequencies, sidebands generated by hopping, and spurious frequencies generated as by-products.

Figure 2.12 is a simplified block diagram of a frequency hopping transmission system. The frequency spectrum of this frequency hopper is shown in Figure 2.13.

Over a period of time, the ideal frequency hopping spectrum would be perfectly rectangular, with transmissions distributed evenly in every available frequency channel. The transmitter should also be designed to transmit, to a degree as close as practical, the same amount of power in every channel.

The received frequency hopping signal is mixed with a locally generated replica, which is offset a fixed amount such that $\{f_1, f_2, \cdots, f_n\} \times \{f_1 + f_{\text{IF}}, f_2 + f_{\text{IF}}, \cdots, f_n + f_{\text{IF}}\}$ produces a constant difference frequency f_{IF} when transmitter and receiver code sequences are in synchronism.

As in a direct sequence system, any signal that is not a replica of the local reference is spread by multiplication with the local reference. Bandwidth of an undesired signal after multiplication with the local reference is again equal to the covariance of the two signals; for example, a CW signal appearing at the frequency hopping receiver's input would be identical to the local reference when translated to the IF frequency. A signal with the same bandwidth as the local reference (but nonsynchronous) would have twice the reference bandwidth at the IF. The IF following the correlator, then, can reject all of the undesired signal power that lies outside its bandwidth. Because this IF bandwidth is only a fraction of the bandwidth of the local reference, we can see that almost all the undesired signal's power is rejected, whereas a desired signal is enhanced by being correlated with the local reference.

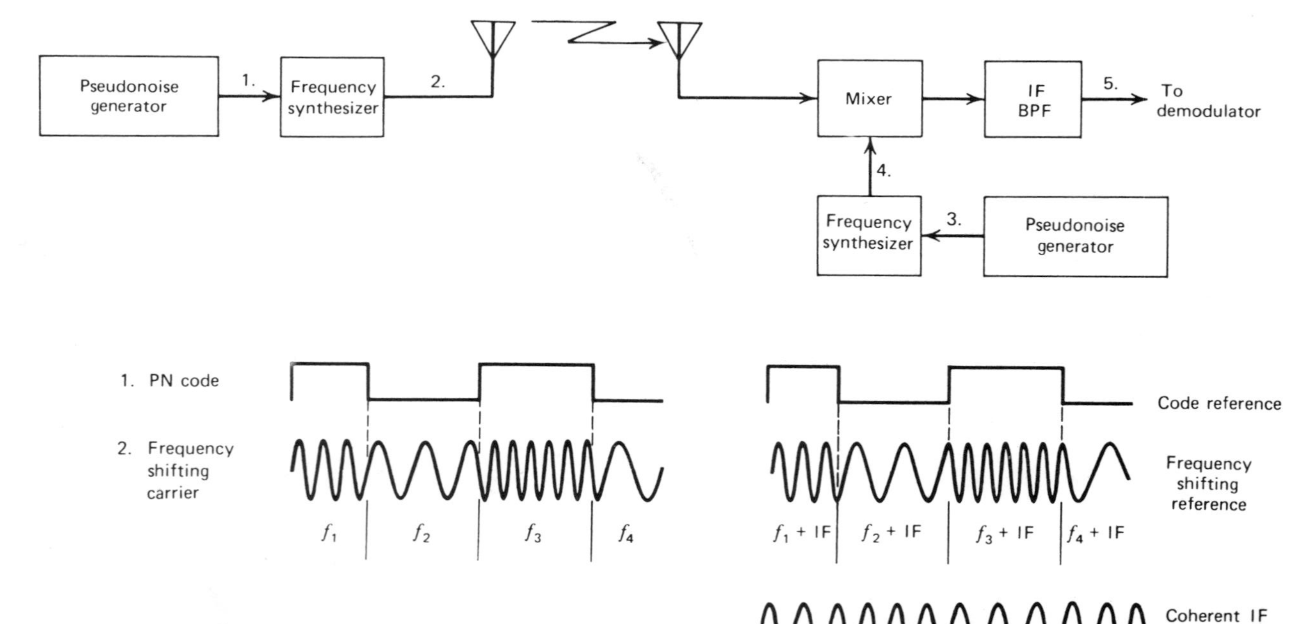

Figure 2.12 Basic frequency hopping system with waveforms.

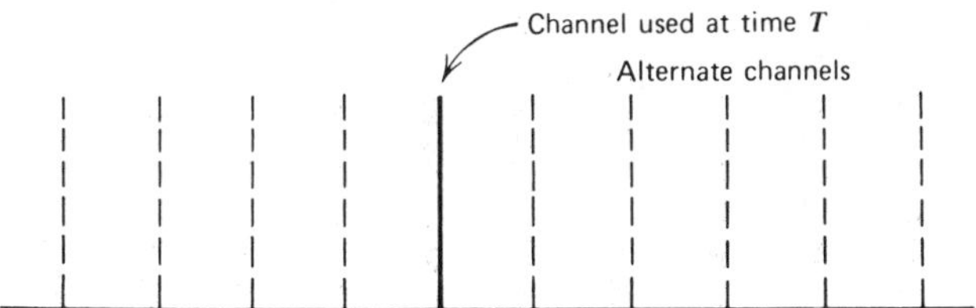

Figure 2.13 Ideal Frequency Hopping signal spectrum.

In the preceding section on direct sequence systems, we saw that a direct sequence system's operation is identical from the standpoint of undesired signal rejection and remapping of the desired signal. From this general viewpoint direct sequence and frequency hopping systems are identical; they are different, however, in the details of their operation.

Process gain for the frequency hopping system (in which channels are contiguous) is the same as that for the direct sequence system; that is,

$$G_p = \frac{\mathrm{BW}_{\mathrm{RF}}}{R_{\mathrm{info}}}$$

If the channels are not contiguous, a better measure of process gain is

$$G_p = \text{the number of available frequency choices} = N$$

which also holds for contiguous channels; for example, a frequency hopping system containing 1000 frequency choices could have 30 dB available process gain.

The only restriction in use of these simple calculations for process gain is that interchannel interference (crosstalk) is neglected. When IF

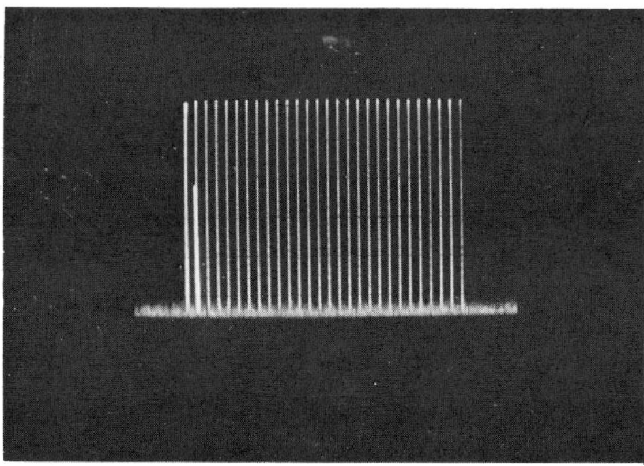

Figure 2.14 Ideal frequency hopper output spectrum.

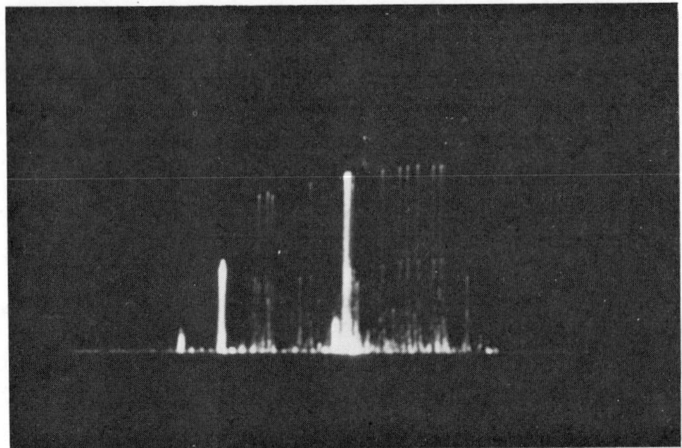

Figure 2.15 Multifrequency hopper spectrum with CW interference.

bandwidth or channel overlap does not provide for unambiguous channel recognition, an allowance must be made for this source of error, which reduces process gain.

Figures 2.14 through 2.17 illustrate actual frequency hopping signals typical of those that might be seen with a spectrum analyzer.

Figure 2.14 shows an ideal frequency hopping spectrum, actually far better than those normally seen from wideband, multifrequency synthesizers. This particular spectrum would be seen from a narrowband frequency hopper (only 28 frequencies are shown), so that the output level is flat over the output band. Compare this spectrum with those of Figures 2.15 and 16.

Figure 2.15 is a time exposure of a several-thousand-frequency hopper output. The bright center trace is a CW carrier equal in power to the

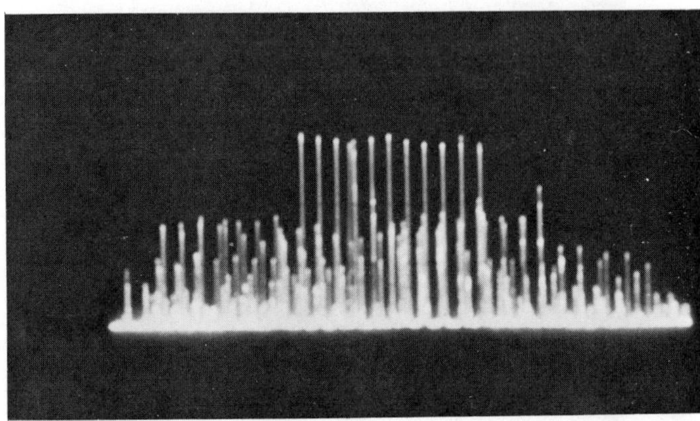

Figure 2.16 Multifrequency hopper spectrum with partial band multitone interference.

frequency hopping signal. Note that the output spectrum of this wider-band synthesizer is not so flat in amplitude as the preceding narrowband hopper.

A wideband frequency hopper, combined with partial band interference, is shown in Figure 2.16. The interference is the larger set of signals at its center. The actual frequency hopping signal is the set of smaller, much-wider-band signals. Here again it is obvious that amplitude linearity across the band is less than perfect, as is spurious suppression.

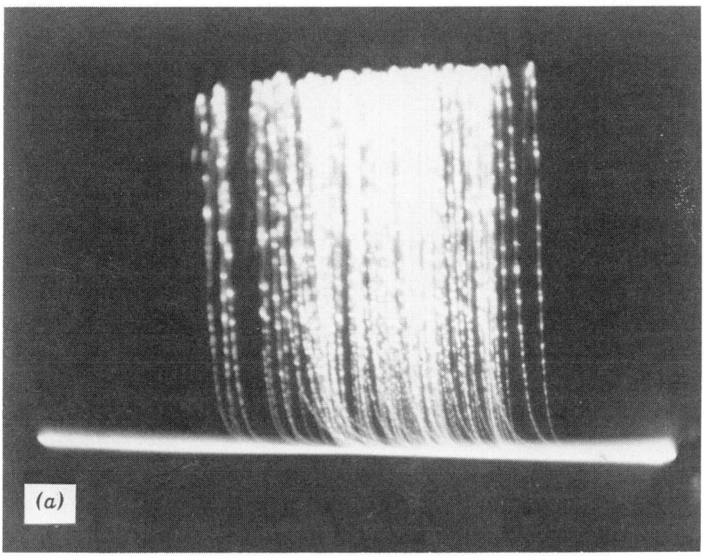

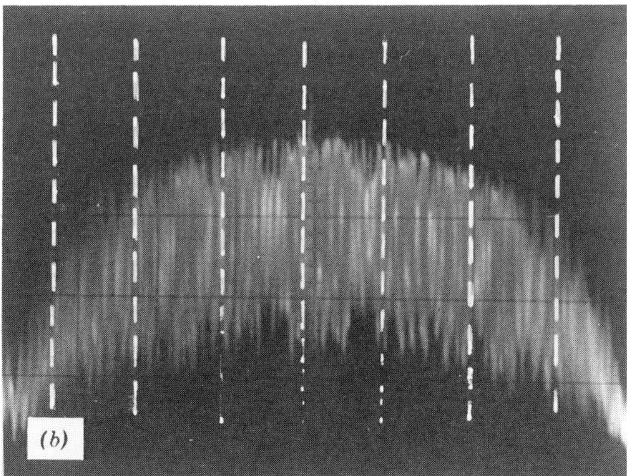

Figure 2.17 Frequency hopping and direct sequence signal comparison. (*a*) Integrated circuit synthesizer at 500 hops/sec, frequency separation 10 kHz. (*b*) Direct sequence signal compared to equivalent-power frequency hopped signal.

Where spurious signal suppression is desired, it is necessary to use pulse shaping of the frequency hopped carrier. Transmission of a rectangular pulse results in (as might be expected) a $(\sin x)/x$ spectrum centered at each frequency cell. This is the origin of most of the spurious signals seen in Figure 2.16.

It is not necessarily obvious that the output spectrum for a frequency hopping system needs to be flat. If bit decisions are made on the basis of signal-power comparison in alternate channels, it becomes important that transmitted signal variation be small. When the mark/space detector has a narrow threshold band, interference must be within that band to be effective. Use of a narrow decision band depends, however, on having a transmitter that has closely controlled amplitude from channel to channel.

Determining Frequency Hopping Rate and Number of Frequencies

The minimum frequency switching rate usable in a frequency hopping system is determined by a number of parameters:

1. The type of information being sent and its rate.
2. The amount of redundancy used, if any.
3. Distance to the nearest potential interferer.

Information in a frequency hopping system may be transmitted in any way available to other systems. Usually, however, some form of digital signal is used, whether the information is a digitized analog signal or data. Assume for the present that some digital rate is prescribed and that frequency hopping has been chosen as the transmission medium. How, then, is the frequency hopping or "chip" rate chosen?

A frequency hopping system must have a large number of frequencies usable on demand. The number required is dependent on the system error rate; for instance, a library of 1000 frequencies could provide good operation when interference or other noise is evenly distributed at every available frequency. For equal distribution of noise in every channel the noise power required to block communications would approach 1000 times desired signal power (in other words, jamming margin would be 30 dB). Unless some form of redundancy that allows for bit decisions based on more than one frequency is used, however, a single narrowband interferer would cause an error rate of 1×10^{-3}, which is generally acceptable for digital data. For a simple frequency-hopping system without any form of transmitted data redundancy, the expected error rate is just J/N, where J equals the number of interferers with power greater than or equal to signal power and N equals the number of frequencies available to the system. We see then that some form of data transmission other than simple binary frequency hopping is required because of the inherent high error rates caused by a small amount of interference.

Error rate for a frequency-hopping system, in which we assume that binary redundant FSK is used (f_a = mark, f_b = space) may be approximated* by the expression for the cumulative binomial expansion:

$$P_e = \sum_{x=r}^{c} \binom{c}{r} p^x q^{c-x}$$

where p = error probability for a single trial = J/N,
J = the number of jammed channels,
N = the number of channels available to the frequency hopper,
q = probability of no error for a single trial = $1 - p$
c = the number of chips (frequencies sent per bit of information),
r = the number of wrong chip decisions necessary to cause a bit error.

A chip decision is defined as occurring any time the interference power in a "space" channel exceeds the power in an intended "mark" channel (or vice versa) by some amount e that is sufficient to cause a decision in favor of what was not intended.

If three or more† frequencies (chips) are sent for each bit of information, we find that performance with interference is greatly enhanced. When the receiver's bit decision is made on the basis of two out of three being correct, a single-channel interferer would cause no more than

$$\binom{3}{2} p^2 q^{3-2} = 3p^2 q^{3-2} = 3p^2 q$$

but $q = 1 - p$, therefore, $3p^2 q = 3(p^2 - p^3)$ errors

*This error-rate approximation is good to the extent that the effects of interference falling into the occupied signal channel are ignored; that is, in addition to the probability J/N of falling into the channel that signifies the opposite of what is intended, there is a probability (also J/N) that an interfering signal will fall into the intended or occupied channel. When this happens, the phase relationship of the two signals occupying the channel determines what occurs. To complicate matters further, there is the probability that an interfering signal will fall in both the intended and opposite channels, in which case, the mark/space decision is based on whether the signal in the opposite channel is greater than the composite of the signals in the intended channel.

†Three or more chips per bit of information allows a majority decision (or some other decision) to be made. Usually an odd number of bits is used. In majority decision systems (in which a decision is based on two of three, three of five, · · ·, trials) an odd number of trials (r) is always better than $r + 1$ or $r - 1$ trials because another trial gives no more information but does increase the chance of error. A three-of-five decision, for instance, allows two errors in five, yet gives a correct decision. Increasing the number of trials to six requires four of the six to be correct and still allows only two errors.

Returning to our 1000-channel system, we see that p is $1/1000$ and $q = 1 - 1/1000 = 0.999$ and the error rate is now

$$3\left(\frac{1}{10^6} - \frac{1}{10^9}\right) \approx 3 \times 10^{-6}$$

which is far better than the 1×10^{-3} error rate given by the simple one-chip-per-bit system.

We have paid for the reduction in errors with a threefold increase in transmission (hop) rate; our frequency synthesizer must now hop three times as fast. Given that the synthesizer can do so, however, it is almost always advantageous to make the three-for-one trade in speed to reduce errors 3000 to 1.

The probability of increasing redundancy (and, with it, the hop rate) to improve bit error rate depends on the system parameters. It is obvious that the more chips sent for a bit, the lower the bit error rate is. The hopping rate required and the RF bandwidth increase in direct proportion. If either the bandwidth allocated or the frequency capability of the frequency synthesizer is limited, then some tradeoff must be made between sending a larger number of chips per bit and reducing the number of frequencies available. As an example, assume that a data source at 1 kbps is to be transmitted and that a 10 MHz RF bandwidth is allowable. The 1 kb data rate suggests that frequency hopping be at least 1 khps, so that the main lobe of the dehopped $[(\sin x)/x]^2$ carrier spectrum is 2 kHz wide. Therefore, if no overlap is allowed (ignoring sidelobes), 10 MHz/2 kHz = 5000 frequency hopping channels are available. Maximum desired error rate is assumed to be 1×10^{-3}, when the interference-to-signal ratio is 100:1.

Now if only one chip is sent per bit of data, $P_s + e$ interference power in 100 of our 5000 available channels (which would equal our desired 100:1 interference power ratio) would have a probability of causing errors once chips are sent for each bit of information, the transmission rate goes to 3000 hps and the received signal bandwidth to 6 kHz. Thus the number of nonoverlapping signal channels available is 10 MHz/6 kHz = 1666 and the 100:1 interference ratio gives us $J/N = 100/1666 = 0.06 = P_e$ for 100 channels containing interference among the 1666 available. The error-rate could then be expected to be

$$\sum_{x=2}^{3} \binom{3}{x} 0.06^x 0.94^{3-x} = 1.2 \times 10^{-2}$$

which still does not reach our error-rate goal.

Observing Figure 2.18, we see that to reach the desired error rate (1×10^{-3}) it is necessary to decrease the ratio J/N or change the decision

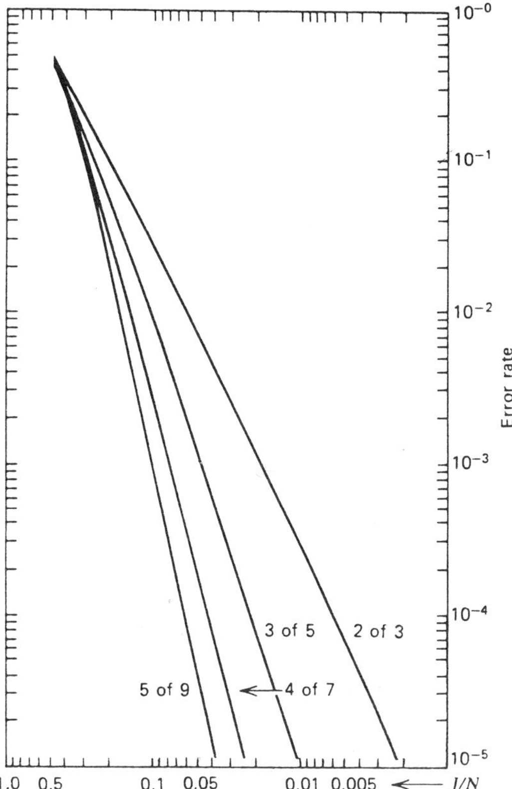

Figure 2.18 Error rate versus fraction of channels jammed (J/N) for various chip decision criteria in multichip transmissions.

criteria. For a two-out-of-three (one error allowable in three) decision at a 10^{-3} error rate we see that J/N can be no greater than 0.019 and that $100/0.019 = 5260$ transmit frequencies are the minimum allowable for 100:1 interference rejection. At three chips per bit of information the receiver dehopped-signal bandwidth is 6 kHz. Therefore, the RF bandwidth would be $6 \times 5260 \times 10^3 = 31.5$ MHz.

Alternatively, if the number of chips is increased to five per information bit, we see that J/N is increased to 0.047 at a 10^{-3} error rate. For the same interference ratio (100:1) we are required to use $100/0.047 = 2130$ channels. Because the contiguous (nonoverlapping) dehopped-signal bandwidth is now 10 kHz, the RF bandwidth must be 21.3 MHz, or approximately one-third less than that required for decisions made on the basis of two of three chips.

Figure 2.19 plots bandwidths required for various data rates, desired jamming margin, and J/N. Note that any kind of bit decision that can be made on the basis of one error allowed in three, or two in five, etc., would

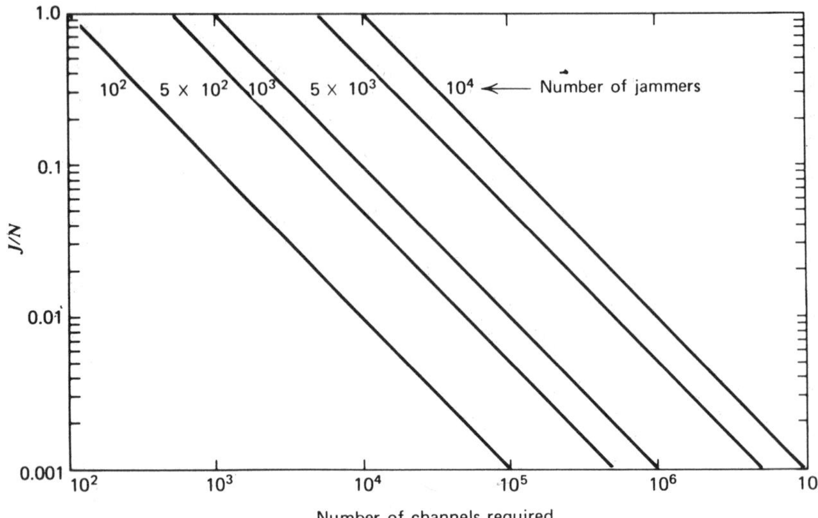

Figure 2.19 Number of channels required versus fraction of channels jammed (J/N) and number of jammers.

give the same result. Here the use of simple majority bit decisions has been implied. Any other form of error allowance would do just as well. It is worthy of mention that there are techniques for forcing three out of three chip errors before causing a bit error.

It is been assumed for the frequency hopping systems discussed so far that contiguous frequency spacing is used; that is, the dehopped signal appearing at the receiver is not allowed to overlap from one channel into another. This is not strictly a true picture. In many systems, depending on the receiver used, transmit spacing can be such that significant overlap occurs. This overlap greatly reduces the RF bandwidth required for the transmitted spread spectrum signal. Figure 2.20 illustrates overlapping channels and the bandwidth savings. Between Figures 2.20*a* and 2.20*b*, there is a doubling of channels in the same bandwidth. Overlapping is such that the center of one channel falls at a null for the adjacent channels (we are assuming noncoherent carrier reception). Here, then, is a technique for allowing the frequency hopping system in our example to meet the RF bandwidth restrictions imposed while keeping chip rate low. The three-chip-per-bit system of the example would have only 15.5 MHz bandwidth, whereas the five-chip-per-bit system would come close to the 10 MHz bandwidth goal with a 10.6 MHz requirement.

One other significant consideration regarding chip rate is the effect of signals which are received at the same frequency as the desired signal but are different in phase. These signals are due to multipath or, even worse, deliberate interference. In most cases, the multipath signal arriving at a receiver is much smaller than the desired signal and is therefore not of

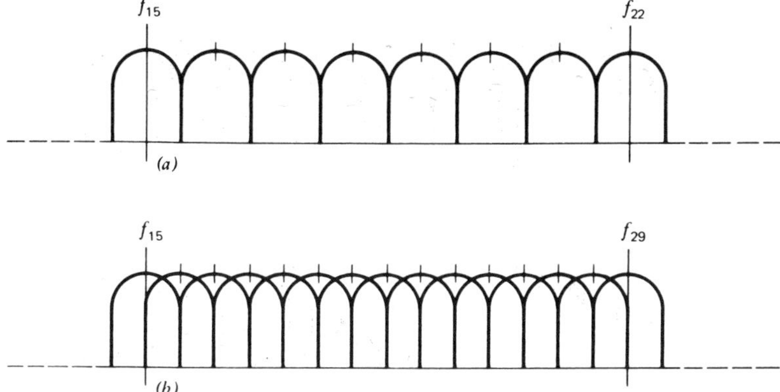

Figure 2.20 Illustration of contiguous versus overlapping channels showing gain in channels per unit bandwidth; (a) contiguous channel spacing; (b) overlapping channel spacing

great consequence. The deliberate interferer who receives a signal from the frequency hopping transmitter, amplifies it, modulates with noise (or, if he or she knows the key, transmits the complement) can be extremely effective with transmitter power similar to the friendly one. Against this kind of threat, the frequency hopper might have no advantage. To combat the threat the frequency hopper must hop at a rate that allows it to skip to another frequency before the interferer can respond to the last one. The desired hopping rate is then greater than $(T_r - T_d)^{-1}$, where T_r is the total propagation time from the frequency hopping transmitter to the interferer and from there to the intended receiver. T_d is direct path delay, illustrated in Figure 2.21.

As we observe Figure 2.21, we see that the minimum required hopping rate is a function of distance to the interfering station and the angle of offset from the direct path. For fixed stations, we can derive a minimum

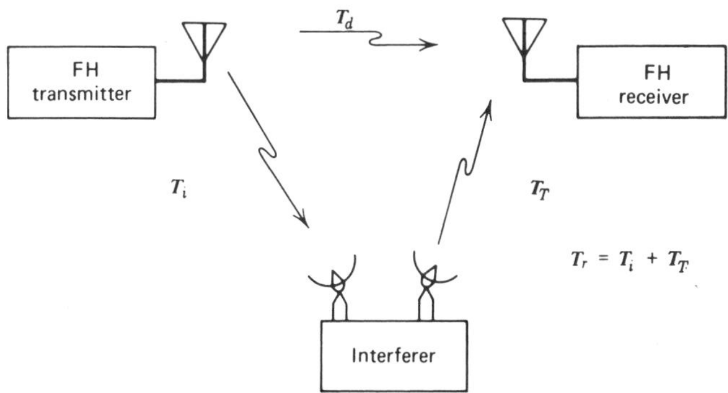

Figure 2.21 Repeater-type interference diagram

hop rate, but for mobile applications the conclusion must be to "make the system hopping rate as fast as possible."

To give an idea of the magnitude of a possible problem assume that the transmitter and desired receiver are 10 miles apart and that the interferer is 8 miles from each of them; $T_r - T_d = 109 - 68 = 41\ \mu s$ and the hopping rate should be at least 24.2 khps. This is a readily achievable hopping rate.

Because of its importance in frequency hopping, let us recap what has been said:

Frequency hopping rate is a function of information rate.

Information error rate cannot be reduced below J/N unless some form of redundant* transmission is used.

The number of frequency channels required is determined by the desired interference rejection capability (jamming margin) and chip error tolerance.

The correlated (dehopped) signal seen in a noncoherent frequency hop receiver is $[(\sin x)/x]^2$ in shape and has main-lobe bandwidth equal to twice the chip rate.

Chip rate is bounded on the low side by multipath and repeater interference considerations. The best rule for variable geometry situations is to hop as fast as possible.

Avoiding the Repeat Jammer

We have mentioned that frequency hopping systems can be faced with the prospect of a jamming signal which follows the signal from frequency to frequency, and can thus have an effect on the desired signal similar to that which a single-frequency jammer would have on a nonhopping signal. It is possible to negate or at least greatly reduce the effects of the repeat jammer by relatively simple techniques, using a noncoherent detector.

One simple approach is to use frequency hopped on–off keying. That is, if a One is intended, let the transmitter send a signal selected from the library by the code. If a zero is intended, however, the transmitter sends nothing. The repeat jammer then has no advantage since:

Transmission of a One provides a signal for the jammer to detect and repeat, but doing so will in most instances only help the receiver to detect the signal and make the correct decision that a One has been sent. The only hope that the jammer has to cause an incorrect decision is to attempt to arrive at the receiver with a signal that is identical to the

*"Redundant" includes error detection and correction coding.

desired signal but opposite in phase, and this is a highly improbable (if not impossible) state to maintain.

When a zero is intended, the desired signal transmitter sends nothing, so the jammer can only make a guess at random as to the correct signal frequency if a One were being sent. In this case, the jammer's probability of causing an error is just J/N.

The repeat jammer is similarly ineffective against other frequency hopping techniques as long as the frequency hopped carrier is unmodulated, and the receiver uses noncoherent detection. Thus in many applications the repeat jammer can generally be ignored as a viable threat to the frequency-hopping system.

Let us turn for a moment to m-ary frequency hopping systems in which a block of $\log_2 m$ coded data bits is sent by transmitting one of a set of m frequencies. (This corresponds in an analog system to one of m levels.) When such transmissions are used, an interfering signal needs only to hit in any one of the $m - 1$ channels not intended for transmission to cause a block error. In an 8-ary system this means that one-seventh the number of jamming signals is required to cause the same probability of a chip error that would be required in a simple binary system. This is a clear loss of 8.5 dB in processing gain. Of course, the reason for using an m-ary transmission format is to reduce chip rate, not to reduce errors or increase jamming margin. Chip-rate reduction, however, is not necessarily a good thing when we are faced with the "look-through" or repeating jammer.

At any rate, for m-ary frequency hopping, only N/m* m-ary channels or sets are available, which not only reduces the processing gain available (to $10 \log N/m$ dB) but increases the probability of chip error per jammed (single) channel. Chip error rates for m-ary transmitters may then be calculated on the basis of J/N_c = number of jammed channels per number of m-ary sets. Character or block error rate for m-ary transmission would be

$$p(\epsilon) = \sum_{a=r}^{m} \binom{m}{a} p^a q^{m-a}$$

which gives the probability distribution for independent trials with probability p of a correct bit and $q = 1 - p$ incorrect bit probability for any given chip. In the case of m-ary signaling $p - J/N_c$ and $q = 1 - J/N_c = (N_c - J)/N_c$. Given a probability of error that is acceptable, we can determine the number of jamming signals that will produce this error.

*N is the number of channels available. For contiguous channels $N = \text{BW}/R$, where BW = bandwidth available, R = chip rate.

Table 2.2 Block Error Rate for *m*-ary Transmission of Three-, Five-, and Seven-Bit Blocks with Various Numbers of Errors Corrected

	Percent of Channels Jammed									
Corrected	5	10	15	20	25	30	35	40	45	50
	Error Rate									
1 in 7	0.04	0.15	0.28	0.42	0.56	0.67	0.76	0.84	0.89	0.93
2 in 7	0.004	0.026	0.073	0.15	0.24	0.35	0.47	0.58	0.68	0.77
3 in 7	0.0002	0.003	0.012	0.033	0.07	0.13	0.2	0.29	0.39	0.5
1 in 5	0.022	0.081	0.16	0.26	0.37	0.47	0.57	0.66	0.74	0.81
2 in 5	0.0012	0.0086	0.027	0.058	0.10	0.16	0.24	0.32	0.41	0.5
1 in 3	0.007	0.03	0.06	0.1	0.16	0.22	0.28	0.35	0.43	0.5

Table 2.2 lists the percentage of jammed channels allowable for various levels of error, correctable in blocks of three, five, and seven bits.

The prime advantage of *m*-ary transmissions in frequency hopping is that they are well adapted to the use of some powerful error-correction techniques without a great increase in the hopping rate being required. A good example is the TATS technique, which will be discussed in Chapter 4 in more detail. That system employs an octal Reed–Solomon error-correcting code in such a way that six-bit input data blocks are converted to seven-octit output symbols. Each octit (a three-bit word) is then transmitted as one of eight contiguous frequency hopping signals. Two advantages accrue from this format: One is that up to three errors in seven symbols can be corrected, and up to four errors in seven symbols can be detected. The second advantage is that the technique (which converts from binary to octal and back again) permits the use of an extremely powerful error-correcting code while increasing the symbol and hopping rates (the symbol rate is the rate at which the system sends encoded information) by a factor of only $\frac{7}{6}$.

2.3 TIME HOPPING

Time hopping, in other words, is the familiar pulse modulation; that is, the code sequence is used to key the transmitter on and off, as in Figure 2.22. Transmitter on and off times are therefore pseudorandom, like the code, which can give an average transmit duty cycle of as much as 50%. This form of signal, as spread spectrum modulation, has found its major application in combination with frequency hopping as in the RACEP systems developed by the Martin Marietta Corp. The fine point of difference separating time frequency and plain frequency hopping is that in frequency hopping systems the transmitted frequency is changed at

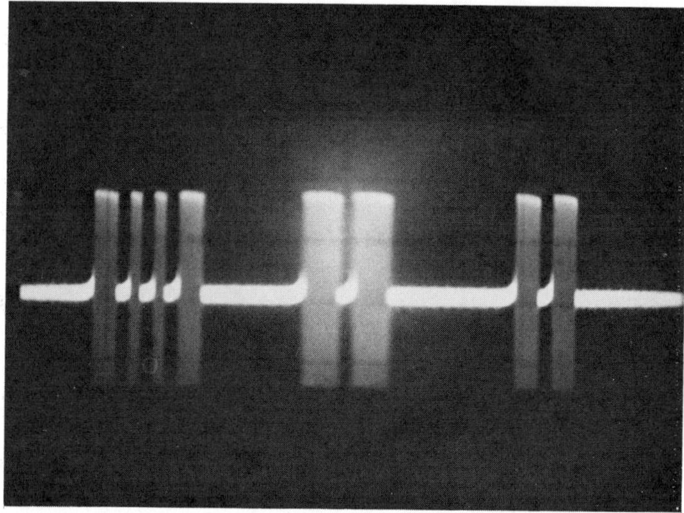

Figure 2.22 Pseudorandom code-keyed (time-hopping) signal waveform.

each code chip time, whereas a time-frequency hopping system may change frequency only at one/zero transitions in the code sequence. Figure 2.23 shows a time-hopping system in block form. The simplicity of the modulator is obvious. Any pulse-modulatable signal source capable of following code waveforms is eligible as a time-hopping modulator.

Time hopping may be used to aid in reducing interference between systems in time-division multiplexing. Stringent timing requirements must be placed on the overall system to ensure minimum overlap between

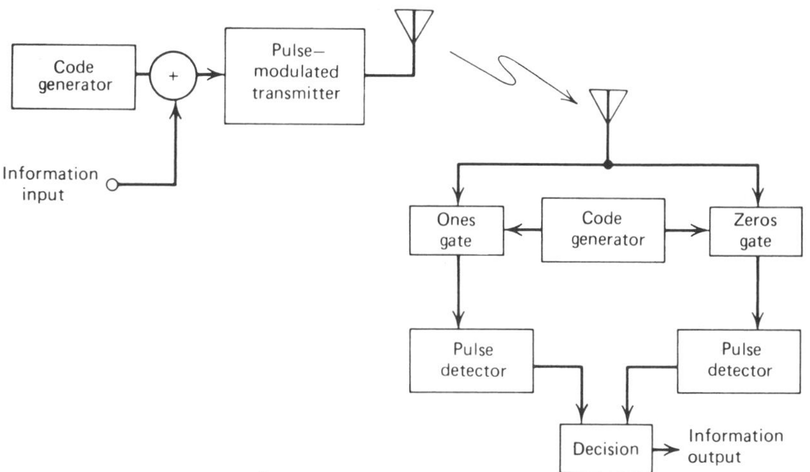

Figure 2.23 Simple time-hopping (pseudorandom pulse) system.

transmitters, however. Also, as in any other coded communications system, the codes must be considered carefully from the standpoint of their cross-correlation properties.

Simple time-hopping modulation offers little in the way of interference rejection because a continuous carrier at the signal center frequency can block communications effectively. The primary advantage offered is in the reduced duty cycle; that is, an interfering transmitter to be really effective would be forced to transmit continuously (assuming the coding used by the time hopper is unknown to the interferer). The power required of the reduced-duty-cycle time hopper would be less than that of the interfering transmitter by a factor equal to the signal duty cycle.

Because of this relative vulnerability to interference, simple time-hopping transmissions should not be used for antijamming unless combined with frequency hopping to prevent single frequency interferers from causing significant losses. For ranging, multiple access, or other special uses time hopping may be especially useful, if only because of the simplicity of generating the transmitted signal.

2.4 PULSED FM (CHIRP) SYSTEMS

Characteristics of Chirp Signals

One type of spread spectrum modulation that does not necessarily employ coding but does use a wider bandwidth than the minimum required so that it can realize processing gain is "chirp" modulation. This form has found its main application in radar but is also applicable to communications. Chirp transmissions are characterized by pulsed RF signals whose frequency varies in some known way during each pulse period. The advantage of these transmissions for radar is that significant power reduction is possible. The receiver used for chirp signals is a matched filter, matched to the angular rate of change of the transmitted frequency-swept signal. Coding is not normally used with this type of matched filter.

The transmitted frequency-swept-signal chirp signal is just that produced by a common laboratory sweep generator because most chirp systems use a linear sweep pattern. Any pattern, suitable to the requirement that a matching receiver filter must be built, is suitable. Here, however, we consider only linearly swept signals. Figure 2.24 illustrates typical waveforms that exist in a chirp system.

Letting $\mu = d\omega / dt$ for the frequency sweep rate, we see that the linearly swept transmitted signal (and the input to the chirp filter) is

$$F(t) = A \cos (\omega_c t + \tfrac{1}{2}\mu t^2)$$

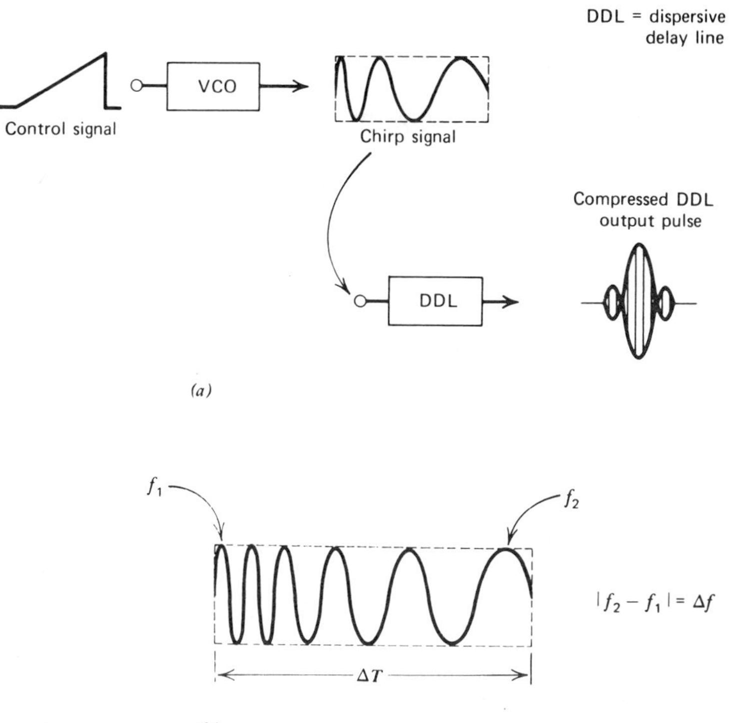

Figure 2.24 Chirp system waveform: (a) typical chirp waveforms; (b) transmitted chirp signal (one sweep).

over the linear range of frequency sweep. For the larger compression ratios of interest here amplitude distribution in our linear FM pulse is quite flat. Cook[210] and Klauder et al.[223] have shown that the chirp signal is distributed according to the Fresnel integral

$$Z(u) = \int_0^u e^{i\pi}\alpha^{1/2}\,d\alpha$$

and have plotted spectral amplitude for compression ratios as high as 130:1. Their plots confirm that ripple decreases as compression ratio increases.

The dispersive filter at the heart of a chirp receiver is a storage and summing device that accumulates the energy received over an interval, assembles it, and releases it in one coherent burst. We may visualize the filter's operation by imagining a transmitter that sends a series of signals over a time period, each of which is tagged in some way. The receiver then sorts the signals according to their tags and assembles them so that they add together within a much shorter time period, reinforcing one another to produce a much stronger signal.

In actual practice, the "tagging" process used by the transmitter is its frequency sweep (which may be in either direction—low to high or high to low), with each increment in frequency denoting a new signal to be sorted and stored by the receiver. A received signal is stored until the entire ensemble of frequencies in one sweep arrives and the summed power at all frequencies is output at one time.

The transmitted signal may be developed in several ways, the simplest of which is application of a linear voltage sweep to a voltage-controlled oscillator. An alternative to the voltage-controlled oscillator is the chirp-generating filter. This is a filter whose impulse response is such that its output is a time series of output frequency components; that is, an impulse with a "continuous" frequency spectrum causes a series of output signals whose time delay is dependent on the frequency component being delayed. If the delay slope dt/df is linear (constant over the range of interest), the output is a linear frequency sweep and the chirp-generating filter can be made to be the exact complement of the receiver's dechirping filter. In fact, if chirp filters were designed as bilateral devices, a chirp-generating filter could also be used to dechirp.

The transmitted chirp signal has two parameters of primary interest. Once these parameters are defined, system performance is also defined, and the rest of the problem is making use of the capability inherent in the chirp signal. These parameters are (1) the frequency sweep $|f_2 - f_1| = \Delta f$ and (2) the amount of time ΔT used to sweep the Δf band. We assume that any amplitude change during ΔT is negligible.

The dispersive delay line or chirp-matched filter compresses a frequency sweep, usually linear, which provides an improvement in output

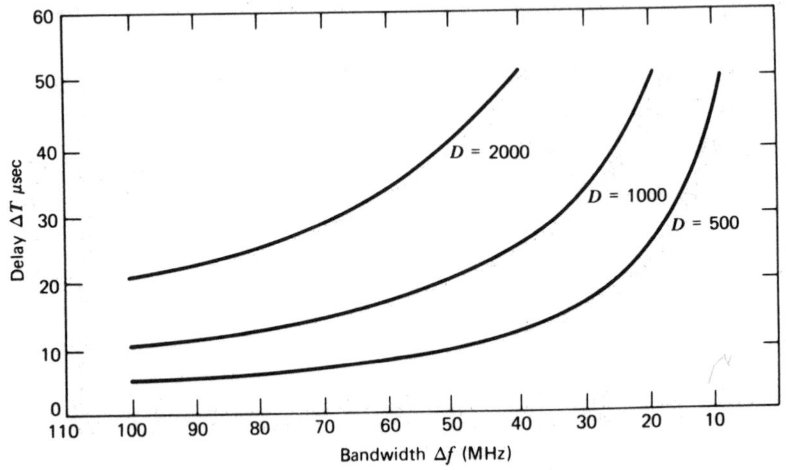

Figure 2.25 Chirp filter TW product curve.

signal (voltage) to noise ratio equal to $\sqrt{D}$, where the compression ratio is

$$D = \Delta T \, \Delta f$$

Δf is defined as the frequency sweep $|f_2 - f_1|$ and ΔT is defined as the time occupied by the sweep in frequency.

All signal parameters are transformed by the factor D between the filter's input and output as given below:

	Input	Output
Frequency sweep	Δf	$\Delta f / D$
Sweep time (signal duration)	ΔT	$\dfrac{\Delta T}{D} = \tau$
Signal (peak) power	$\hat{P}_i$	$DP_i = \hat{P}_0$
Time resolution	ΔT	$\pm\dfrac{\Delta T}{D} = \pm\dfrac{1}{\Delta f}$
Frequency resolution	Δf	$\pm\dfrac{\Delta f}{D} = \pm\dfrac{1}{\Delta T}$

Chirp systems are not the only ones that employ matched filters for reception. Matched filters made up of delay elements matched to a coded signal are also used. These are discussed in Chapter 6.

Figure 2.25 illustrates typical delay and bandwidth parameters available in contemporary delay lines of the dispersive type.

2.5 HYBRID FORMS

In addition to the more usual forms of spread spectrum modulation, there are the hybrid combinations of modulation that offer certain advantages over, or at least extend the usefulness of, the direct sequence and frequency hopping techniques. The most often used combination (or hybrid) spread spectrum signals are made up of (1) simultaneous frequency hopping and direct sequence modulations, (2) simultaneous time and frequency hopping, and (3) simultaneous time-hopping and direct sequence modulations.

The advantage in combining two spread spectrum modulation methods is usually that characteristics can be provided which are not available from a single modulation method. Implementation is not necessarily increased in difficulty by the same factor, however; that is, a combined frequency hopping and direct sequence transmitter might be built up of readily constructed code sequence generators and frequency synthesizers whereas a straightforward direct sequence modulator or frequency hopper could not be constructed to do the same job.

The specific construction of the various subsystems is not discussed here because it is covered elsewhere. The emphasis here is on describing the hybrid techniques in use and reasons for that use.

Frequency Hopped/Direct Sequence Modulation

As suggested by its title, frequency hopped/direct sequence (FH/DS) modulation consists of a direct sequence modulated signal whose center frequency hops periodically. Figure 2.26 illustrates the frequency spectrum from such a modulator. The spread spectrum signal shown is made up of a number of spread spectrum signals. A direct sequence signal covering a part of the band appears instantaneously, and the entire signal appears in other parts of the band as dictated by a frequency hopping pattern.

Hybrid FH/DS signals are used for various reasons:

To extend spectrum spreading capability.

For multiple access and discrete address.

For multiplexing.

When the maximum in reliable sequence generator clock speed has already been obtained or a limit in the number of frequency hop channels has been reached, a hybrid modulator can be especially valuable. Say, for example, that a RF spectrum 1 GHz wide is wanted. The system designer could specify a 1136 Mcps code sequence generator to support a direct sequence transmitter, a frequency synthesizer with 200,000 5-kHz-spaced outputs for frequency hopping, or a hybrid combination. By combining the direct sequence and frequency hopping approaches we could construct a signal that has a 1 GHz spread spectrum bandwidth by using a 114 Mcps code sequence generator and a frequency synthesizer with only twenty 50-MHz-spaced frequencies available. It is obvious that the difficulty of building the hybrid modulator is significantly lower than that of either of the other approaches. Both the 114 Mcps code sequence generator and the 20-frequency synthesizer could be readily constructed today, which is much more than can be said for a 1136 Mcps code sequence generator or a 200,000 frequency, 1 GHz bandwidth frequency synthesizer.

Hybrid FH/DS transmitters are straightforward superimpositions of direct sequence modulation on a frequency hopping carrier, as shown in

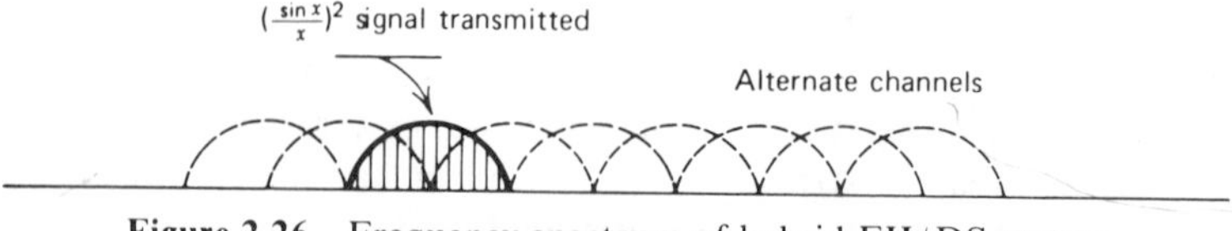

Figure 2.26 Frequency spectrum of hybrid FH/DS system.

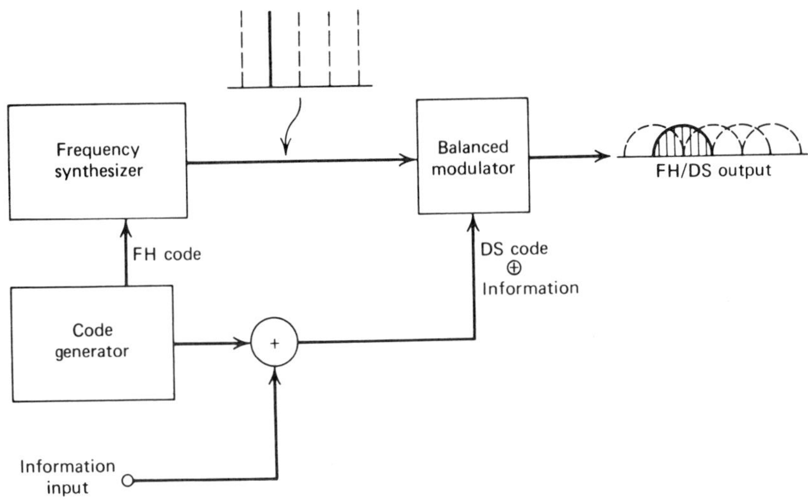

Figure 2.27 Hybrid FH/DS modulator.

Figure 2.27. This modulator differs from a simple direct sequence modulator mainly in that the carrier frequency is varying (hopped) rather than being at a constant frequency, as for simple DS modulation. Note also that the same code sequence generator can supply code data both to the frequency synthesizer to program its hopping pattern and to the balanced modulator for direct modulation.

Synchronism between the FH and DS code patterns means that whenever a given sector of the DS code pattern is transmitted the frequency channel will be the same. It should be noted that the DS code rate is normally much faster than the rate of frequency hopping. Therefore, many chips of DS modulation occur in a single frequency channel. Also, the number of frequency channels available is usually much smaller than the number of code chips so that in the course of a complete code length all the frequency channels are used many times. The pattern of their use, however, is as random (or pseudorandom) as the code itself.

The correlator used to strip off the spread spectrum modulation in a receiver, before baseband demodulation, is, for FH/DS hybrids, again a superimposition of a direct sequence correlator on a frequency hopping correlator; that is, the local reference signal becomes a hybrid FH/DS signal, which is then multiplied with all received input signals. Figure 2.28 shows a typical FH/DS receiver configuration in which the local reference generator is essentially a replica of the transmitting modulator, with two exceptions: (1) the local reference center frequency is offset an amount equal to the IF and (2) the DS code is unmodified by baseband input.

The process gain for FH/DS hybrids is a composite of that furnished

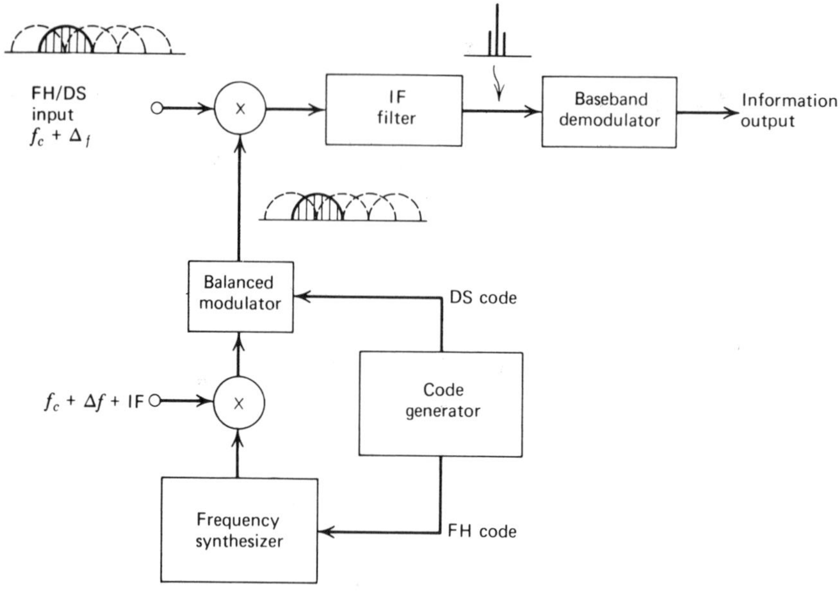

Figure 2.28 Hybrid FH/DS receiver.

separately by the frequency hopping and direct sequence modulations, with certain restrictions; that is, from an overall view the process gain is a function of

$$\frac{\text{BW}_{\text{RF}}}{R_{\text{info}}} = \text{FH/DS process gain}$$

When viewed with respect to particular signals, however, we see that this does not hold true when predicting jamming margin; for a hybrid FH/DS system, jamming margin is not necessarily related to process gain by the relation (see Chapter 1)

$$\text{jamming margin} = G_p - \left[L_{\text{sys}} + \left(\frac{S}{N} \right)_{\text{out}} \right]$$

although from an overall standpoint the relation holds. The reason is that in any particular segment of the overall input spectrum the FH/DS receiver can reject only that amount of interference commensurate with the spread spectrum bandwidth of the partial-band direct sequence signal; that is, if we refer to out 1 GHz bandwidth example, in which the DS signal bandwidth was 100 MHz, the process gain in that segment is limited by the 100 MHz DS bandwidth. This is true because an interfering signal (say a CW carrier) whose power falls wholly within one sector of the 1 GHz band is spready by only the bandwidth of the DS signal whenever the signal hops into that sector. This means than when the signal lies in that one sector, a narrowband interferer can cause a loss of performance, with significantly less power than is required to interfere with the entire FH/DS signal structure.

The reason for pointing out this possible loss in partial-band jamming margin is that when the separated direct sequence channels are used as access channels for multiplexing more than one DS signal on a frequency hopping carrier it is possible for a single frequency interfering signal to interfere with only one channel, with no significant effect on the others. To prevent one user (or channel) from being completely lost, while spreading the effect of interference across all channels, we need to change the order of channel use periodically so that each user employs all available frequencies equally. In this way, even though an interfering signal causes complete loss of one frequency channel, each of the multiplexed signals would be affected only a part of the time equal to

$$\frac{\text{number of jammers}}{\text{number of channels}}$$

Thus, if our 20-channel system had an interfering signal in one channel, it would be affected 1/20th of the time, and if all users shared that channel, a particular user would be affected 1/400th of the time. It is reasonable to say, then, that process gain in decibels for a properly designed FH/DS hybrid system is the sum of the process gains produced by two overlaid spread spectrum modulations; that is

$$\begin{aligned} G_{p(\mathrm{FH/DS})} &= G_{p(\mathrm{FH})} + G_{p(\mathrm{DS})} \\ &= 10\log(\text{number of channels}) + 10\log\frac{\mathrm{BW}_{\mathrm{DS}}}{R_{\mathrm{info}}} \end{aligned}$$

Comparing the results of our 1-GHz spread spectrum example, we have process gain for a straightforward direct sequence system (assuming a 5-kbps information rate):

$$G_p = 10\log\frac{1\times10^9}{5\times10^3} = 53\ \mathrm{dB}$$

for a straightforward frequency hopper with 200,000 frequencies,

$$G_p = 53\ \mathrm{dB}$$

for the FH/DS system,

$$G_{p(\mathrm{FH/DS})} = 10\log 20 + 10\log\frac{1\times10^8}{5\times10^3} = 56\ \mathrm{dB}$$

Process gain is actually 3 dB higher for the FH/DS system because of overlap allowed in the twenty 100-MHz-wide channels. (The same degree of overlap in a straightforward frequency hopper would produce 400,000 channels for a 56-dB process gain.) We emphasize that jamming margin may be significantly less than the simple FH or DS system, however.

Time-Frequency Hopping

Time-frequency hopping modulation has found its greatest application in those systems in which a large number of users with widely variable distances or transmitted power are to operate simultaneously in a single link. Such systems tend to employ simple coding, primarily as an addressing medium, rather than to spread the spectrum specifically. The general tendency is to design for the equivalent of a wireless telephone switching system in which random access and discrete address are the prime operational goals. For such uses time-frequency hopping is well adapted; it offers one of the few (and perhaps only) viable solutions to the near–far problem.

Consider for a moment two receiving and transmitting links whose receivers and transmitters are spaced as in Figure 2.29 and in which the receiver for each link is positioned so that the other link transmitter causes interference. The problem in this system lies in the difference in distance from a receiver to its desired transmitter and to the nearby transmitter.

The antijamming capacity of spread spectrum systems aids in such a situation but unfortunately is not enough under most conditions to overcome the disparity seen in nearby and remote signals. A close transmitter, for instance (say within a mile), would normally produce signals attenuated by 40-plus dB (the plus is a function of transmitting frequency), whereas another transmitter 20 miles away would produce

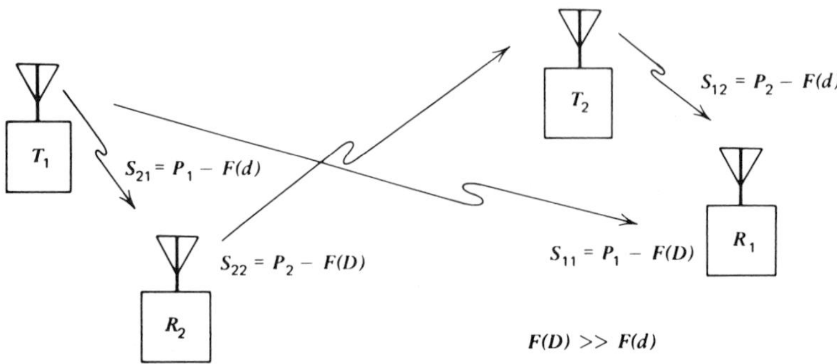

Figure 2.29 Two-link communication system illustrating the near-far problem.

signals attenuated by 63-plus dB. To operate in such an environment a receiver would need process gain in the order of 35 dB with no allowance for fading. Processing alone is not really the answer to near–far reception difficulties. A much better solution is to time all transmissions so that the desired and undesired transmitters are never transmitting at the same time. Even better, with time-frequency hopping the two transmitters can be programmed to transmit on different frequencies as well as at different times. Many links can then be operated at once if their time slots and frequency hopping channels are properly synchronized.

Time-Hopping Direct Sequence

When direct sequence transmission is used and code division multiplexing does not permit sufficient access to the link, time-hopping has proved to be a useful way of adding time-division multiplexing (TDM) to aid in traffic control. The high degree of time synchronization between direct sequence transmit–receive terminals because of their code correlation requirements leads ideally to time-hopping; that is, because a direct sequence receiver must align its local reference code within a fraction of a *pn* code chip time, it already has timing good enough to support TDM operation.

All that is required to add time-hopping TDM to a direct sequence system is on–off switching and control. For time-hopping the on and off decision can be easily derived from the same code sequence generator used to derive the spectrum spreading code. An easily implemented keying control would be an n-input gate which senses some previously chosen shift register state and turns the transmitter on or off at the appearance of that state. Figure 2.30 illustrates a time-hopping direct sequence (TH/DS) transmitter. The receiver for this system would operate similarly, except that the control signal would be used to block or enable the receiver front end synchronously with the expected transmit signal.

When selecting the code state for time control, we choose a shift register state that occurs repeatedly and detect the occurrence of that state. For an n-stage maximal-connected shift register generator there are $2^n - 1$ unique states, each occurring once per repetition of the m-sequence. By choosing a number of register stages $(n - r)$ and sensing an all-ones condition in those stages, we could expect to see that condition 2^r times a code sequence and the distribution of that occurrence would be (pseudo-) random.

On the average, the state sensed would occur at a rate

$$\frac{R_c 2^r}{2^n - 1} = \text{average state occurrence for } (n - r) \text{ stage vector}$$

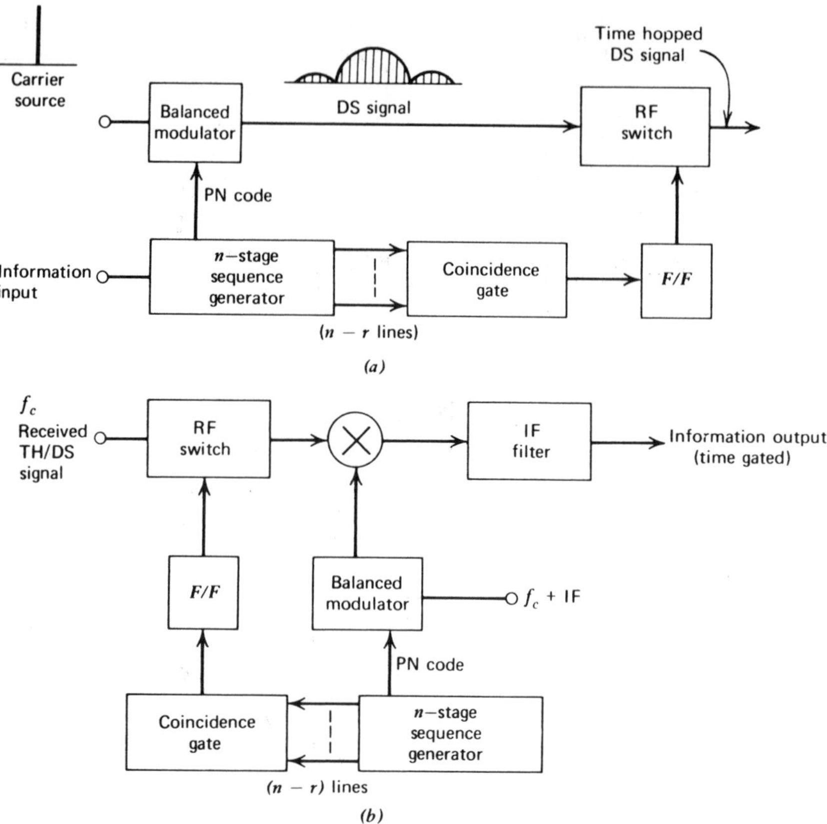

Figure 2.30 TH/DS system block diagrams: (a) time-hopping/direct sequence transmitter; (b) TH/DS receiver.

and the direct sequence "on" periods would occur at an average rate

$$\frac{R_c 2^r}{2^{n+1} - 2} = \text{average pulse occurrence for TH/DS transmission}$$

where R_c = code clock rate,
n = shift register length,
r = register stages not sensed.

We could then select the average frequency of occurrence of direct sequence output, with the actual output occurring at pseudorandom intervals for a pseudorandom duration, as dictated by the code. This pseudorandom time division can permit more channel users, improve operation over simple code division multiplex, and allow more accesses. Again, as in all other applications of codes for signal discrimination, code cross-correlation should be well defined before a TH/DS operation is attempted.

PROBLEMS

1. Why employ a balanced mixer to generate a direct sequence spread spectrum signal?
2. Remembering that a direct sequence signal has a $[(\sin x)/x]^2$ power spectrum, what is the signal's 3-dB bandwidth? What is the peak power level of the first sidelobe compared with that of the main lobe?
3. What percentage of total signal power is contained in the main lobe of a direct sequence signal?
4. Compare the power densities that would exist at a point 5 MHz removed from the center frequency of a 20 MHz wide (a) frequency hopped signal, and (b) direct sequence signal, BPSK, QPSK, classic MSK.
5. What number of channels is required by a frequency hopper whose maximum bit error rate is 0.01 and which must have a 30-dB jamming margin?
6. What RF bandwidth would this frequency hopper employ if its frequency cells were contiguous, its data rate were 1 kpbs, and three chips were transmitted per bit?
7. For the same 30-dB jamming margin, 0.01 bit error rate, and 1 kbps data, what RF bandwidth would be required by a direct-sequence system?
8. The process gain expected from a hybrid FH/DS system with a 20 Mcps code rate, 100 frequency cells, and 3 kbps data rate would be how many decibels?
9. Compare the demodulator bandwidth required on coherent and non-coherent frequency hopping systems.
10. Compare demodulator bandwidth for one- and three-chip-per-bit frequency hopping systems.
11. What would be the effect of a Doppler frequency shift on an upchirp–downchirp system?
12. Compare multipath sensitivity for direct sequence and frequency hopping systems.
13. How might time/frequency hopping be employed to overcome near–far and transmitter imbalance problems?
14. Determine the minimum hop rate that should be used for a fixed receiver when an interferer is 10 miles away and the desired signal transmitter is 100 miles away at a 45° angle from the desired signal. The interferer is a repeater.
15. Compare power density for a chirp signal with that of direct sequence and frequency hopping signals.

3

CODING FOR COMMUNICATIONS AND RANGING

It is intended in this chapter to discuss the codes used in communications and ranging systems—those that act as noiselike (but deterministic) carriers for the information being transmitted. Error-correcting codes are not emphasized since they are more suited to the task of bit- or word-error correction in the information stream. The code sequences of interest are of much greater length than those considered in the usual areas of coding for information transfer, since they are intended for bandwidth spreading and not for the direct transfer of information.

This chapter does not, and indeed cannot, cover the entire field of coding for communications, even for spread spectrum communications alone. Volumes of information and speculation have been written on this subject by specialists in the coding field, and it is suggested that those who wish to explore it consult some of the works listed in the Bibliography under coding techniques.

The importance of the code sequence to a spread spectrum communications or ranging system is difficult to overemphasize, for the type of code used, its length, and its chip rate set bounds on the capability of the system that can be changed only by changing the code.

The purpose of this chapter will be well served if only enough of the rudiments of code selection for communications and ranging systems are taken into account to give a feel for the scope of the code's influence. The emphasis is placed on a description of desirable characteristics and their effects rather than on the design of codes with these characteristics.

Other code types, especially those of the nonlinear variety, are given only minimum attention here—not because of their lack of importance but because of the lack of a requirement for their use when less than true message security is desired. In some communications or ranging systems this security is not required, and in others, security is provided by processing linear codes nonlinearly. Furthermore, nonlinear encoding, in addition to not being critical to the subject at hand—that of providing codes suitable for signal bandwidth spreading in communications and ranging systems—is beyond the scope of this book. It must be stressed, however, that though the linear codes are suitable for interference rejection, ranging, and other spread spectrum applications they are not usable to secure a transmission system. The linear codes are easily decipherable once a short sequential set of chips $(2n + 1)$ from the sequence is known. This has been shown by Birdsall and Ristenbatt,[34] among others. Appendix 5 describes one simple approach to linear code breaking.

If the reader feels that the greatest emphasis is placed on linear codes, or to be more specific, on maximal linear codes, he or she is correct. The maximal linear code sequences (often called *m*-sequences or *pn* codes) are unexcelled for general use in communications and ranging. (Other codes can do no better than equal their performance.) Therefore, it is proper that they and some of their applications and variations be given adequate exposure. It is also proper that we point out here that the most advanced systems ever constructed employ linear codes internal to themselves.

Figure 3.1 illustrates linear and nonlinear code generation on a block diagram basis. To be sure that the point is understood, a spread spectrum system is not a secure system unless the codes used* are cryptographically secure, and although they are useful in spread spectrum systems, linear codes are not secure.

Having (hopefully) dispatched any misconceptions about security, let us define the concepts used in coding for nonsecure spread spectrum systems. It is in order to point out, however, that in general a nonsecure spread spectrum system may be made secure simply by replacing its nonsecure code sequence with a secure code sequence.

To summarize the properties of the codes for use in spread spectrum systems, in which we are interested:

Protection against interference. Coding enables a bandwidth trade, for processing gain against interfering signals.

Provision for privacy. Coding enables protection of signals from eavesdropping to the degree that the codes themselves are secured.

Noise-effect reduction. Error-detection and -correction codes can reduce the effects of noise and interference.

(Note that coding is necessary to provide all of the properties claimed for spread spectrum systems.)

*To be exact, if the band-spreading code is not a secure code, the overall system could still be secure if the information itself were encoded by a cryptographically secure technique. The point here is that encoding in some form is necessary but not necessarily sufficient for security purposes.

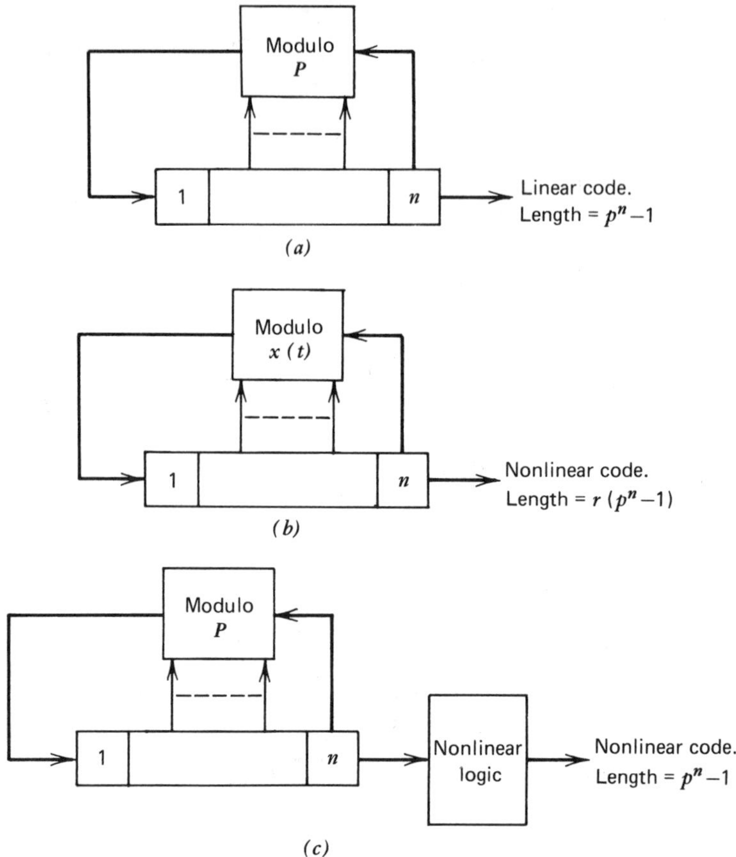

Figure 3.1 Linear and nonlinear generator configurations: (*a*) linear code generator for p-ary shift register (each stage has p possible states); (*b*) nonlinear code generator using p-ary shift register (there are $p^n - 1$ possible states for each feedback condition); (*c*) nonlinear code generator based on linear generator.

3.1 MAXIMAL SEQUENCES

Maximal codes are, by definition, the longest codes that can be generated by a given shift register or a delay element of a given length. In binary shift register sequence generators, which are the only type considered here, the maximum length sequence is $2^n - 1$ chips, where n is the number of stages in the shift register. A shift register sequence generator consists of a shift register working in conjunction with appropriate logic, which feeds back a logical combination of the state of two or more of its stages to its input. The output of a sequence generator, and the contents of its n stages at any sample (clock) time, is a function of the outputs of the stages fed back at the preceding sample time. Feedback connections have been tabulated for maximal code generators from 3 to 100 stages, so that some sequences of any length from 7 through $2^{36} - 1$ chips are readily available. A number of feedback connections are listed in Table 3.6.

Properties held by all maximal code sequences are briefly these:

1. The number of ones in a sequence equals the number of zeros within one chip. For a 1023-chip code there are 512 ones and 511 zeros. Consider a code implementation in which a one is represented by a positive voltage, $+V$, and a zero by a negative voltage, $-V$. The amount of offset over the code length is proportional to the inverse of the code length, or $V/(2^n - 1)$. Similarly, when a code sequence biphase modulates a carrier, the residual carrier component is down by a factor $(2^n - 1)^{-1}$. Thus we see that not only is the modulator important in carrier suppression but the codes used must be capable of supporting the amount of suppression required; for example, when carrier suppression is 30 dB, the shortest code usable is 1000 chips. (For most practical purposes much longer codes are usually required.)
2. The statistical distribution of ones and zeros is well defined and always the same. Relative positions of the runs vary from code sequence to code sequence but the number of each run length does not.
3. Autocorrelation of a maximal linear code is such that for all values of phase shift the correlation value is -1, except for the 0 ± 1 chip phase shift area, in which correlation varies linearly from the -1 value to $2^n - 1$ (the sequence length). A 1023-chip maximal code $(2^{10} - 1)$, therefore, has a peak to minimum autocorrelation value of 1024, a range of 30.1 dB. (It must be realized, however, that this value for autocorrelation is valid only for integrating over the entire sequence length.)
4. A modulo-2 addition of a maximal linear code with a phase-shifted replica of itself results in another replica with a phase shift different from either of the originals.
5. Every possible state, or n-tuple, of a given n-stage generator exists at some time during the generation of a complete code cycle. Each state exists for one and only one clock interval. The exception is that the all-zeros state does not normally occur and cannot be allowed to occur.

Each of these properties is especially useful in a communications or ranging system. Therefore, let us engage in further discussion and explanation of each one and what each means in these systems.

The fact that an almost equal number of ones and zeros exists in any maximal linear code is important in that it allows the DC component in a code or in a code-modulated signal to be neglected. [It is noted that whether a code is $2^3 - 1$ (from a three-stage generator) or $2^{99} - 1$ (from a 99-stage generator) the total number of ones always exceeds the total number of zeros by the same amount—one.] To be exact, the number of ones in any linear maximal code is

$$\frac{2^n}{2} = \text{number of ones}$$

and the number of zeros is

$$\frac{2^n}{2} - 1 = \text{number of zeros}$$

where n is, as usual, the number of stages in the code generator, and the code length is $2^n - 1$ chips.

When modulating a carrier with a code sequence, one–zero balance can limit the degree of carrier suppression obtainable because carrier suppression is dependent on the symmetry of the modulating signal. Therefore, the longer the code sequence, the less the effect of the code sequence on carrier balance.

Run-Length Distribution

Freymodsson[311] has shown that there are exactly $2^{n-(p+2)}$ runs of length p for both ones and zeros in every maximal code sequence (except that there is only one run containing n ones and one containing $n - 1$ zeros; there are no runs of zeros of length n or ones of length $n - 1$). A run is defined as a series of ones or zeros grouped consecutively (Figure 3-2).

An example of the distribution of runs is shown in Table 3.1 for $2^7 - 1$ chip m-sequences. The pattern exhibited here is typical of all m-sequences; that is, the number of runs of each length is a decreasing power of 2 as the run length increases. Therefore, as might be suspected, the number of runs of length p is equal to the sum of the number of runs of all lengths $p + r$, where $1 \leqslant r \leqslant n - p$.

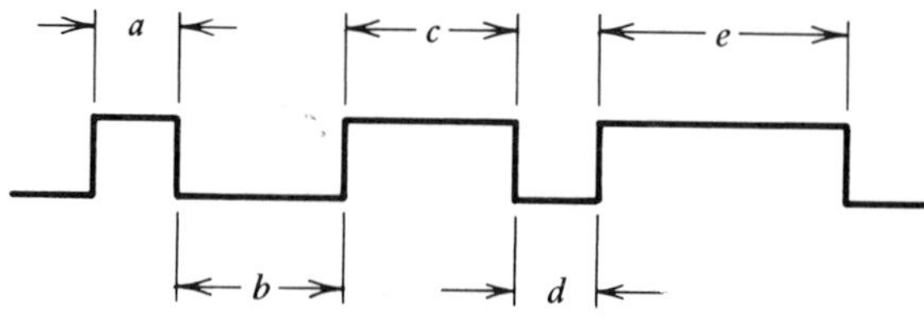

a Run of length one one
b Run of length two zeros
c Run of length two ones
d Run of length one zero
e Run of length three ones

Figure 3.2 Illustration of runs.

Table 3.1 Distribution of Runs for a $2^7 - 1$ Chip m-Sequence

	Number of Runs		
Run Length (Chips)	Ones	Zeros	Number of Chips Included
1	16	16	32
2	8	8	32
3	4	4	24
4	2	2	16
5	1	1	10
6	0	1	6
7	1	0	7
			127 Total

Comparing the structure of m-sequences with truly random noise, White[335] confirms Tausworthe's theory[329] that for r-tuples such that $r \leqslant n$, taken as samples from the output sequence of a maximally connected generator, the distribution of chips is statistically independent. This was done by comparing the actual distribution of ones in samples of length $r \leqslant n$ with the binomial probability of occurrence of ones that would be evidenced by a truly random stream. Thus it was shown that for periods less than $n \times R_c$ the output of a maximal linear generator is random for all practical purposes. When randomness is a governing criteria, we should, as might be supposed even without analysis, increase the generator length (and thereby the code length.)

Even though the randomness properties can be demonstrated, the maximal linear sequences are deterministic. (Certainly, no proof is required because m-sequences repeat at intervals of $2^n - 1$ chips.) Each repetition exhibits the same one–zero distribution. As sample size is increased, the one–zero distribution becomes less and less random. Finally, when sample size equals $2^n - 1$, the number of ones per sample is $2^n/2$ and the number of zeros is $2^n/2 - 1$.

Autocorrelation

Autocorrelation in this context refers to the degree of correspondence between a code and a phase-shifted replica of itself. Autocorrelation plots show the number of agreements minus disagreements for the overall length of the two codes being compared, as the codes assume every shift in the field of interest. Such a plot can be generated easily with a pair of code generators or a computer simulation.

The plot of code correlations for the linear maximal sequences is two-valued, with a peak only at the zero shift point. This is an invaluable property because it allows the receiver to discriminate between signals on a yes–no basis.

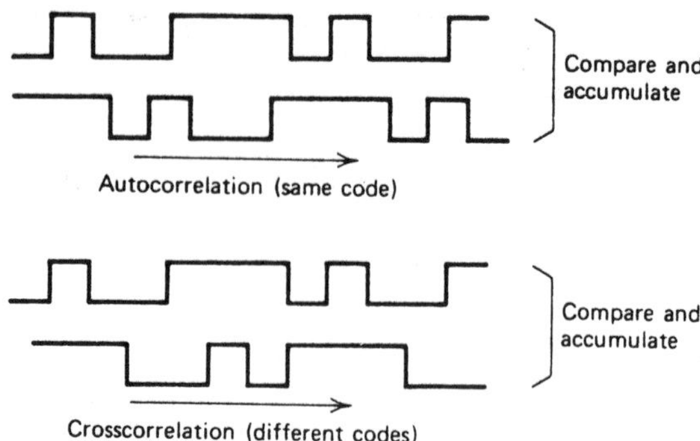

Figure 3.3 Illustration of autocorrelation and cross-correlation of binary code sequence (ψ = agreements − disagreements).

Cross-correlation of two codes is of similar importance. Cross-correlation, however, is the measure of agreement between two different codes. Again, it is plotted as a function of phase shift. Unfortunately, cross-correlation is not so well behaved as autocorrelation, and when large numbers of transmitters, using different codes, are to share a frequency band, the code sequences must be carefully chosen to avoid interference between users. The effect of a high degree of correlation between an undesired code received and a receiver reference is an increase in the receiver's false alarm rate and, under extreme circumstances, false recognition of synch by the receiver (Figure 3.3).

An even worse possible effect of poorly chosen codes with high cross-correlation is that a jammer might transmit a code from the set being used. This in turn could cause every receiver within range of the jammer to be affected by partially correlating with their reference codes and thereby causing false synchronization.

Linear Addition Properties

Maximal linear code sequences possess a particularly interesting combinatorial property. This property, which allows generation of any desired code phase (e.g., 500 chips delay or any other desired delay up to $2^n - 1$ chips), is valuable any time a different code phase is required. One of its uses is in operating multiple correlators to reduce effective synchronization time. Another treats the phase-shifted sequence as a different sequence, with both the normal output and the shift-and-add generated output for separate communication links. This is possible because of the well-defined autocorrelation. An example of a typical system employing the same code for modulation, by all transmitters, is the global positioning system (GPS). In GPS, each transmitter sends the same code (the code having a period of one week) but each transmitter's signal is offset in time, which prevents a receiver from synchronizing to more than one signal at once. The autocorrelation of the code is useful in this case to permit good effective signal orthogonality.

Another valuable property of *m*-sequences is the way in which two (or more) sequences add. When two *m*-sequences of different lengths, say $2^n - 1$ and $2^p - 1$, are linearly (modulo-2) added, the result is a composite sequence with a length $(2^n - 1)(2^p - 1)$. This composite is not maximal but may be a segment of a longer maximal sequence. The primary application of these composite sequences has been in the JPL[219] ranging technique.

Possibly the most valuable linear addition property is that the addition of two *m*-sequences, each of length *r*, produces a composite sequence also of length *r* but not maximal. The composite sequence itself, however, is different for each combination of delay between the two *m*-sequences. Therefore, a pair of sequence generators of length *r* can generate *r* nonmaximal linear codes, each *r* chips long. More important, if the component *m*-sequences are properly chosen, the set of *r* composite sequences will have low and equal autocorrelation; for example, a pair of 10-stage shift register generators (see Figure 3.4) would be capable of generating 1023 different 1023-chip nonmaximal linear codes in addition to the two basic linear maximal codes. For each change in the feedback logic to either of the two shift registers a new set of 1023 codes would be produced.

A disadvantage of the shift and add property is that the linear maximal codes are also predictable by anyone who knows the current code state, so that future operation can be anticipated.

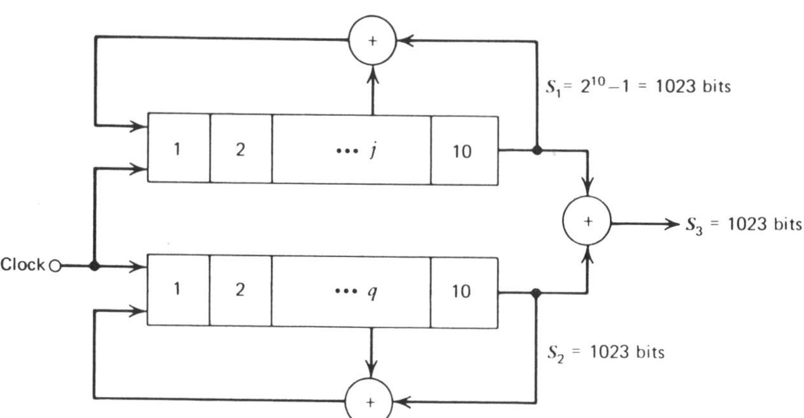

Figure 3.4 Shift register generator for 1023-chip nonmaximal codes.

Table 3.2 Binary-to-Decimal Conversion of Code n-Tuples (Four-Stage Register) Showing Pseudorandom Ordering

State	4-Tuple				Base-10 Equivalent
1	1	1	1	1	15
2	0	1	1	1	7
3	0	0	1	1	3
4	0	0	0	1	1
5	1	0	0	0	8
6	0	1	0	0	4
7	0	0	1	0	2
8	1	0	0	1	9
9	1	1	0	0	12
10	0	1	1	0	6
11	1	0	1	1	11
12	0	1	0	1	5
13	1	0	1	0	10
14	1	1	0	1	13
15	1	1	1	0	14
1	1	1	1	1	15

State Exhaustion

The number of states possible for a set of n elements, each capable of r discernible states, is r^n. A binary shift register generator with maximally connected feedback goes through $2^n - 1$ states in generating a $(2^n - 1)$-chip m-sequence. These states are n-tuples that may be employed to control a processor such as a frequency synthesizer or a Monte Carlo test generator. Table 3.2 lists a set of 15 binary 4-tuples produced by a four-stage shift register generator with its corresponding base-10 conversion. This pseudorandom ordering of n-tuples is typical of all maximal linear sequences. The particular ordering depends on the feedback used. Note that no n-tuple corresponds to zero. Selection of one particular n-tuple or code vector permits monitoring the code generation process to ensure proper operation and to use as a lower rate clock or as a code phase measuring mark.

3.2 LINEAR CODE GENERATOR CONFIGURATION

A linear code sequence generator can be made up of any set of delay elements in conjunction with linear combining elements in a feedback path such that the number of states the generator can assume is a function of the length (in time) of the delay elements and the particular

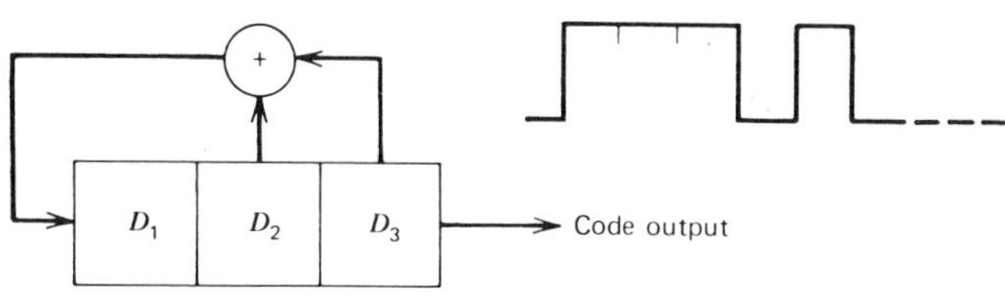

Figure 3.5 Typical code sequence generator (simple type).

combination of feedback. Typical implementations of code sequence generators have often used delay lines or even lengths of coaxial cable, but the most common contemporary technique uses digital circuits (flip-flops) in a shift register configuration.

Figure 3.5 illustrates the general form of a simple linear generator. Outputs from the last delay stage D_n* and an intermediate stage D_j are combined in a modulo-2 adder and fed back to the input of the first delay element. The truth table in Table 3.2 illustrates the action of feedback in generating all of the possible states of a register. In this particular case the code sequence generated, 1110010, is cyclic (repetitive) with a total period $2^n - 1$ times the period of a single delay element. This is the longest code sequence that can be generated by a given number of stages of delay; that is, for n delay stages the longest sequence that can be generated is $2^n - 1$, and this sequence is entitled "maximal" or, to be more specific, "linear maximal" when the generator is linear.

An equivalent and much preferable sequence generator configuration places the feedback adders between stages, as in Figure 3.6. That the same code sequence can be generated with the same number of delay stages and modulo-2 adders, as in the simple generator, is shown by Figure 3.7.

Often a large number of code sequences is needed for applications such as code-division multiplexing. Under these conditions multiple feedback points are necessary, since the maximum number of codes of any length available from a set of delay elements using single-tap feedback would be only $n - 1$, or one less than the number of delay elements. Unfortunately,

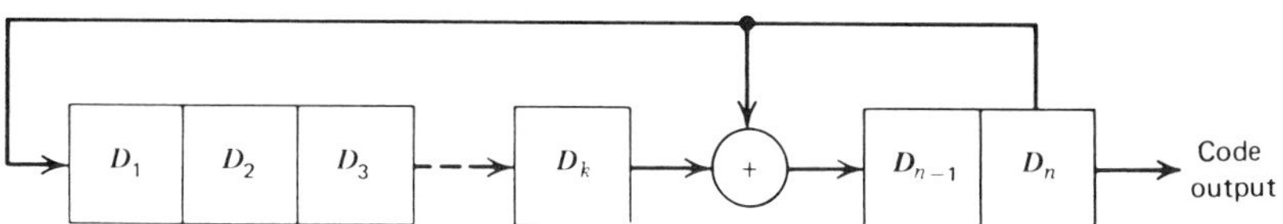

Figure 3.6 Alternate sequence generator configuration (modulator type).

*The notation D_n signifies delay element (such as flip-flop or delay line) number n; that is the nth delay element, where the convention used is to count from left to right.

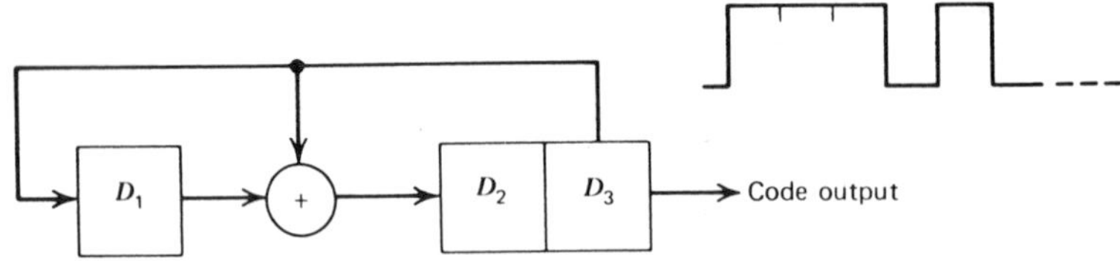

Figure 3.7 Code sequence generator equivalent to that of Figure 3.5.

many of the $n - 1$ codes may be very short cyclic sequences, or the generator may halt operation altogether by entering zeros in all elements in some configurations.

Using all of the possible linear combinations of feedback taps for an n-stage register, there are $[\phi(2^n - 1)]/n$ maximal sequences* that can be generated. [Here, $\phi(2^n - 1)$ is an Euler number—the number of positive integers including 1 that are relatively prime to and less than $2^n - 1$.] As an example of the use of this formula let us observe a five-stage register, for which $2^n - 1$ is 31. Because 31 is itself prime, all positive integers less than 31 are relatively prime to it. The five-stage register then has maximal linear sequences available:

$$\frac{\phi(2^n - 1)}{n} = \frac{30}{5} = 6$$

Table 3.3 lists the number of Mersenne† prime length codes available for register lengths 2 through 89.

When more than one feedback point is used, the alternate generator configuration of Figure 3.6 is of most value. Consider Figure 3.8, which shows a multitap sequence generator of the simple type. In this configuration, the delay inherent in the feedback loop is additive and consists of the sum of all the propagation delays in the path from last stage to first stage. Therefore, the limitation set by the feedback delay path sets the maximum useful speed of the sequence generator. Figure 3.9 extends the idea of Figure 3.5 to illustrate a sequence generator with a modulo-2 adder between each delay stage and its neighbor. This configuration reduces the delays inherent in the feedback path so that the total delay is

*Here and in the remainder of the chapter, the feedback points or taps referred to do not include the last stage feedback because the last stage must always be used, otherwise the generator produces codes of a length that is a function of the last stage in the feedback network. Thus, when single-tap feedback is mentioned, two stages are actually supplying feedback—the last delay element and one intermediate point. Notation signifying feedback includes the final stage, however, and thus specifies the register. As an example (13.9.5.1) specifies a 13-stage shift register with feedback (to modulo-2 adders) from stages 13.9.5 and 2.

†Mersenne prime codes are defined as those codes for which the code length $L = 2^n - 1$ is a prime number.

Table 3.3 Number of Maximal Sequences Available from Register Lengths 3 through 31, 61, and 89

n	Number of Codes	Prime Factors of $2^n - 1$
3	2	7
4	4	3;5
5	6	31
6	4	3;3;7
7	18	127
8	16	3;5;17
9	48	7;73
10	60	3;11;31
11	176	23;89
12	96	3;3;5;7;13
13	630	8,191
14	756	3;43;127
15	1,800	7;31;151
16	2,048	3;5;17;257
17	7,710	131,071
18	1,728	3;3;3;7;19;73
19	27,594	524,287
20	19,200	3;5;5;11;31;41
21	72,576	7;7;127;137
22	120,032	3;23;89;683
23	356,960	47;178,481
24	184,320	3;3;5;7;13;17;241
25	1,296,000	31;601;1,801
26	1,719,900	3;2,731;8,191
27	4,260,864	7;73;262,657
28	4,741,632	3;5;29;43;113;127
29	18,407,808	233;1,103;2,089
30	11,880,000	3;3;7;11;31;151;331
31	69,273,666	2,147,483,647
61	31,800,705,069,076,960	2,305,843,009,213,693,951
89	6,954,719,320,827,979,072,466,990	618,970,019,642,690,137,449,562,111

never greater than that of a single-tap sequence generator. Another important, though less evident, property of the configuration in Figure 3.9 is that it lends itself more easily to implementation, being readily divisible into units consisting of a delay element (flip-flop) and a modulo-2 adder. For this reason sequence generators of this type have been dubbed "modular," and shift register sequence generators made up in this way are called "modular shift register generators" or MSRGs.

In recent years, first with the advent of high-speed switching transistors and then integrated circuits (ICs), it has become progressively simpler to design high-chip-rate code generators. One particular TTL integrated

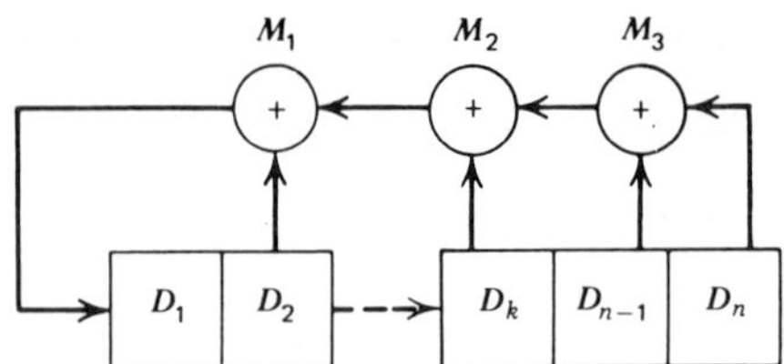

Figure 3.8 Multiple-tap simple sequence generator (SSRG).

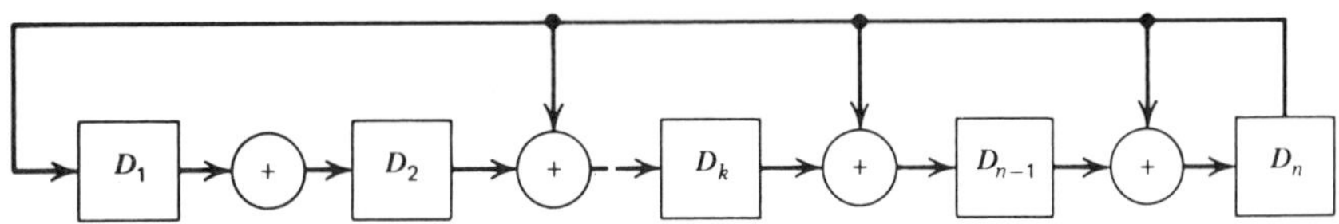

Figure 3.9 Modular multiple-tap sequence generator (MSRG).

circuit, whose block diagram is shown in Figure 3.10, has been specifically designed to act as a flip-flop and gated modulo-2 adder module in MSRGs with code chip rates up to or higher than 20 Mcps. It is expected that entire code generators, with all stages selectable for feedback, will be available as single-chip monolithic integrated circuits. Already, shift registers are available with up to 100 chips storage and need only have feedback added to perform as sequence generators.

One particularly applicable integrated circuit capable of acting as an MSRG is the MC8504 developed by Motorola. This IC includes four D

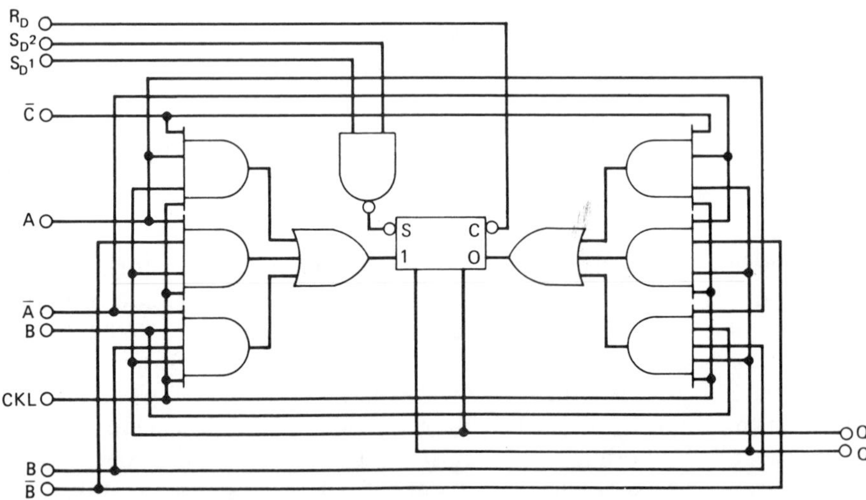

Figure 3.10 Single package flip-flop and gated modulo-2 adder for code sequence generation.

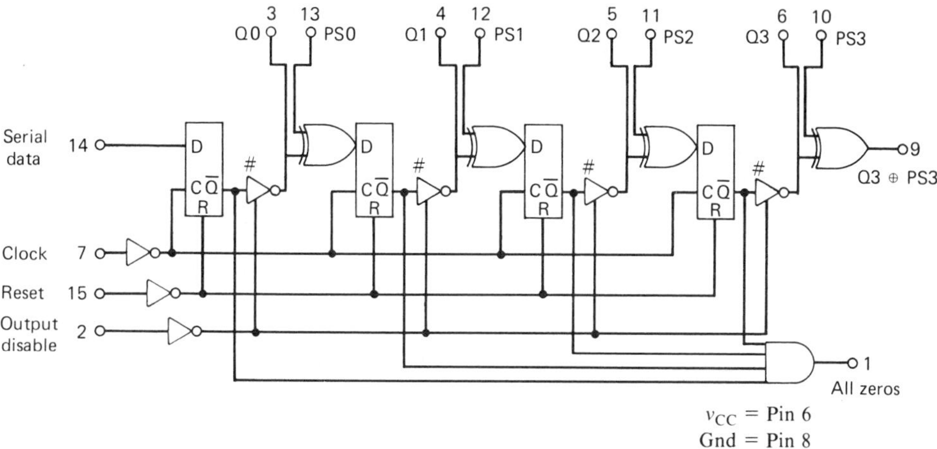

Figure 3.11 MC8504 integrated circuit.

flip-flops and four modulo-two adders, plus a gate that can be used to sense the nonallowable all zeros condition. For longer generators, the output of one MC8504 can be cascaded as an input to the next, and code generators that operate to 17 Mcps can be readily constructed. Figure 3.11 is a block diagram of the MC8504, and Figure 3.12 shows a typical 13-stage MSRG constructed with these devices. Note that the internal all-zeros gate is used in this case to sense an all-zeros hangup condition (and to correct) as well as to generate an n-1 zeros output. (The n-1 zeros

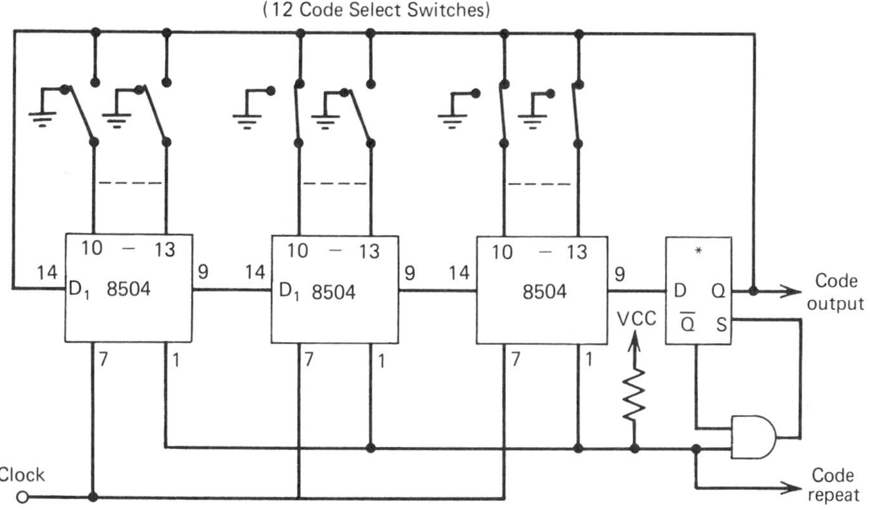

Figure 3.12 Thirteen-stage MSRG design for MC8504. (*½ 74 74. Power and grounds not shown.)

state is equivalent to the all-ones state for test or other purposes, since it also occurs once and only once in each complete cycle of a linear maximal code.)

A shift register configuration worthy of mention here is the "toggle" shift register generator. The toggle SRG is a special type of code sequence generator that takes advantage of the logic included in the generally available integrated circuit *J-K* flip-flops; that is, toggle-type code sequence generators do not employ modulo-2 adders that are separate from the shift register stages themselves. This configuration is advantageous in that it provides for code generation at the highest rates that the flip-flops are capable of operating without their being slowed down by external logic. On the minus side, however, not all codes normally available for a given shift register length are possible with the toggle configuration.

Table 3.4 lists polynomials for linear maximal codes from toggle

Table 3.4 Toggle Registers which are Maximal[a]

Stages	Number of Toggles	Associated Polynomial
3	1	$1 + x + x^3$
3	2	$1 + x^2 + x^3$
4	1	$1 + x + x^4$
5	2	$1 + x^2 + x^5$
5	3	$1 + x + x^2 + x^3 + x^5$
6	1	$1 + x + x^6$
6	5	$1 + x + x^4 + x^5 + x^6$
7	1	$1 + x + x^7$
7	3	$1 + x + x^2 + x^3 + x^7$
7	4	$1 + x^4 + x^7$
7	6	$1 + x^2 + x^4 + x^6 + x^7$
9	4	$1 + x^4 + x^9$
9	5	$1 + x + x^4 + x^5 + x^9$
10	7	$1 + x^2 + x^3 + x^4 + x^5 + x^6 + x^7 + x^{10}$
11	2	$1 + x^2 + x^{11}$
15	1	$1 + x + x^{15}$
15	4	$1 + x^4 + x^{15}$
15	8	$1 + x^8 + x^{15}$
15	11	$1 + x + x^2 + x^3 + x^8 + x^9 + x^{10} + x^{11} + x^{15}$
17	3	$1 + x + x^2 + x^3 + x^{17}$
17	5	$1 + x + x^4 + x^5 + x^{17}$
17	6	$1 + x^2 + x^4 + x^6 + x^{17}$
17	11	$1 + x + x^2 + x^3 + x^8 + x^9 + x^{10} + x^{11} + x^{17}$
17	12	$1 + x^4 + x^8 + x^{12} + x^{17}$
17	14	$1 + x^2 + x^4 + x^6 + x^8 + x^{10} + x^{12} + x^{14} + x^{17}$
18	11	$1 + x + x^2 + x^3 + x^8 + x^9 + x^{10} + x^{11} + x^{18}$

[a]We note that there are no maximal toggle registers of 8, 12, 13, 14, 16, or 19 stages.

registers up through a length of 18 stages (degree 18). There are no available linear maximal toggle codes of degree 8, 12, 13, 14, 16, or 19.

Any binary code may be represented by a polynomial, where the degree of the polynomial is equal to the number of stages in the generating register. The terms in the polynomial (i.e., $x, x^2, x^4, \ldots, x^n$) represent the stages in the register, and the coefficients that are 0 or 1 determine which stages are included in the feedback network. (A 1 is always included as the x^0 stage, with coefficient 1.) As an example, $1 + x + x^2 + x^4 + x^7$ is a seven-stage code generator with feedback from stages 1, 2, and 4, with of course, 7.

3.3 AUTOCORRELATION AND CROSS-CORRELATION OF CODES

The correlation properties of the code sequences used in spread spectrum communications depend on code type, length, chip rate, and even the chip-by-chip structure of the particular code being used. Both autocorrelation and cross-correlation are of interest in communication system design.

Autocorrelation, in general, is defined as the integral

$$\psi(\tau) = \int_{-\infty}^{\infty} f(t)\, f(t - r)\, dt,$$

which is a measure of the similarity between a signal and a phase-shifted replica of itself. An autocorrelation function is a plot of autocorrelation over all phase shifts $(t - r)$ of the signal, where Δt is in one-chip intervals.

Autocorrelation is of most interest in choosing code sequences that give the least probability of a false synchronization. In a communications system designed for maximum sensitivity it is no mean task to discriminate between correlation peaks in a poorly chosen code. Therefore, the designer should investigate the code he or she uses carefully, even if that code is one of the relatively safe *m*-sequences. Statements such as "our extremely long 127-bit chip code sequence assures noiselike properties," which have been observed in the literature on spread spectrum systems exhibit a lack of investigation.

Cross-correlation is of interest in several areas such as (a) code division multiple access systems (or any code addressed system) in which receiver response to any signal other than the proper addressing sequence is not allowable, and (b) antijamming systems that may employ codes with extremely low cross-correlation as well as unambiguous autocorrelation.

Cross-correlation is the measure of similarity between two different code sequences. The only difference between auto- and cross-correlation is that in the general convolution integral for autocorrelation a different term is substituted:

$$\psi_{(\text{cross})} = \int_{-\infty}^{\infty} f(t)\, g(t - r)\, dt$$

Cross-correlation for different code sequences can be tabulated by generating a comparison table and curve of agreements minus disagreements, just as in autocorrelation (Figure 3.16).

It is well, at this point, to introduce a term for the property of a code sequence, pair of sequences, or a sequence and other signal that determines a receiver's ability to recognize the proper point of code synchronization. This property is called the index of discrimination (ID) and denotes the difference in correlation between the fully correlated (perfectly synchronized) code and the peak of minor autocorrelation or of cross-correlations. A particular code will then have separate ID values for autocorrelation and cross-correlation with noncoded signals. The higher the ID value, the better the code.

Code sequence autocorrelation and cross-correlation are expressed as the number of agreements minus the number of disagreements when the code or codes are compared chip by chip. The following example shows autocorrelation for all shifts of a three-stage shift register generator, generating a seven-chip maximal linear code:

$$S = S_1 S_2 S_3,\ S_3 \oplus S_2 \rightarrow S_1,\ L(S) \rightarrow R(S)$$

Reference sequence: 1110010

Shift	Sequence	Agreements (A)	Disagreements (D)	$A - D$
1	0111001	3	4	−1
2	1011100	3	4	−1
3	0101110	3	4	−1
4	0010111	3	4	−1
5	1001011	3	4	−1
6	1100101	3	4	−1
0	1110010	7	0	7

Note that the net correlation $A - D$ is -1 for all except zero-shift or synchronous condition and $2^n - 1 = 7$ for the zero-shift condition. This is typical of all m-sequences.

In the region between the zero and plus or minus one chip shifts, correlation increases linearly so that the autocorrelation function for an m-sequence is triangular as shown in Figure 3.13. This characteristic autocorrelation is used to great advantage in communication and ranging

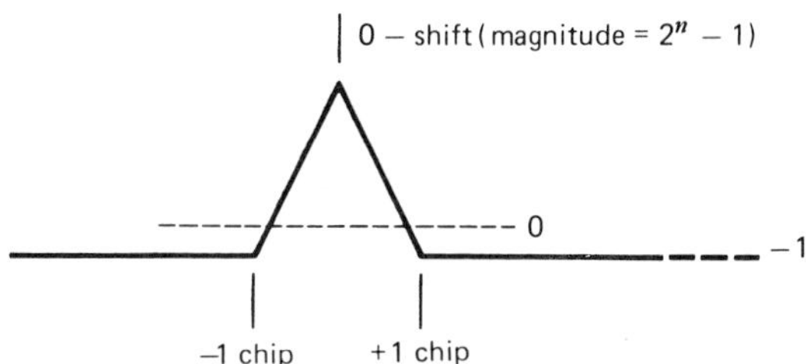

Figure 3.13 *m*-Sequence autocorrelation function.

systems. Two communicators may operate simultaneously, for instance, if their codes are phase-shifted more than one chip. In a ranging system a range measurement is ensured of being accurate within one chip by using the correlation peak as the marker for measurement. This may be accomplished by setting the correlation detector in such a way that it recognizes the level associated with ±1-chip synchronization and does not recognize the lower level.

When codes other than *m*-sequences are used, autocorrelation properties may be markedly different from those of the *m*-sequences. Figure 3.14 illustrates a typical autocorrelation function for a nonmaximal code. The minor correlation peaks are dependent on the actual code used and are caused by partial correlations of the code with a phase-shifted replica of itself. When such minor correlations occur, a receiving system's ability to synchronize may be impaired because it must discriminate between the major (±1 chip) and minor correlation peaks, and the margin of discrimination is reduced.

For purposes of illustration, let us consider the five-stage shift register generator shown in Figure 3.15. If feedback is taken from stages five and three, the code sequence output is

$$\cdots 1111100011011101010000100101100 \cdots \quad \text{(31 chips)}$$

The autocorrelation of this sequence is shown in Figure 3.13; its maximum value is $2^n - 1 = 31$ and its $\text{ID}_{\text{auto}} = 32$. This ID value is, as

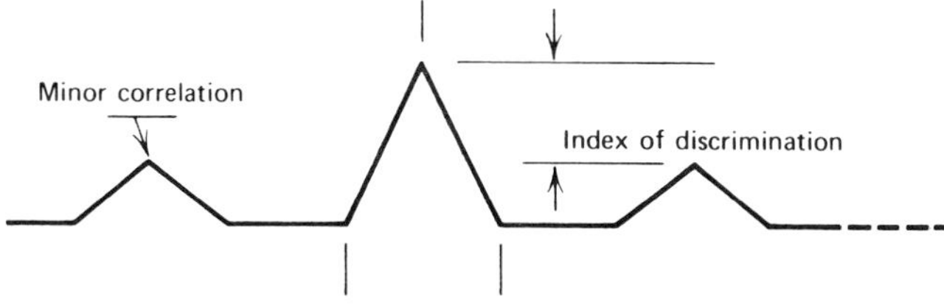

Figure 3.14 Typical nonmaximal code autocorrelation function.

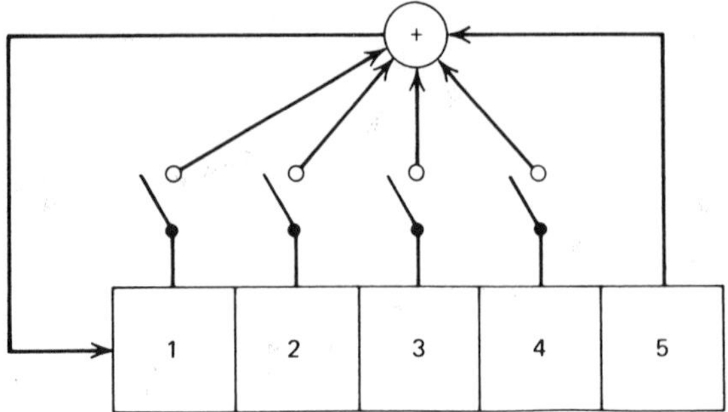

Figure 3.15 Variable tap five-stage SRG.

expected, typical of all linear maximal sequences (of which this is an example) for which ID_{auto} is always equal to 2^n.

Now if we modify the feedback to come from stages five and four, one possible output sequence is only 21 bits long:

$$\cdots 111110000100011001010 \cdots$$

This is an example of a nonmaximal linear sequence that is less than $2^n - 1$ chips long. There are two other nonmaximal linear sequences available from this same feedback configuration whose lengths are seven and three chips:

$$\cdots 1001110 \cdots \quad \text{and} \quad \cdots 101 \cdots$$

The initial start vector contained in the register determines which of the sequences is generated. For this reason greater care is necessary when nonmaximal sequences are used, both to ensure that the initial start vector is correct (or at least is one of the allowable states) and that noise does not cause the register to go to a state outside the desired set. (In such a case the output code could suddenly change from one sequence to another.)

The three sets (not counting the images) of sequence generator states for the available nonmaximal sequences are

	Set 1 $Q_1 Q_2 Q_3 Q_4 Q_5$	Set 2 $Q_1 Q_2 Q_3 Q_4 Q_5$	Set 3 $Q_1 Q_2 Q_3 Q_4 Q_5$
Initial condition	1 1 1 1 1	1 1 0 0 1	0 1 1 0 1
Next states	0 1 1 1 1	1 1 1 0 0	1 0 1 1 0
	0 0 1 1 1	0 1 1 1 0	1 1 0 1 1
	0 0 0 1 1	1 0 1 1 1	0 1 1 0 1
	0 0 0 0 1	0 1 0 1 1	(3-chip cycle)

(*Continued*)	Set 1 $Q_1 Q_2 Q_3 Q_4 Q_5$	Set 2 $Q_1 Q_2 Q_3 Q_4 Q_5$	Set 3 $Q_1 Q_2 Q_3 Q_4 Q_5$
	1 0 0 0 0	0 0 1 0 1	
	0 1 0 0 0	1 0 0 1 0	
	0 0 1 0 0	1 1 0 0 1	
	0 0 0 1 0	(7-chip cycle)	
	1 0 0 0 1		
	1 1 0 0 0		
	0 1 1 0 0		
	0 0 1 1 0		
	1 0 0 1 1		
	0 1 0 0 1		
	1 0 1 0 0		
	0 1 0 1 0		
	1 0 1 0 1		
	1 1 0 1 0		
	1 1 1 0 1		
	1 1 1 1 0		
Starting point (21-chip cycle)	1 1 1 1 1		

Note that a total set of only 31 $(2^5 - 1)$ states exists in all of these nonmaximal sequences—the same number that exists in a single maximal sequence. It is typical of linear sequence generators that for every feedback point that produces a subset of length $(2^n - 1) - k$ there are one or more other nonmaximal feedback connections whose subsets (in combination with the original set) have a total length k. Nonmaximal sequences often have high minor autocorrelation peaks. For this reason, the use of nonmaximal codes or even sectors of maximal codes for communications should be approached with caution.

Code sequences available from the five-stage generator of Figure 3.15 are the following:

Feedback	Sequence	Length
[5, 3]	· · · 1111100011011101010000100101100 · · ·	31
[5, 2]	· · · 1111100110100100001010111011000 · · ·	31
[5, 4, 3, 2]	· · · 1111100100110000101101010001110 · · ·	31
[5, 3, 2, 1]	· · · 1111101110001010110100001100100 · · ·	31
[5, 4, 3, 1]	· · · 1111101000100101011000011100110 · · ·	31
[5, 4, 2, 1]	· · · 1111101100111000011010100100010 · · ·	31
[5, 4]	· · · 111110000100011001010 · · ·	21
[5, 1]	· · · 111110101001100010000 · · ·	21

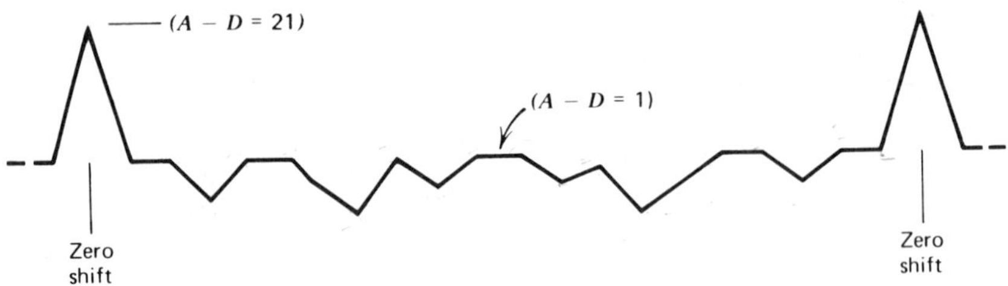

Figure 3.16 Autocorrelation for 21-chip maximal code [5, 4].

Six of these sequences are maximal ($2^5 - 1$) in length, whereas two are nonmaximal. Observation of pairs ([5, 4, 3, 2], [5, 3, 2, 1]), ([5, 4, 3, 1], [5, 4, 2, 1]), and ([5, 4], [5, 1]) will show that they are paired inverses. None of these 31-chip codes is useful as the spectrum spreading element for a practical system because of their short length, but they are listed here to provide models. Let us examine these sequences for their autocorrelation and cross-correlation properties; autocorrelation of all six maximal codes is the same (i.e., equal to −1 for all except the 0 ± 1 chip shift). The zero shift produces an autocorrelation value of 31 for all of them.

Autocorrelation of one of the 21-chip nonmaximals is shown in Figure 3.16 (for sequence [5, 4]), which is also typical for the other 21-chip sequence (although backward). The ID_{auto} value for the 21-chip sequences is 19, which could cause a reduction of more than 40% in a receiver's synchronization capability below that for the 31-chip maximal code.

Even the linear maximal sequences are not immune to cross-correlation problems, though they are, in general, the best available. It is also of some interest to note, even when the codes used exhibit excellent cross-correlation properties when averaged over their entire length, that short-term cross-correlations, which are quite effective in disrupting communications, can (and do) occur.

Note that we have restricted our considerations in this chapter to integration over a long period ($-\infty$ to ∞). This is essentially (for our simple case) the same as integrating over the code length, for the codes spoken of here repeat at intervals of $2^n - 1$ chips. We hasten to point out that integration (as in a synchronization detector) over a period less than that of the code used allows short-term correlations; that is, a short pattern occurring in two different codes or twice in the same code could appear as a legitimate code synchronization when the integration period does not significantly exceed the pattern period.

Figures 3.17 and 3.18 are given as illustrations of cross-correlation and autocorrelation for maximal sequences. The autocorrelation curve of the [5, 3] code shows a zero-shift correlation value of 31. For the [5, 3] and [5, 2] codes cross-correlated, however, the peak value is 11, which gives an index of discrimination of 20, or 35% less than the autocorrelation value.

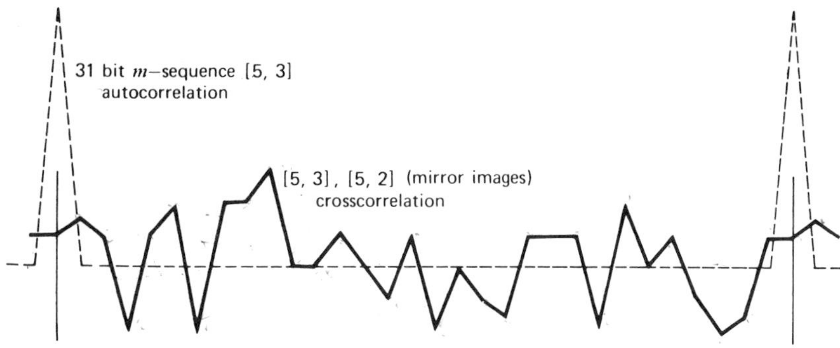

Shift		Agreements (A)		Disagreements (D)		$A - D$	
0	17	17	15	14	16	3	−1
1	18	18	13	13	18	5	−5
2	19	17	12	14	19	3	−7
3	20	11	17	20	14	−9	3
4	21	17	17	14	14	3	3
5	22	19	17	12	14	7	3
6	23	11	11	20	20	−9	−9
7	24	19	19	12	12	7	7
8	25	19	15	12	16	7	−1
9	26	21	17	10	14	11	3
10	27	15	13	16	18	−1	−5
11	28	15	11	16	20	−1	−9
12	29	17	12	14	19	3	−7
13	30	15	17	16	14	−1	3
14	31	13	17	18	14	−5	3
15		17		14		3	
16		11		20		−9	

Figure 3.17 Comparative autocorrelation and cross-correlation for 31-chip mirror image m-sequences.

The [5, 3] and [5, 2] codes are images; that is, one is the same as the other, but generated in reverse order. Cross-correlation of the [5, 3] and [5, 4, 3, 2] codes is lower than that for the image codes, but is still such that the peak cross-correlation value is seven, a value that occurs at 10 different shift positions.

The significant point is that these particular pairs of code sequences are not capable of operating in the same link if the transmitted power from either transmitter exceeds the other enough to raise the peak cross-correlation to a value near peak autocorrelation. Of course, such short codes should not be used, but the comparison is reasonably representative of operation even with much longer sequences used in code division multiplexing or other multiple-access applications.

Judge[222] has considered code division multiplexing by using "quasi-

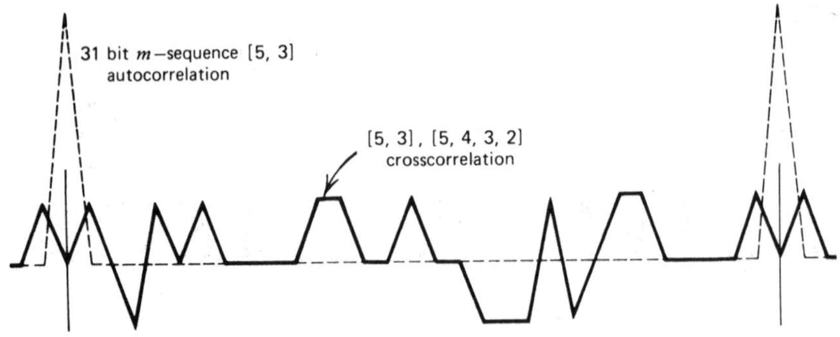

Shift		Agreements (A)		Disagreements (D)		$A-D$	
0	17	15	15	16	16	−1	−1
1	18	19	11	12	20	7	−9
2	19	15	11	16	20	−1	−9
3	20	11	11	20	20	−9	−9
4	21	19	19	12	12	7	7
5	22	15	11	16	20	−1	−9
6	23	19	15	12	16	7	−1
7	24	15	19	16	12	−1	7
8	25	15	19	16	12	−1	7
9	26	15	15	16	16	−1	−1
10	27	15	15	16	16	−1	−1
11	28	19	15	12	16	7	−1
12	29	19	15	12	16	7	−1
13	30	15	19	16	12	−1	7
14	31	15	15	16	16	−1	−1
15		19		12		7	
16		15		26		−1	

Figure 3.18 Comparative autocorrelation and cross-correlation for 31 chip *m*-sequences (not images).

orthogonal binary functions" (linear maximal sequences) and states that for two equal-power signals multiplexed together signal to noise is

$$\frac{S}{N_{(2)}} = \frac{S}{\left(K_1^2 + \dfrac{T_0}{T}\right)^{1/2}}$$

in each receiver. For b signals

$$\frac{S}{N_{(b)}} = \frac{S}{\left[b\left(K_1^2 + \dfrac{T_0}{T}\right)\right]^{1/2}}$$

where T = the crosscorrelation integration period,
T_0 = the code bit period,
K_1 = value of DC correlation.

Judge's result shows that some Mersenne prime sequences exhibit cross-correlation values superior to others, sometimes even for nonprime sequences longer than prime sequences.

The composite code sequences already mentioned are of great utility when cross-correlation is a prime consideration. Their real advantage lies in that for every code in a set of $2^n - 1$ codes, each of length $2^n - 1$, cross-correlation values are well defined, and a system can be designed to operate within this definition. Unless some similar coding scheme is used, each code employed must be separately analyzed with respect to every other code to ensure noninterfering operation.

3.4 COMPOSITE CODES

Though unwavering in the assertion that for communications and ranging the maximal linear sequences cannot be bettered, we must admit that composite code sequences generated by a combination of linear maximal sequences should be considered. Composite codes constructed in this way have special properties that are most advantageous under proper circumstances; for instance, the JPL ranging codes[219] and the Gold[312–314] codes, though constructed from maximal sequences, are not maximal. The JPL ranging codes have special correlation properties that permit rapid synchronization, whereas the Gold codes allow construction of families of $2^n - 1$ codes from pairs of n-stage shift registers in which all codes have well-defined correlation characteristics.

Gold Code Sequence Generators

Gold code sequence generators are useful because of the large number of codes they supply, although they require only one pair of feedback tap sets. A bonus awarded on the basis of the use of these codes is that few sets of feedback taps are needed. Thus the possibility of using a pair of single-tap feedback, simple SRG's while retaining the capacity to generate a large number of codes is present. The single-tap simple SRG is the fastest configuration possible. Thus the Gold code sequences are potentially available at rates equal to the capability of the fastest simple SRG.

The Gold codes are generated by modulo-2 addition of a pair of maximal linear sequences as shown in Figure 3.19. The code sequences are added chip-by-chip by synchronous clocking. The codes themselves

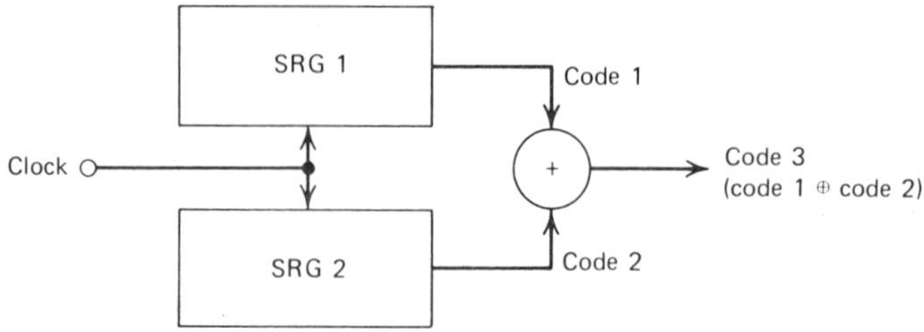

Figure 3.19 Gold code sequence generator configuration.

are the same length. Thus the two code generators maintain the same phase relationship, and the codes generated are the same length as the two base codes, which are added together but are nonmaximal. Figure 3.20 is a specific example.

The shift and add property of maximal sequences tells us that any maximal sequence added to a phase-shifted replica of itself (any integral number of bits) produces a different phase shift as an output. Here the same operation is performed, with the new sequence having the same length as those being added, and nonmaximal. Furthermore, every change in phase position between the two generators causes a new sequence to be generated. To show this advantage, consider the following example.

Given a five-stage sequence generator, we choose a set of feedback taps from Table 3.7. Note that there are only six feedback sets available for the five-stage register and half are images of the other half. If more than six 31-chip codes are needed, we cannot get them from our five-stage register.

The solution, as stated, is to use two five-stage sequence generators connected in the Gold configuration, as shown in Figure 3.20. This figure

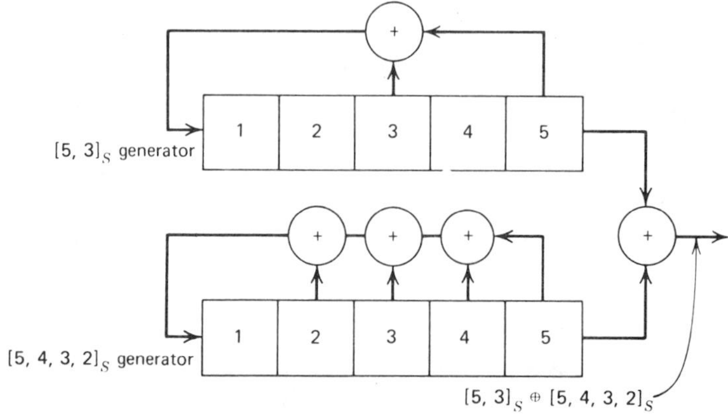

Figure 3.20 Illustration of Gold code generation with $[5, 3]_S$ and $[5, 4, 3, 2]_S$ generators.

also shows the modulo-2-combined Gold codes produced by combining the two output maximal codes with different initial offsets; that is, the two code generators are started with initial conditions offset by various amounts to give different output codes. The all-ones vector is set into both registers as an initial condition. In addition, one- and five-chip shifts (from all-ones vector) are also shown in initial conditions:

Zero-shift combination:

1 1 1 1 1 0 0 0 1 1 0 1 1 1 0 1 0 1 0 0 0 0 1 0 0 1 0 1 1 0 0
1 1 1 1 1 0 0 1 0 0 1 1 0 0 0 0 1 0 1 1 0 1 0 1 0 0 0 1 1 1 0

---

0 0 0 0 0 0 0 1 1 1 1 0 1 1 0 1 1 1 1 1 0 1 1 1 0 1 0 0 0 1 0

One-chip-shift combination:

1 1 1 1 1 0 0 0 1 1 0 1 1 1 0 1 0 1 0 0 0 0 1 0 0 1 0 1 1 0 0
1 1 1 1 0 0 1 0 0 1 1 0 0 0 0 1 0 1 1 0 1 0 1 0 0 0 1 1 1 0 1

---

0 0 0 0 1 0 1 0 1 0 1 1 1 1 0 0 0 0 1 0 1 0 0 0 0 1 1 0 0 0 1

Five-chip-shift combination:

1 1 1 1 1 0 0 0 1 1 0 1 1 1 0 1 0 1 0 0 0 0 1 0 0 1 0 1 1 0 0
0 0 1 0 0 1 1 0 0 0 0 1 0 1 1 0 1 0 1 0 0 0 1 1 1 0 1 1 1 1 1

---

1 1 0 1 1 1 1 0 1 1 0 0 1 0 1 1 1 1 1 0 0 0 0 1 1 1 1 0 0 1 1

Any shift in initial conditions from zero to 30 chips can be used. (A 31-chip shift is the same as the zero shift.) Thus, from this Gold sequence generator, 33 maximal-length codes are available. Extending this demonstration, we can show that any two-register Gold code generator of length n can generate $2^n - 1$ maximal-length sequences (length $2^n - 1$) plus the two maximal base sequences. A multiple-register Gold code generator can generate $(2^n - 1)^r$ nonmaximal sequences of length $2^n - 1$ plus r maximal sequences of the same length where r is the number of registers and n is register length.

In addition to their advantage in generating large numbers of codes, the Gold codes may be chosen so that over a set of codes available from a given generator the cross-correlation between the codes is uniform and bounded. Thus the Gold codes are attractive for applications in which a number of code-division-multiplexed signals are to be used. The same guarantee of bounded cross-correlation is impossible for maximal sequences to the same length.

Gold has presented a method[312] for choosing the linear maximal codes used as components to Gold sequences that gives a set of sequences, each of whose members has cross-correlation, and autocorrelation side lobes, bounded by $|\theta(r)| \leqslant 2^{(n+1)/2} + 1$ for n odd, and by $|\theta(r)| \leqslant 2^{(n+2)/2} - 1$ for n even.

An equivalent result is given by Anderson[32] for the Gold codes; that is, Anderson's expression for the cross-correlation bound is

$$|\theta(\tau)|_G \leqslant \left(\frac{\sqrt{2}\sqrt{1 + 1/L} + 1/\sqrt{2}}{\sqrt{L}}\right)^{1/2}$$

It is apparent from this expression that as $L \to \infty$, $|\theta(\tau)| \to \sqrt{2}/\sqrt{L}$. Convergence is sufficiently rapid that for any code sequence length of interest $|\theta(\tau)| = \sqrt{2}/\sqrt{L}\%$.

Notice that one expression gives cross-correlation in chips, whereas the other gives a percentage of maximum correlation. By normalizing maximum correlation to one, $2^{(n+1)/2} + 1 \approx \sqrt{2L}/\sqrt{L}$ for large L.

Anderson also states that the cross-correlation function for maximal sequences is bounded by

$$|\theta(\tau)| \leqslant \left(\frac{1 + 1/L - 1/L^2}{L}\right)^{1/2}$$

Now, as $L \to \infty$, $|\theta(\tau)| \to 1/\sqrt{L}$. For a given value of L the Gold codes exhibit cross-correlation that is $(\sqrt{2}/\sqrt{L})/(1/\sqrt{L}) = \sqrt{2}$ greater than maximal length sequences of the same length.

Appendix 7 describes an algorithm for Gold code sequence design.

Syncopated-Register Generators

A technique that multiplexes two or more slower generators to generate high-rate sequences was used in complex form as early as 1958 by H. G. Posthumus. The method is similar to that used in Gold code sequence generators in that two separate sequences are modulo-2 added to produce a composite output. The difference lies in that separate clocks, phase-shifted by $360/P$ degrees, each at a rate R/P, are used. (Here R is the desired output code chip rate and P is the number of registers used or the rate reduction desired.) Figure 3.21 shows a two-register generator of the syncopated type, which would allow clock rate reduction by a factor of 2. Timing for this code generator is also shown. Note that the chip rate of code 1 $\oplus$ code 3 is twice that of either code 1 or code 2 before their combination.

Codes generated by syncopation are subject to the same correlation requirements as others used for communications. Unfortunately, work in this specific area is unknown, if it exists, so that the designer who must resort to syncopation techniques must also spend some time in code analysis of correlation properties.

There is no difficulty in extending the syncopation technique. Five

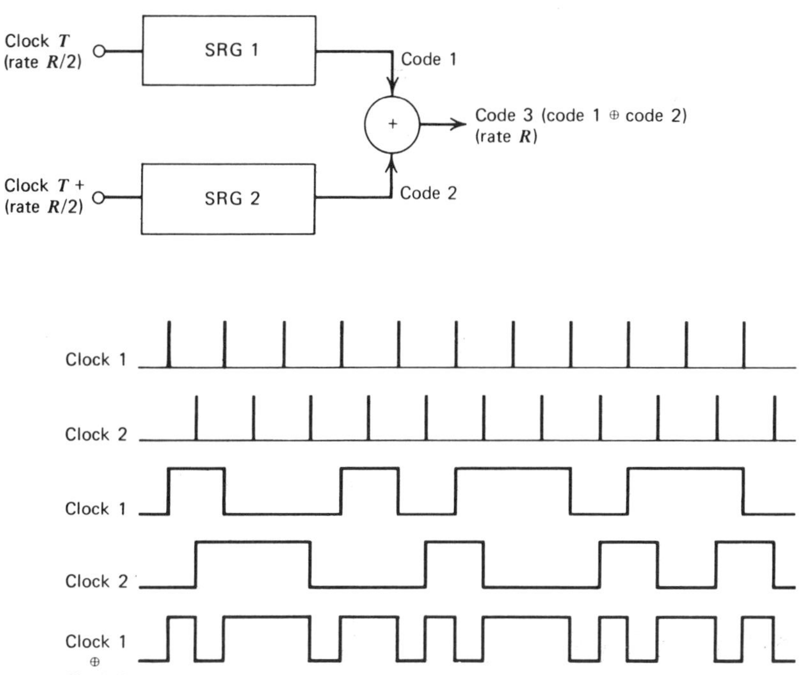

Figure 3.21 Block diagram of syncopated code generator for clock rate reduction by a factor of 2.

registers, for instance, each operating at 50 Mcps, could be used to generate a 250-Mcps code sequence, with a five-phase clock. Theoretically, the syncopation technique could achieve a rate multiplication as great as desired. The rapid increase in hardware needed places a limit on what is practical, however. The overall length of any number of syncopated codes is the product of the lengths of all the composite codes; for example, a pair of codes $(2^r - 1)$ and $(2^s - 1)$ chips long produces a composite code $(2^r - 1)(2^s - 1)$ chips in overall length. We assume here that $r \neq s$. When $r = s$, the composite code is $2^r - 1$ chips long.

Syncopation techniques, though useful for increasing the speed of marginal designs (e.g., increasing low-speed MOS large-scale integrated circuits to give effective 10-Mcps or higher output speeds), will probably find less use in the near future because of the high-speed capability of new integrated circuits, which can operate directly at logic rates that in the past only techniques like syncopation could reach.

JPL Ranging Codes

The JPL (Jet Propulsion Laboratory) ranging codes are constructed by modulo-2 addition of two or more maximal linear sequences whose

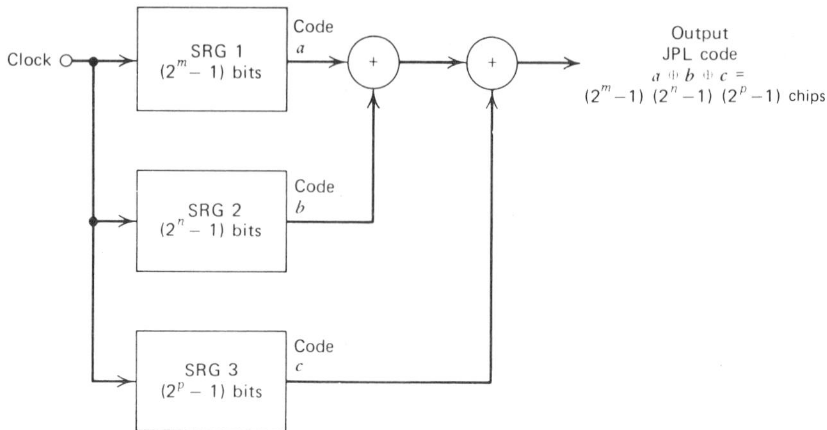

Figure 3.22 Typical JPL code sequence generator configuration.

lengths are relatively prime to one another. Often a clock term is also modulo-2 added to the other sequences (the clock is a short maximal sequence in itself). There are several advantages to such a technique: (a) very long codes useful for unambiguous ranging over long ranges are available; (b) these long codes are generated by a relatively small number of shift register stages; and (c) synchronization of a receiver can be accomplished by separate operations on the component codes. This can greatly reduce the time required for synchronization.

Figure 3.22 shows a typical JPL code generator configuration, which has three basic maximal-length shift register generators, each with a different number of stages. This configuration is identical to the Gold and syncopated code generators except for the difference in the individual code lengths. The JPL codes and their application are discussed further in Chapter 6, and details of their selection and construction may be found in Reference 219.

3.5 CHIP RATE AND CODE LENGTH

Code chip rates in spread spectrum systems affect their systems in many ways. The most obvious is in a BPSK direct sequence system, in which the transmission bandwidth is a direct function of the code chip rate (i.e., main lope null–null RF bandwidth is twice the code chip rate). Code repetition rate is also a function of the chip rate; that is, code repetition rate is simply

$$R_{\text{rep}} = \frac{\text{clock rate in chips per second}}{\text{code length in chips}}$$

Table 3.5 Code Sequence Periods for Various *m*-Sequence Lengths, 1-Mcps Rate

Register Length n	Sequence Length	Sequence Period
7[a]	127	1.27×10^{-4} sec
8	255	2.55×10^{-4} sec
9	511	5.11×10^{-4} sec
10	1023	1.023×10^{-3} sec
11	2047	2.047×10^{-3} sec
12	4095	4.095×10^{-3} sec
13[a]	8191	$9{,}191 \times 10^{-3}$ sec
17[a]	131071	1.31×10^{-1} sec
19[a]	524287	5.24×10^{-1} sec
23	8388607	8.388 sec
27	134217727	13.421 sec
31[a]	2147483647	35.8 min
43	879609302207	101.7 days
61[a]	2305843009213693951	7.3×10^{4} yr
89[a]	618970019642690137449562111	1.95×10^{9} yr

[a]Mersenne prime generators. All codes of length $2^n - 1$.

This repetition rate determines the line spacing in the RF output spectrum and is an important consideration in a system design.

One criterion for selecting code repetition rate is that the period of the code must exceed the length of any mission in which it is to be used. In most aircraft, for instance, an eight-hour code period would exceed the flight capability. Table 3.5 lists the various code lengths for a 1-Mcps chip rate to acquaint the reader with the range of code lengths. Other considerations that bear on the choice of code rate and length are the relationship of the repetition rate to the information baseband and use of the system for ranging.

It is advisable that a direct sequence system's code repetition rate* be adjusted by choosing a satisfactory code length so that it will not lie in the information band. Otherwise, unnecessary noise will be passed into the information demodulators, especially under jammed conditions. In frequency hopping systems the equivalent of having a code repetition rate that falls within the baseband is to have a chip rate that falls there. Both conditions should be avoided whenever possible.

Code chip rates for frequency hoppers usually bear an integral relationship to the information sent; that is, 1 to p frequency chips are sent for each bit of enclosed information. Sending multiple information bits at a single frequency is more conducive to errors than multiple chip/bit

*Repetition rate is usually adjusted by choosing code length, for chip rate must be chosen to give a particular RF bandwidth in a direct-sequence System.

Table 3.6 Chip Rates for Various Basic Range Resolutions

Basic Accuracy	Nautical Miles	Statute Miles
0.1 mile	1.61875 Mcps (608 ft/chip)	1.86333 Mcps (528 ft/chip)
0.05 mile	3.23750 Mcps (304 ft/chip)	3.72666 Mcps (264 ft/chip)
0.02 mile	8.09375 Mcps (121.6 ft/chip)	9.19665 Mcps (105.6 ft/chip)

transmission, although a lower-frequency hopping rate is possible and the synthesizer may be simpler to build.

When ranging is important, a properly chosen code rate can ease the measurement problem and sometimes even improve resolution. If the chip rate is chosen in such a way that an integral number of code chips is accumulated for each mile of delay (propagation time), a simple count of code offset can be used to measure range without elaborate correction. Clock rates that are integral multiples of 161,875 Hz produce code at rates integrally related to the speed of propagation of the RF signal. This relationship may be derived as follows:

Speed of propagation of light (in a vacuum) is

$$c = 2.997926 \times 10^8 \text{ m/sec}$$

There are 1.852×10^3 m/nautical mile; therefore, for 1-chip per mile resolution, clock rate should be (setting clock wavelength = 1 n mi):

$$\frac{c}{R} = \frac{2.997925 \times 10^8}{1.852 \times 10^3} = 161{,}875 \text{ Hz}$$

The clock rate for a 1-statute mile resolution would be 186,333 Hz. Table 3.6 lists the rates that should be used for various range resolutions (numbers of chips per mile) for nautical and statute mile measurements.

3.6 CHOOSING A LINEAR CODE

Among the biggest problems in code generation are (1) finding feedback logic that gives the desired code length and (2) checking the code once a sequence generator has been constructed to ensure that it is operating properly. It is of little use to know how to construct a shift register generator unless one also knows the feedback connections necessary to generate a useful code sequence. Of course, it is possible to find a set of feedback connections experimentally, but this requires not only constructing a shift register generator but also taking the time to check the length of codes generated until one chances on a maximal connection.

Happily, tables of feedback connections and tables of irreducible polynomials have been generated,[326,320,335] which make the job much easier. Table 3.7 was compiled from information given in many sources,

Table 3.7 Feedback Connections for Linear *m*-Sequences

Number of Stages	Code Length	Maximal Taps
2[a]	3	[2, 1]
3[a]	7	[3, 1]
4	15	[4, 1]
5[a]	31	[5, 2] [5, 4, 3, 2] [5, 4, 2, 1]
6	63	[6, 1] [6, 5, 2, 1] [6, 5, 3, 2]
7[a]	127	[7, 1] [7, 3] [7, 3, 2, 1] [7, 4, 3, 2] [7, 6, 4, 2] [7, 6, 3, 1] [7, 6, 5, 2] [7, 6, 5, 4, 2, 1] [7, 5, 4, 3, 2, 1]
8	255	[8, 4, 3, 2] [8, 6, 5, 3] [8, 6, 5, 2] [8, 5, 3, 1] [8, 6, 5, 1] [8, 7, 6, 1] [8, 7, 6, 5, 2, 1] [8, 6, 4, 3, 2, 1]
9	511	[9, 4] [9, 6, 4, 3] [9, 8, 5, 4] [9, 8, 4, 1] [9, 5, 3, 2] [9, 8, 6, 5] [9, 8, 7, 2] [9, 6, 5, 4, 2, 1] [9, 7, 6, 4, 3, 1] [9, 8, 7, 6, 5, 3]
10	1023	[10, 3] [10, 8, 3, 2] [10, 4, 3, 1] [10, 8, 5, 1] [10, 8, 5, 4] [10, 9, 4, 1] [10, 8, 4, 3] [10, 5, 3, 2] [10, 5, 2, 1] [10, 9, 4, 2]
11	2047	[11, 1] [11, 8, 5, 2] [11, 7, 3, 2] [11, 5, 3, 5] [11, 10, 3, 2] [11, 6, 5, 1] [11, 5, 3, 1] [11, 9, 4, 1] [11, 8, 6, 2] [11, 9, 8, 3]
12	4095	[12, 6, 4, 1] [12, 9, 3, 2] [12, 11, 10, 5, 2, 1] [12, 11, 6, 4, 2, 1] [12, 11, 9, 7, 6, 5] [12, 11, 9, 5, 3, 1] [12, 11, 9, 8, 7, 4] [12, 11, 9, 7, 6, 5] [12, 9, 8, 3, 2, 1] [12, 10, 9, 8, 6, 2]
13[a]	8191	[13, 4, 3, 1] [13, 10, 9, 7, 5, 4] [13, 11, 8, 7, 4, 1] [13, 12, 8, 7, 6, 5] [13, 9, 8, 7, 5, 1] [13, 12, 6, 5, 4, 3] [13, 12, 11, 9, 5, 3] [13, 12, 11, 5, 2, 1] [13, 12, 9, 8, 4, 2] [13, 8, 7, 4, 3, 2]
14	16, 383	[14, 12, 2, 1] [14, 13, 4, 2] [14, 13, 11, 9] [14, 10, 6, 1] [14, 11, 6, 1] [14, 12, 11, 1] [14, 6, 4, 2] [14, 11, 9, 6, 5, 2] [14, 13, 6, 5, 3, 1] [14, 13, 12, 8, 4, 1] [14, 8, 7, 6, 4, 2] [14, 10, 6, 5, 4, 1] [14, 13, 12, 7, 6, 3] [14, 13, 11, 10, 8, 3]
15	32, 767	[15, 13, 10, 9] [15, 13, 10, 1] [15, 14, 9, 2] [15, 1] [15, 9, 4, 1] [15, 12, 3, 1] [15, 10, 5, 4] [15, 10, 5, 4, 3, 2] [15, 11, 7, 6, 2, 1] [15, 7, 6, 3, 2, 1] [15, 10, 9, 8, 5, 3] [15, 12, 5, 4, 3, 2] [15, 10, 9, 7, 5, 3] [15, 13, 12, 10] [15, 13, 10, 2] [15, 12, 9, 1] [15, 14, 12, 2] [15, 13, 9, 6] [15, 7, 4, 1] [15, 4] [15, 13, 7, 4]

Table 3.7 ***(Continued)***

Number of Stages	Code Length	Maximal Taps
16	65, 535	[16, 12, 3, 1] [16, 12, 9, 6] [16, 9, 4, 3] [16, 12, 7, 2] [16, 10, 7, 6] [16, 15, 7, 2] [16, 9, 5, 2] [16, 13, 9, 6] [16, 15, 4, 2] [16, 15, 9, 4]
17[a]	131, 071	[17, 3] [17, 3, 2, 1] [17, 7, 4, 3] [17, 16, 3, 1] [17, 12, 6, 3, 2, 1] [17, 8, 7, 6, 4, 3] [17, 11, 8, 6, 4, 2] [17, 9, 8, 6, 4, 1] [17, 16, 14, 10, 3, 2] [17, 12, 11, 8, 5, 2]
18	262, 143	[18, 7] [18, 10, 7, 5] [18, 13, 11, 9, 8, 7, 6, 3] [18, 17, 16, 15, 10, 9, 8, 7] [18, 15, 12, 11, 9, 8, 7, 6]
19[a]	524, 287	[19, 5, 2, 1] [19, 13, 8, 5, 4, 3] [19, 12, 10, 9, 7, 3] [19, 17, 15, 14, 13, 12, 6, 1] [19, 17, 15, 14, 13, 9, 8, 4, 2, 1] [19, 16, 13, 11, 19, 9, 4, 1] [19, 9, 8, 7, 6, 3] [19, 16, 15, 13, 12, 9, 5, 4, 2, 1] [19, 18, 15, 14, 11, 10, 8, 5, 3, 2] [19, 18, 17, 16, 12, 7, 6, 5, 3, 1]
20	1, 048, 575	[20, 3] [20, 9, 5, 3] [20, 19, 4, 3] [20, 11, 8, 6, 3, 2] [20, 17, 14, 10, 7, 4, 3, 2]
21	2, 097, 151	[21, 2] [21, 14, 7, 2] [21, 13, 5, 2] [21, 14, 7, 6, 3, 2] [21, 8, 7, 4, 3, 2] [21, 10, 6, 4, 3, 2] [21, 15, 10, 9, 5, 4, 3, 2] [21, 14, 12, 7, 6, 4, 3, 2] [21, 20, 19, 18, 5, 4, 3, 2]
22	4, 194, 303	[22, 1] [22, 9, 5, 1] [22, 20, 18, 16, 6, 4, 2, 1] [22, 19, 16, 13, 10, 7, 4, 1] [22, 17, 9, 7, 2, 1] [22, 17, 13, 12, 8, 7, 2, 1] [22, 14, 13, 12, 7, 3, 2, 1]
23	8, 388, 607	[23, 5] [23, 17, 11, 5] [23, 5, 4, 1] [23, 12, 5, 4] [23, 21, 7, 5] [23, 16, 13, 6, 5, 3] [23, 11, 10, 7, 6, 5] [23, 15, 10, 9, 7, 5, 4, 3] [23, 17, 11, 9, 8, 5, 4, 1] [23, 18, 16, 13, 11, 8, 5, 2]
24	16, 777, 215	[24, 7, 2] [24, 4, 3, 1] [24, 22, 20, 18, 16, 14, 11, 9, 8, 7, 5, 4] [24, 21, 19, 18, 17, 16, 15, 14, 13, 10, 9, 5, 4, 1]
25	33, 554, 431	[25, 3] [25, 3, 2, 1] [25, 20, 5, 3] [25, 12, 5, 4] [25, 17, 10, 3, 2, 1] [25, 23, 21, 19, 9, 7, 5, 3] [25, 18, 12, 11, 6, 5, 4] [25, 20, 16, 11, 5, 3, 2, 1] [25, 12, 11, 8, 7, 6, 4, 3]
26	67, 108, 863	[26, 6, 2, 1] [26, 22, 21, 16, 12, 11, 10, 8, 5, 4, 3, 1]
27	134, 217, 727	[27, 5, 2, 1] [27, 18, 11, 10, 9, 5, 4, 3]
28	268, 435, 455	[28, 3] [28, 13, 11, 9, 5, 3] [28, 22, 11, 10, 4, 3] [28, 24, 20, 16, 12, 8, 4, 3, 2, 1]

Table 3.7 (*Continued*)

Number of Stages	Code Length	Maximal Taps
29	536, 870, 911	[29, 2] [29, 20, 11, 2] [29, 13, 7, 2] [29, 21, 5, 2] [29, 26, 5, 2] [29, 19, 16, 6, 3, 2] [29, 18, 14, 6, 3, 2]
30	1, 073, 74, 1, 823	[30, 23, 2, 1] [30, 6, 4, 1] [30, 24, 20, 16, 14, 13, 11, 7, 2, 1]
31[a]	2, 147, 483, 647	[31, 29, 21, 17] [31, 28, 19, 15] [31, 3] [31, 3, 2, 1] [31, 13, 8, 3] [31, 21, 12, 3, 2, 1] [31, 20, 18, 7, 5, 3] [31, 30, 29, 25] [31, 28, 24 10] [31, 20, 15, 5, 4, 3] [31, 16, 8, 4, 3, 2]
32	4, 294, 967, 295	[32, 22, 2, 1] [32, 7, 5, 3, 2, 1] [32, 28, 19, 18, 16, 14, 11, 10, 9, 6, 5, 1]
33	8, 589,934, 591	[33, 13] [33, 22, 13, 11] [33, 26, 14, 10] [33, 6, 4, 1] [33, 22, 16, 13, 11, 8]
61[a]	2, 305, 843, 009, 213, 693, 951	[61, 5, 2, 1]
89[a]	618, 970, 019, 642, 690, 137, 449, 562, 112	[89, 6, 5, 3]

[a]Mersenne prime length generator.

and cross checked to ensure accuracy as far as was practical. An example of the use of the table follows.

After selecting a given code length, say 127 chips, we select a seven-stage shift register generator from the table, together with a set of maximal taps for a seven-stage register. For a minimal number of feedback taps we choose [7, 1]. Generators implemented with this feedback appear in Figure 3.21, which shows both SSRG and MSRG configurations. SSRG configurations are denoted $[n, r, q, p]_s$.

For modular shift register generators, the $[n, r, q, p]_m$ notation would be used, and this signifies an n-stage register with feedback from the nth stage modulo-2 added with the output of the rth, qth, and pth stages and fed to the inputs of the $r + 1$st, $q + 1$st stages, respectively.

The two configurations in Figure 3.23 illustrate a point—that in order to generate an identical code sequence, simple shift register and modular shift register generators must be implemented differently. A second example will perhaps reinforce the point and better explain the use of Table 3.7.

Suppose now that another 127-chip code sequence is needed. This time we choose feedback set [7, 3, 2, 1] and construct either $[7, 3, 2, 1]_s$ or $[7, 6, 5, 4]_m$. Again the two configurations shown are equivalent.

As an example of the equivalence of a $[n, p]_s$ and a $[n, n - p]_m$

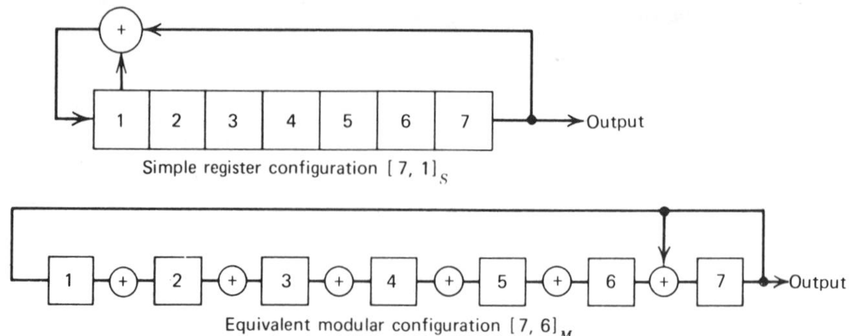

Figure 3.23 Feedback configurations for 127-chip linear *m*-sequences.

generator, we generate the pair of subsequences shown in Table 3.8. Observing the first 15 states of stage one in the $[7, 1]_s$ register and of stage seven in the $[7, 6]_m$ register, we see that both generate an identical subsequence of 15 chips:

$$101010100110011 \cdots$$

For the seven-stage register this is $2n - 1$ chips. It can be readily shown that a sequence of $2n + 1$ chips from a linear sequence generator can be used to derive the feedback in the sequence generator. (See Appendix 6.)

Table 3.8 Comparison of States in $[7, 1]_s$ and $[7, 6]_m$ Sequence Generators

	$[7, 1]_s$	$[7, 6]_m$
State 1	1 1 1 1 1 1 1	1 1 1 1 1 1 1
2	0 1 1 1 1 1 1	1 1 1 1 1 1 0
3	1 0 1 1 1 1 1	0 1 1 1 1 1 1
4	0 1 0 1 1 1 1	1 0 1 1 1 1 0
5	1 0 1 0 1 1 1	0 1 0 1 1 1 1
6	0 1 0 1 0 1 1	1 0 1 0 1 1 0
7	1 0 1 0 1 0 1	0 1 0 1 0 1 1
8	0 1 0 1 0 1 0	1 0 1 0 1 0 0
9	0 0 1 0 1 0 1	0 1 0 1 0 1 0
10	1 0 0 1 0 1 0	0 0 1 0 1 0 1
11	1 1 0 0 1 0 1	1 0 0 1 0 1 1
12	0 1 1 0 0 1 0	0 1 0 0 1 0 0
13	0 0 1 1 0 0 1	0 0 1 0 0 1 0
14	1 0 0 1 1 0 0	0 0 0 1 0 0 1
15	1 1 0 0 1 1 0	1 0 0 0 1 0 1
Stage	1 2 3 4 5 6 7	1 2 3 4 5 6 7

Therefore, because both configurations produce an identical subsequence for $2n + 1$ chips and $2n + 1$ chips is sufficient to establish feedback for a linear sequence generator, the two configurations must be equivalent.

Finally, we must point out that for every set of $[n, \cdots, p]_s$ feedback taps listed in Table 3.6 there is an image set; that is, there is a set of feedback taps that generates an identical code reversed in time sequence. Furthermore, this mirror image of a feedback set $[n, r, \cdots, p]_s$ is $[n, n-r, \cdots, n-p]_s$.

Recalling that $[n, p]_s$ is equivalent to $[n, n-p]_m$, we see that $[n, p]_m$ is the mirror image of $[n, p]_s$, $[n, p]_m$ is equivalent to $[n, n-p]_s$, and $[n, p]_s$ is the mirror image of $[n, p]_m$.

To give a numerical example we select a nine-stage, 511-chip code feedback set $[9, 8, 5, 4]_8$ from our table. If we want the mirror image, we convert to

$$\begin{aligned}[9, 9\text{-}8, 9\text{-}5, 9\text{-}4]_s &= [9, 1, 4, 5]_s \\ &= [9, 5, 4, 1]_s\end{aligned}$$

The same conversion holds for a modular shift register; the modular feedback set would be

$$[9, 5, 4, 1]_m$$

Here the code sequence would not be a mirror image but would be generated by the modular generator as an identical sequence to $[9, 8, 5, 4]_s$.

Caution. No repeating maximal code sequence is produced when two feedback taps (or any even number) besides the nth stage are used. This is obvious if we consider that any three ones (or other odd number) of ones modulo-2 added together produces a one. Thus the all-ones vector could not be permitted and a maximal sequence cannot be generated. If nonmaximal sequences are wanted, however, even numbers of feedback taps may be acceptable.

3.7 GENERATING HIGH RATE CODES

High chip rate code sequences are desirable for many applications. Notable for the purposes of this book are those applications in which high code rates are advantageous for spreading a signal over wide frequency bands. Such is the case when a high data rate signal (wide baseband bandwidth) is to be expanded or when more interference rejection margin is required. Data to be transmitted are often in the megabits-per-second range, and it is obvious that to gain any significant advantage from band spreading the code sequences used must be in the hundreds of megachips range.

Shift register sequence generators have been constructed from most of the known delay elements by those striving for faster code rates, smaller size, or some parameter of interest. Tunnel diodes, delay lines (coaxial acoustic, stripline, etc.), silicon-controlled rectifiers, transistors, vacuum tubes, and even relays have been used. Recent developments in microelectronics have produced logic elements with such flexibility, speed, and general utility that it is now difficult to justify generating code sequences by any other method. Three-hundred-megachip operation has been exceeded by shift registers made up of microelectronic elements. There is little reason to doubt that future circuits and subsystems will include complete sequence generators capable of 100-Mcps clock rates in the same size presently occupied by a single flip-flop.

For the purposes of this book the import of microelectronics is such that we have not (for digital circuits) even considered the use of discrete elements (i.e., transistors). Instead, flip-flops, gates, and other elements are treated as blocks and are assembled to make up the desired code sequence generators.

In a preceding section, we discussed the advantages of modular shift register connections for high-speed code generation. Here we consider modular, simple, and partially modular construction (i.e., with modulo-2 adders inserted between some stages rather than at every stage). Toggle registers are not specifically considered because they generate codes at the operating rate of the flip-flops themselves without special provision for high-speed capability; they are special cases of the modular SRG's and they lack the repertoire of codes available that either the simple or modular generators are able to produce.

Figure 3.24 is a diagram of a seven-stage simple shift register generator made up of ECL elements capable of generating 127-chip ($2^7 - 1$) codes well in excess of 200 Mcps, depending on the layout of the circuit. Figure 3.25 shows a sector of the 127-chip code at a 150-Mcps rate. In high-speed sequence generators layout is extremely critical, because the propagation time in wiring approaches that in the circuit elements. Extending the length of the generator in Figure 3.24 to a useful length and redesigning to provide a larger number of code sequences could slow down the code rate significantly.

Emitter-coupled logic elements that have been used to construct code sequence generators capable of operating to well in excess of 200 Mcps are available. Some instances are known in which simple-type ECL generators have operated beyond 300 Mcps. Figure 3.26 shows a 200-Mcps 15-chip code sequence generated by a four-stage MECL III generator.

The maximum chip rate of a code generator is determined not only by the rate of its shift register stages but by any delay in its feedback network.

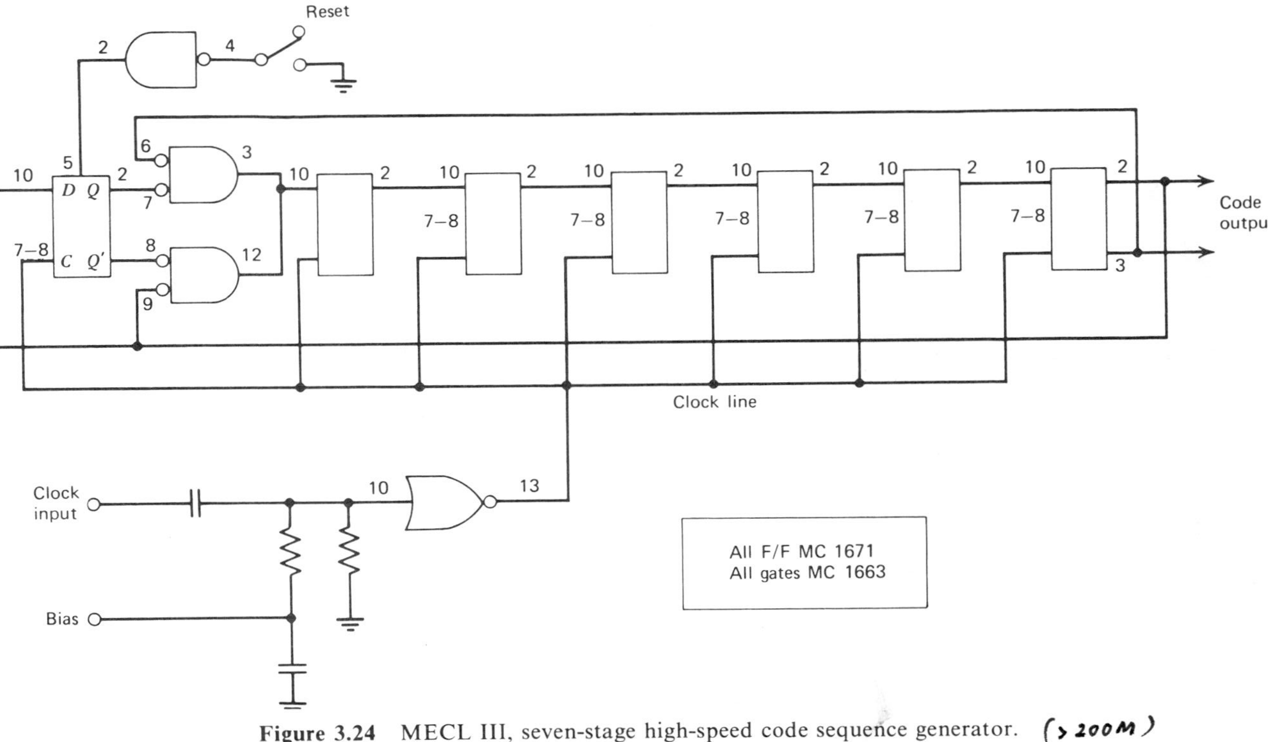

Figure 3.24 MECL III, seven-stage high-speed code sequence generator. (> 200M)

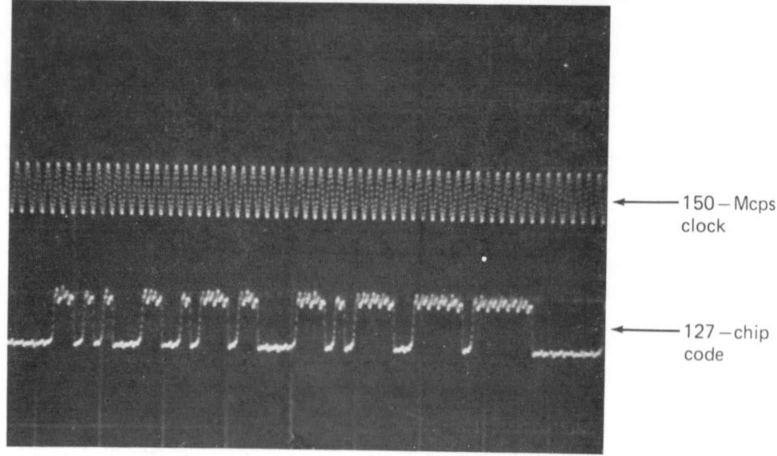

Figure 3.25 127-chip ($2^7 - 1$) code sequence at 150 Mcps.

Because the feedback network output is the next-state information for some of the register stages, both the flip-flops used as feedback points and all modulo-2 adders must complete their operation before the next shift time; that is, the maximum code rate for a shift register generator is the rate at which

$$R_{\text{max}} \approx \frac{1}{\tau_R + \tau_M} = \frac{1}{\tau_K}$$

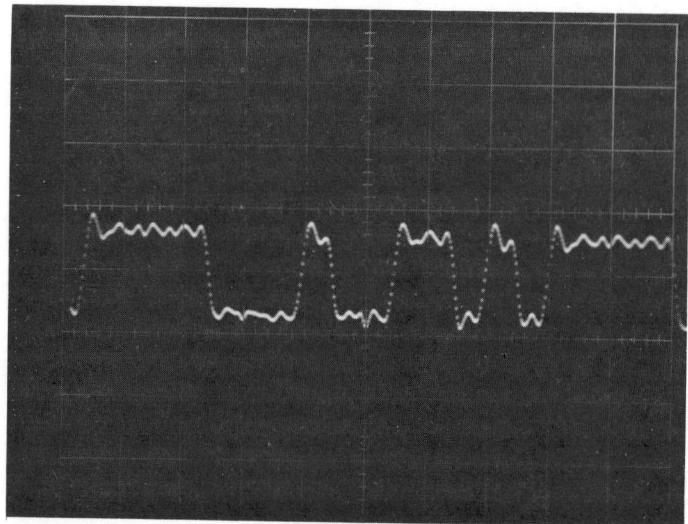

Figure 3.26 15-chip ($2^4 - 1$) code sequences at 200 Mcps.

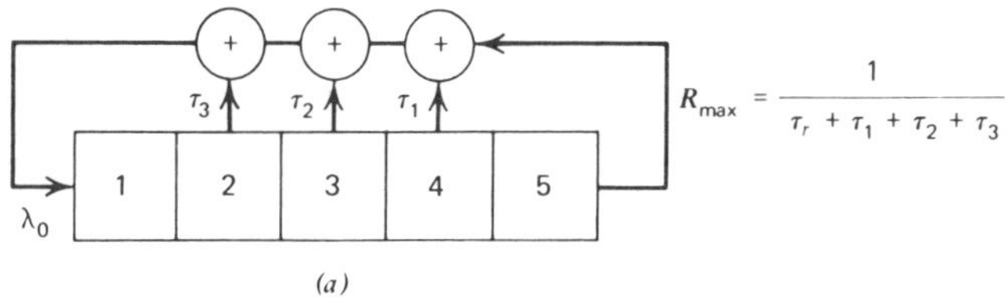

(*a*)

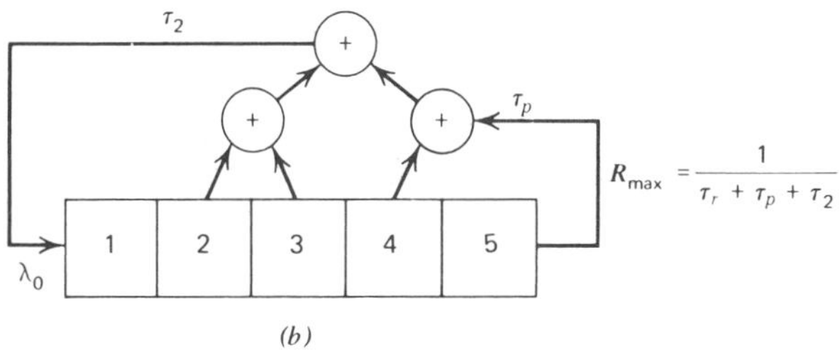

(*b*)

Figure 3.27 Comparative shift rates for two simple SRG configurations: (*a*) series connection; (*b*) series–parallel construction.

and τ_R = time for a shift register stage to reach its next state,
τ_M = propagation time through the feedback network,
τ_K = period of the clock pulse.

The simple SRG with one modulo-2 adder in the feedback network is the fastest possible configuration.* The modular SRG in its simplest form is equivalent to it. When it is necessary to use three, five, or more (always odd) feedback taps, however, the modular register quickly becomes the fastest. (Observation of the table of codes given in this chapter will show that single-tap SRG's are the exception rather than the rule.) The multitap simple SRG again has

$$R_{\max} \approx \frac{1}{\tau_R + \tau_M}$$

but $\tau_M = \tau_{M_1} + \tau_{M_2} + \cdots + \tau_{MK}$. Proper arrangement of the feedback network can speed up the operation of the simple SRG by allowing parallel–series addition, as in Figure 3.27, in which the feedback network of configuration (*b*) is the equivalent of (*a*), but only two levels of logic

*This includes the toggle registers. Toggle SRG's are no faster than simple registers with single-layer feedback logic. After all, a toggle register is a single-layer feedback configuration in which the logic is a part of the delay elements because of their construction.

are in series in the network. Writing the feedback equations

$$\lambda_a^0 = (\lambda^5 \oplus \lambda^4) \oplus \lambda^3 \oplus \lambda^2 = \lambda^5 \oplus \lambda^4 \oplus \lambda^3 \oplus \lambda^2$$
$$\lambda_b^0 = (\lambda^5 \oplus \lambda^4) \oplus (\lambda^3 \oplus \lambda^2) = \lambda^5 \oplus \lambda^4 \oplus \lambda^3 \oplus \lambda^2$$
$$\lambda_a^0 = \lambda_b^0$$

shows their equivalence.

A modular SRG has a maximum operating speed that is a function of the switching time of one flip-flop stage plus one modulo-2 adder, no matter how many feedback points are used. The equivalence of modular and simple SRG's has been shown (also see Appendix 3). A parallel–series feedback configuration is not available, so that one modulo-2 adder is required for each feedback tap; however, the modular SRG is inherently capable of higher code rates in multitap configurations.

Because of the criticality of direct high-speed code generators, several techniques for multiplexing codes have been worked out with less-complex SRG configurations. Two of these composite code generators, which offer advantages for high-speed operation, are the Gold and syncopated coders.

A BPSK-modulated signal using the seven-stage high-speed MECL III code generator in Figure 3.24 is shown in Figure 3.28. Here the code rate is 100 Mcps and the main lobe null-to-null bandwidth is 200 MHz. A system employing this RF bandwidth and having a 10 kpbs information rate would have an inherent process gain of 17,600 or 42 dB.

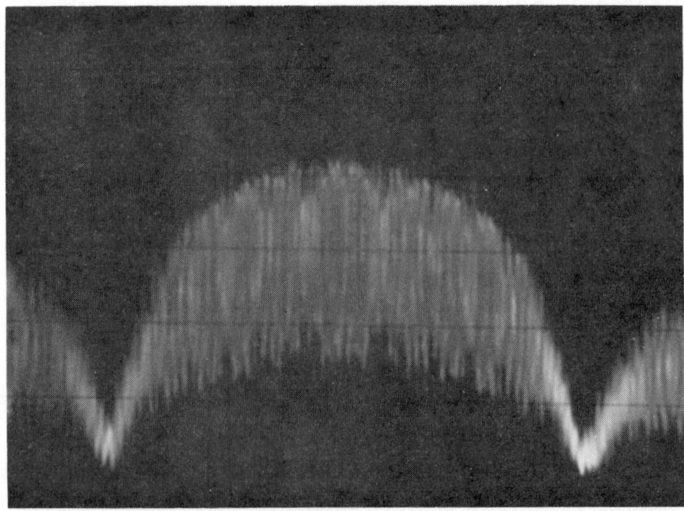

Figure 3.28 Biphase-modulated 300-MHz carrier modulated by a 200-Mcps code. (Code length 127 chips = $2^7 - 1$.)

We must reemphasize here that the 127-chip code shown is not capable of supporting a high process gain. A 42 dB process gain system would require a much longer code. To show the inadequacy of a short code such as this it is necessary only to compare Figures 2.7 and 3.28. These spectra were photographed on two different days with different code rates, yet they are identical. This illustrates the lack of randomness in a short sequence and reinforces the conclusion that just any code will not do. The code sequence and the code rate must be fitted to the task at hand.

3.8 SOFTWARE-IMPLEMENTED CODES

It is, of course, quite practical to use a more general purpose machine such as a calculator or a computer to generate codes. In some cases, it may not be practical to do so because of other loading, but where machine time is available the flow diagrams shown in Figures 3.29 and 3.30 illustrate techniques for generating maximal and Gold codes, respectively.

The simpler of the two diagrams, for maximal codes, requires only that a number of memory locations equal to the degree of the code be set aside (one location per register stage.) Then, with all locations set to one, count the number of ones in the stages from which feedback is taken. If there is

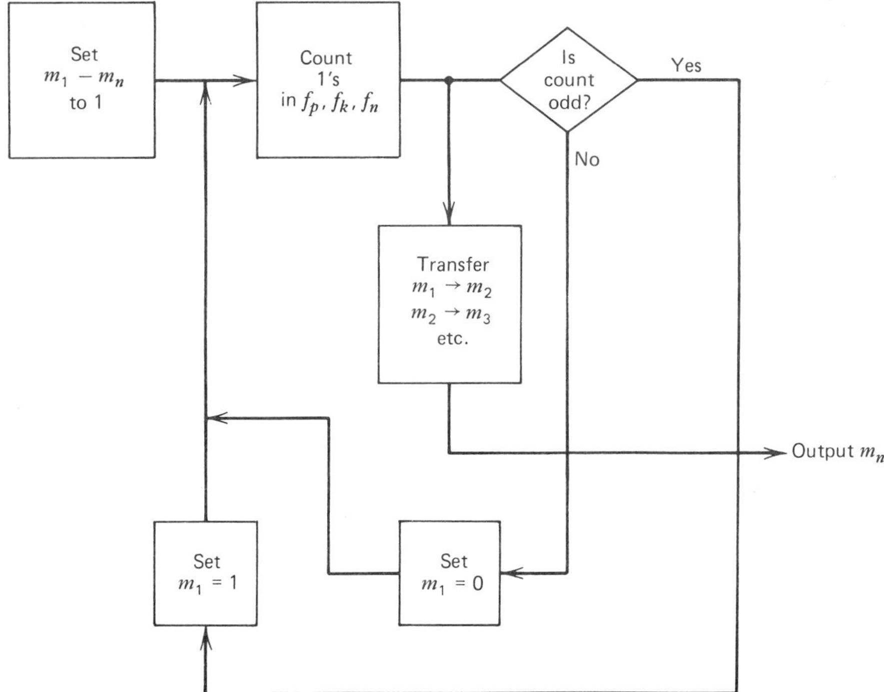

Figure 3.29 Flow diagram for m-sequence generator.

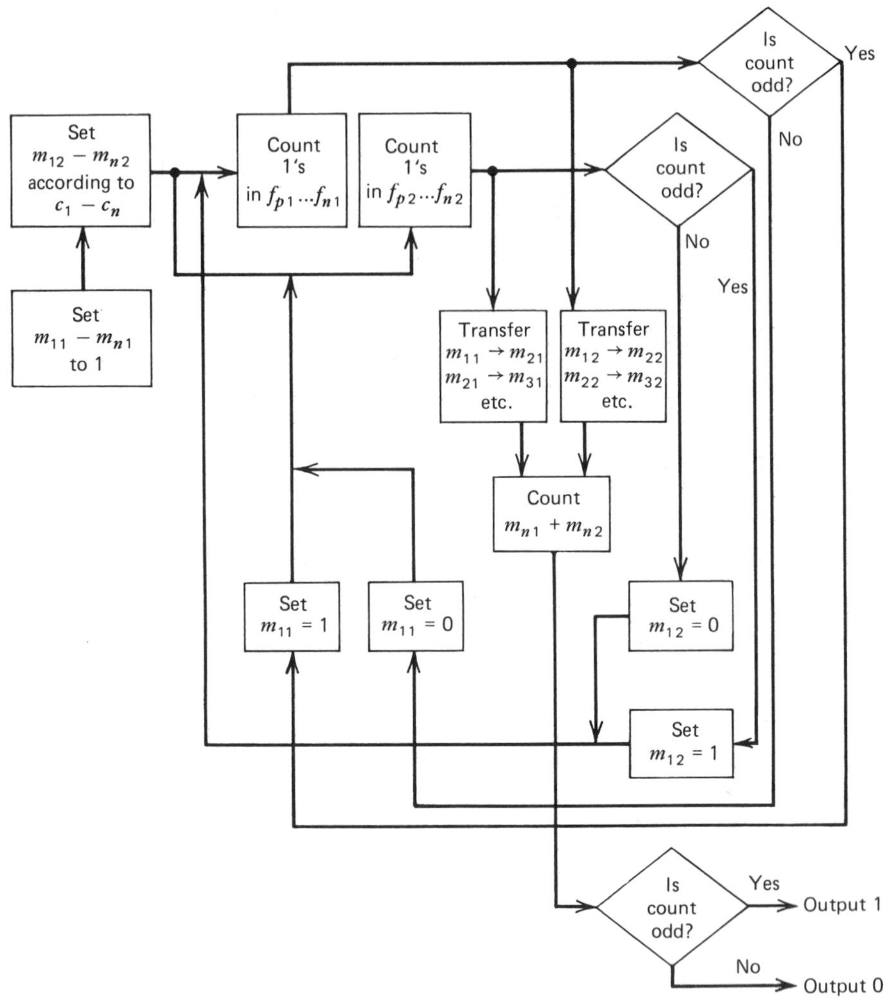

Figure 3.30 Flow diagram for Gold sequence generator.

an odd number of ones, set the first stage to one and shift all to the right. If the number of ones is even, set the first stage to zero and again shift to the right. Repetition of this procedure will produce a code at a rate equal to the algorithm repetion rate.

Figure 3.30 shows a dual maximal sequence generator configuration capable of generating Gold codes.

3.9 CODE SELECTION AND SIGNAL SPECTRA

The following series of photographs illustrates some of the effects of code selection on direct sequence (BPSK and QPSK) spectra. Both "good"

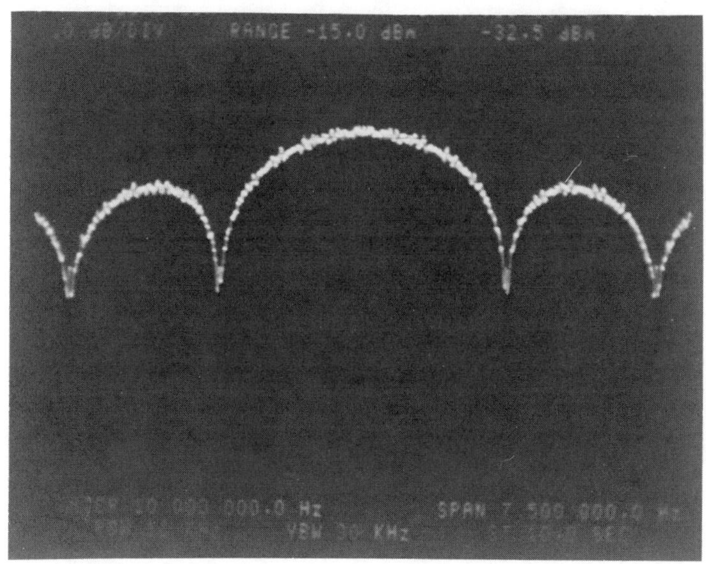

Figure 3.31 BPSK-modulated carrier, 8191-chip linear maximal code. (Note symmetry around center frequency.)

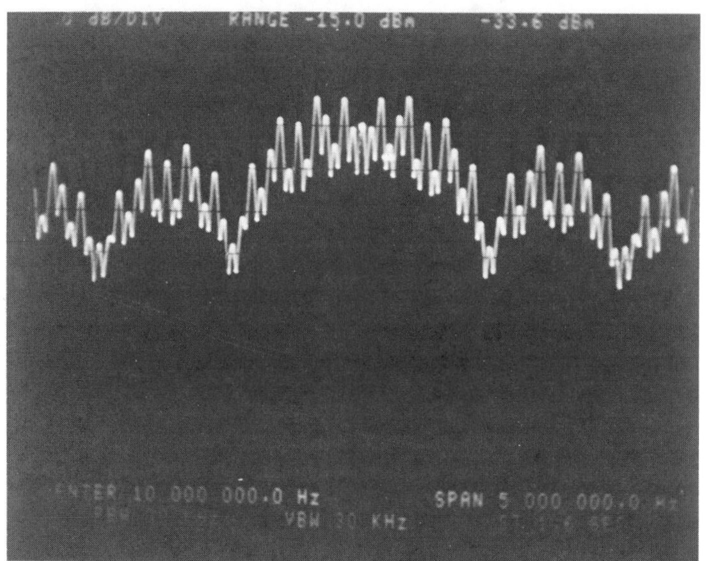

Figure 3.32 BPSK modulation by 14-chip nonmaximal code from 13-stage generator (period 14 μsec).

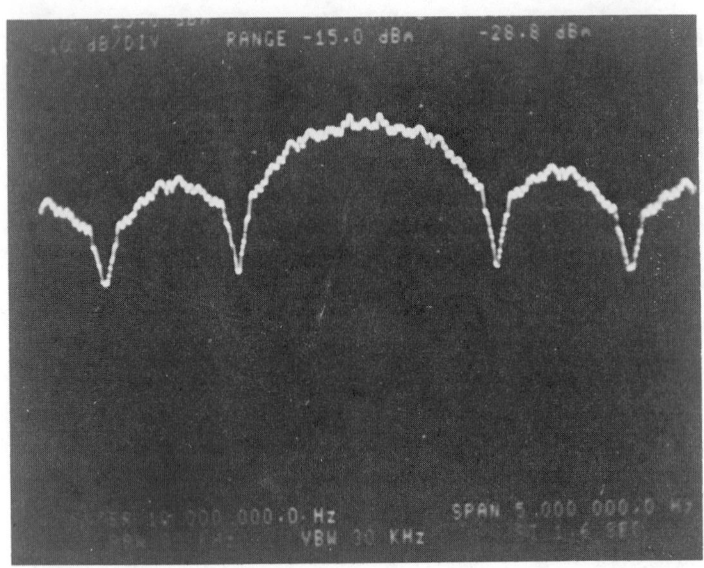

Figure 3.33 BPSK modulation by 105-chip nonmaximal code from 13-stage generator (period 105 μsec).

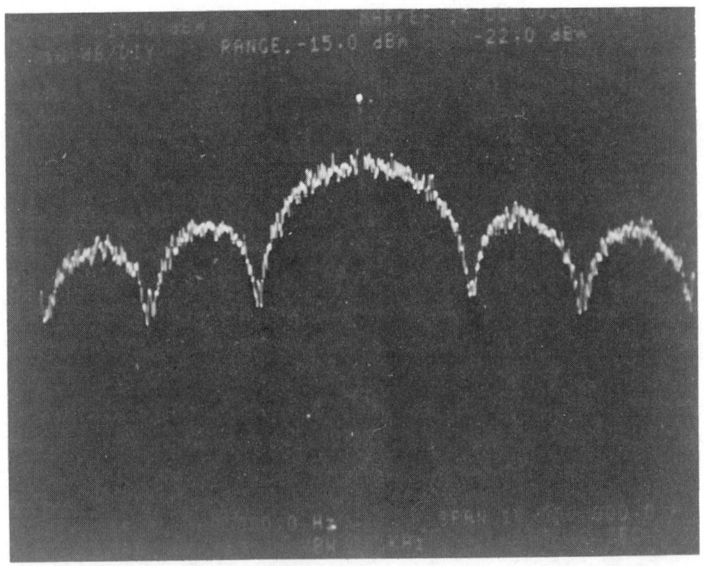

Figure 3.34 QPSK modulation by same code as in Figure 3.31. Codes taken from stages 13 and 4. Marker added at center frequency.

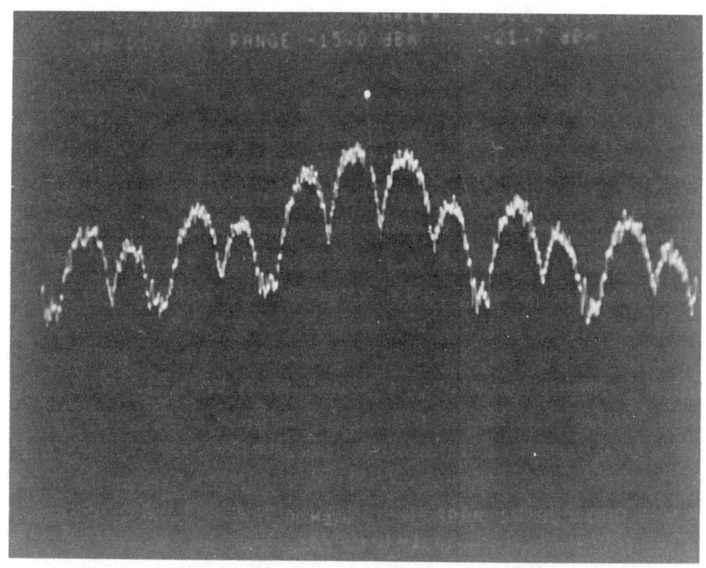

Figure 3.35 QPSK modulation by codes taken from stage 13 and 11 of 13-stage generator.

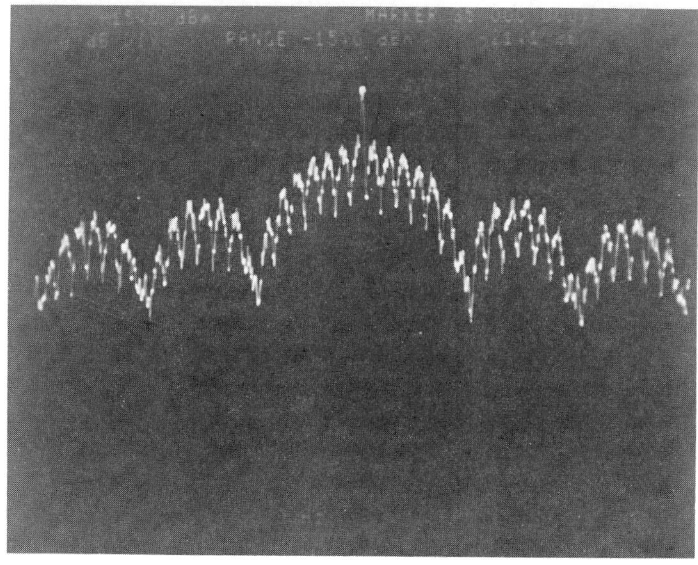

Figure 3.36 QPSK modulation by codes taken from stages 13 and 6 of 13-stage generator.

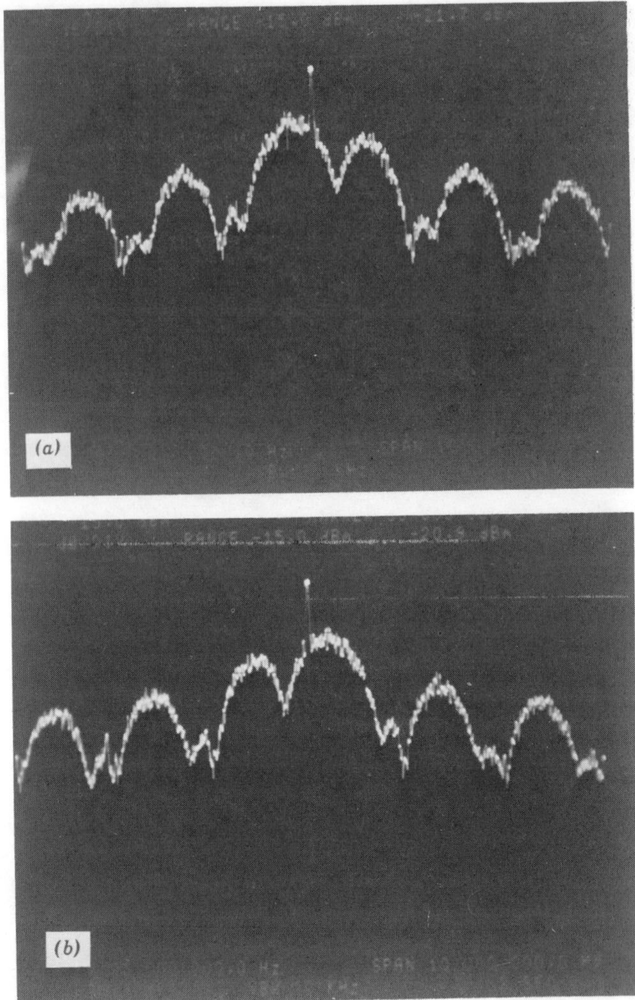

Figure 3.37 QPSK modulation showing complementary asymmetry when codes are taken from $[n, n - 1]$ and $[n, 1]$ stages: (*a*) codes taken from 13 and 12; (*b*) codes taken from 13 and 1.

and "bad" spectra are shown, so that the reader can recognize the effects of generator failure and other phenomena. The first photograph, Figure 3.31, is of the spectrum of a BPSK-modulated carrier, where the code length is $2^{13} - 1$ (8191 chips). The code used is $[13, 4, 3, 1]_m$. A point of interest in Figure 3.31 is the precise symmetry of the spectrum around the center frequency, which shows that BPSK direct sequence signals are no different from others. That is, they generate symmetrical upper and lower sidebands.

An illustration of the possible effects of failure in a feedback element is given in Figures 3.32 and 3.33. If a feedback element fails, or a connection is wrong, a shift register generator may output a much shorter code than the linear maximal code desired, and this can cause the modulated spectrum to be quite different from that which is expected.

For quadriphase modulation, where two code sequences are required, it is possible to take both sequences from the same generator. Some care must be exercised when doing so, however, as is shown in Figures 3.34 through 3.37. The first of the QPSK-modulated carrier spectrum photographs is shown in Figure 3.34. Here the codes driving the QPSK modulator are taken from stages 13 and 4. The spectrum produced has the same shape as the previous BPSK spectrum (the same code generator connections are also used). The QPSK spectrum no longer displays the precise symmetry of the BPSK spectrum however.

The asymmetry of the QPSK spectrum is dramatically illustrated in Figures 3.35 through 3.37. In these photographs, the codes used have been taken from stage 13 together with various other taps on the generator. In none of these is the modulation symmetrical around the center frequency. Note, however, that the number of minor lobes in the one-sided main-lobe bandwidth is equal to the number of stages separating the code output points on the shift register. This illustrates the effects of the strong correlation between the codes, which are delayed replicas of one another. In Figure 3.37, the two individually unsymmetrical but mirror-image replica spectra are those produced by QPSK modulating the carrier with codes separated in time by ± 1 chip time (i.e., in Figure 3.36*a*, stage 12 is one chip time ahead of stage 13, and in Figure 3.36*b*, stage 13 is one chip time ahead of stage 1).

3.10. CODE BASEBAND SPECTRA

The code baseband spectrum can be observed, and is seen to have exactly the same one-sided distribution of frequency components as the two-sided, modulated carrier. A spectrum analyzer having sufficient capability can show not only the overall envelope of the frequency components, but also the fine structure of the code spectrum. Figure 3.38 shows the baseband spectrum of 8191 chip code $[13, 4, 3, 1]_m$ at approximately a 1.5-Mcps rate. This spectrum has lobes from 0 to 20 MHz (and beyond), and code imbalance which is shown by the even harmonic content exhibited by the stationary signals at the nulls.

Fine structure of this same code's spectrum is shown in Figure 3.39. Figure 3.39*a* shows the main lobe between 0 and 10,000 Hz, with a 100-Hz resolution bandwidth. This structure is expanded further in Figure 3.39*b*, with resolution bandwidth of 30 Hz and frequency span of 0–1 kHz. The lines shown are separated by the repetition rate of the code itself. In this case $(1.5 \times 10^6)/8191 = 183$ Hz. This demonstrates very aptly that the

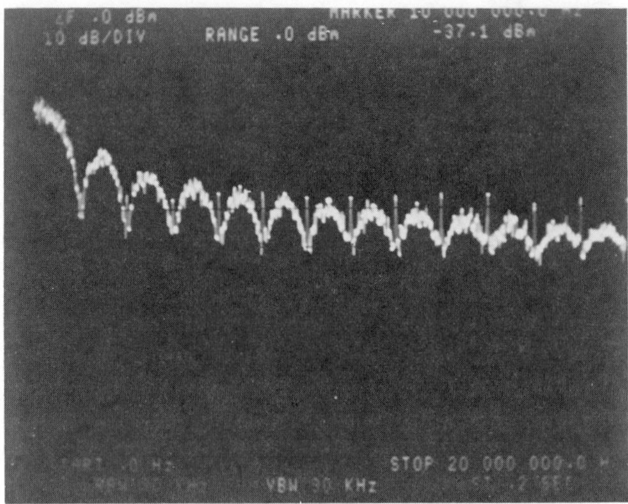

Figure 3.38 Baseband code spectrum of 8191 chip code at 1.5 Mcps.

codes used in spread spectrum systems are noiselike to a casual observer but are deterministic and decidedly not Gaussian.

3.11 ERROR DETECTION AND CORRECTION CODES

Error detection and correction (EDAC) codes are mandatory for use in frequency hopping systems in order to overcome the high error rates induced by partial band jamming and the efficiency (to the jammer) of such a jamming strategy. The code's use, however, must be carefully defined, as error detection and correction codes do not necessarily work well under all conditions. That is, many EDAC codes have thresholds that must be exceeded before satisfactory performance is achieved.

In direct sequence systems, error detection and correction coding may not be advisable in some applications. This is because of the effect of the coding overhead, which increases the signal's apparent data transmission rate and may increase the jamming threshold level. Some demodulators (CVSD, for example) can operate to bit error rates of approximately 1×10^{-2} without EDAC, and since the threshold of EDAC decoders is often near 1×10^{-2}, it may not be worthwhile to include a complex coding and decoding scheme in a system.

Table 3.9 gives results for a rate $\frac{1}{2}$ convolutional code with Viterbi decoding at various bit error rates between 1×10^{-2} and 1×10^{-6}. At these bit error rates, the improvement (or coding gain) seen varies between 2.2 and 5.3 dB. The problem for spread spectrum systems is that unless RF bandwidth is increased when an error-correcting code is added, the doubled transmission rate halves the available processing gain. Since the

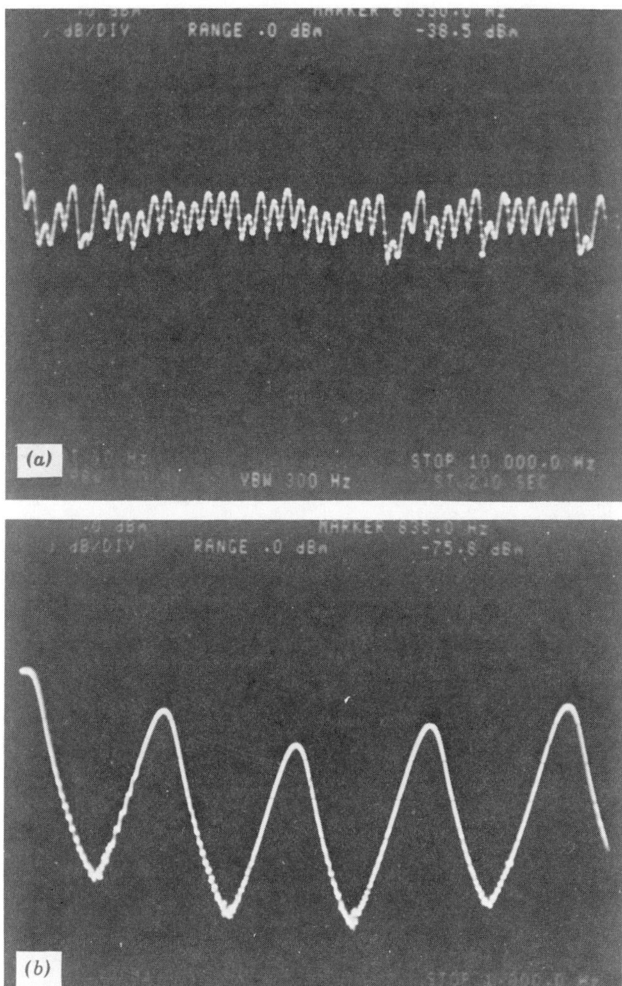

Figure 3.39 Expanded baseband spectra of 8191 chip code at 1.5-Mcps rate: (*a*) span, 0–10 kHz; resolution BW = 100 Hz. (*b*) Span, 0–1 kHz; resolution BW = 30 Hz.

processing gain (and jamming margin) are reduced by 3 dB by doubling information rate while holding RF bandwidth constant, any EDAC technique employed must have coding gain of 3 dB or more, or it is worse than useless. Table 3.8 shows that the net coding gain, for a spread spectrum system using rate $\frac{1}{2}$ convolutional codes and Viterbi decoding, is a loss of 0.8 dB when bit error rate is 1×10^{-2} and only 0.8 dB when bit error rate is 1×10^{-3}. Owing to the complexity of adding such coders and decoders, the modest net of 2.3-dB gain when bit error rate is 10^{-6} may not be enough to justify their use in spread spectrum applications. Nevertheless, coding for error detection and correction has found

Table 3.9 Measured Performance of Rate $\frac{1}{2}$, Constraint Length 7, Convolutional Code/Viterbi Decoder

Bit Error Rate	Eb/NO (dB) Without Coding	Eb/NO (dB) With Coding	Coding Gain (dB)	Loss in Jamming Margin	Net
10^{-2}	4.2	2.0	2.2	3.0	−0.8
10^{-3}	6.8	3.0	3.8	3.0	0.8
10^{-4}	8.4	3.8	4.6	3.0	1.6
10^{-5}	9.6	4.5	5.1	3.0	2.1
10^{-6}	10.5	5.2	5.3	3.0	2.3

application in some spread spectrum systems. Coding that has been employed is basically:

Frequency Hopping (or Frequency Hopping/ Direct Sequence)	Direct Sequence
BCH	Golay
Reed–Solomon	BCH
Convolutional	
Redundant Transmission	

PROBLEMS

1. Given a shift register of 23 stages, what is the longest code sequence that can be generated with a single linear feedback network?
2. How many maximal linear code sequences are available from a 17-stage sequence generator?
3. What code sequence would be produced by a code sequence generator having the equation $D_0 = D_4 \oplus D_2$ and initial conditions $D_1 = 1$, $D_2 = 1$, $D_3 = 0$, $D_4 = 1$?
4. For a Gold code sequence generator how many nonmaximal codes are available if the basic registers employed are each 11 stages long?
5. What are the maximum and minimum autocorrelation values for a $[5, 2]_s$ code?
6. What code sequences could be produced by a $[7, 2]_s$ generator?
7. Draw block diagrams for $[9, 6, 4, 3]_s$ and $[9, 8, 4, 1]_m$ sequence generators.
8. Construct a cross-correlation plot of a $[5, 3]_s$ and a $[5, 4, 3, 2]_s$ code.

9. If a $[17, 3]_s$ code sequence generator is operated at a 5-Mcps clock rate, its repetition period would be how many seconds? Line spacing in a direct-sequence spectrum produced by modulation with this code would be how many hertz?
10. How many runs of three ones would be expected in a $2^{13} - 1$ chip m-sequence?
11. How many runs of three zeros would be expected in the same sequence?
12. What alternate Gold configurations might be employed if at least one thousand 4095-chip code sequences are needed?
13. What is the shortest generator that could be used to provide the needed sequences if all must be maximal?
14. Show that the shift-and-add operation is closed for any $2^5 - 1$ chip m-sequence.
15. Given a maximal shift register sequence generator, show that the all-zero stage cannot be allowed.
16. Show the effect of an even number of feedback taps (in addition to the last stage) on a shift register sequence generator.
17. How may one synchronize to a pseudorandom code sequence to check its correctness?

4

MODULATION AND MODULATORS: GENERATING THE WIDEBAND SIGNAL

The wideband signals characteristic of spread spectrum systems are generated in different ways in the basic techniques of interest here. In a direct sequence system the encoding signal is used to modulate a carrier, usually by phase-shift keying at the code rate; that is, for a "one" in the code pattern, one carrier phase is transmitted and for a "zero" a different carrier phase is sent. More complex phase-shift mappings are also used (such as quadriphase), but in each there is a one-to-one correspondence between the carrier phase transmitted and the code reference or references. It is of some interest to note that balanced modulation* is often used, because this suppresses carrier components and helps to hide the signal. In addition, no power is wasted in transmitting a carrier that would contribute nothing to interference rejection or information transfer.

Frequency hopping systems generate their wideband signal spectra in a different manner from direct sequence systems, although the modulation is still dictated by a code sequence. The relation of the code sequence to the RF output is such that it commands the frequency of a carrier signal rather than directly modulating the carrier; that is, for a given code state the system transmits at one frequency and for any other code state transmits at a different frequency. In each state the code commands the system output frequency. Therefore, the carrier "hops" from one frequency to another in synchronism with code state changes.

*Balanced modulation multiplies a carrier $A \cos w_c t$ with a modulation term $B \cos w_m t$. The resulting signal is $AB \cos w_c \cos w_m t = AB \cos w_r + w_m t$, for which there is theoretically no carrier term. In practice carrier suppression for such modulators is between 20 and 60 dB.

Typically, a frequency hopping system may have a few thousand available frequencies to choose from and a code sequence generator with at least that number of states; for example, a system with a frequency synthesizer capable of outputing any one of 10,000 frequencies separated by 1 kHz would have an RF bandwidth of 10 MHz. Its code generator should then be able to generate a code sequence at least 10,000 chips long so that each of the frequencies available can be used. A 14-stage shift register sequence generator could be used for this purpose because it can generate sequences up to $2^{14} - 1 = 16{,}383$ chips. The wideband signal spectrum from a frequency hopper is not generated by carrier modulation like the direct sequence signal. Instantaneously, the system has only one output rather than a set of symmetrically distributed outputs.

Information is normally embedded in the code sequence that determines frequency for a frequency hopper or spectrum spreading for a direct sequence system. Therefore, it does not matter which type of spread spectrum system one is dealing with, from the standpoint of information transmission, if the information to be sent is a part of the code sequence. The same code might be used for a direct sequence biphase or quadriphase modulator or at a lower rate to determine the frequencies over which a frequency hopper spreads its output.

The point is that both direct-sequence and frequency-hopping systems generate wideband signals under the control of a code sequence generator. In one case, the code is used as direct carrier modulation (direct sequence) and in the other (frequency hopping) the code commands the carrier frequency. This chapter shows spectrum spreading subsystems and circuits typical of those used in practical systems and discusses the problems most often encountered.

4.1 BALANCED MODULATION

An important tool in any suppressed carrier system is the balanced modulator used to generate the transmitted signal. Balanced modulation is employed in many applications in which either one or both of the signals applied to a modulator must be suppressed at its output. Balanced modulation for carrier suppression is at least 60 years old; a patent for this purpose was filed by J. R. Carson in 1915.

Here we intend to show the specific application of balanced modulation in direct sequence spread spectrum systems. We also illustrate the effects of various circuit problems on the output signal.

Almost as many kinds of balanced-modulator (or balanced mixer, depending mainly on who is using it) circuits exists as there are applications. Recent advances in semiconductor diodes (in particular, hot carrier diodes and wideband transformers) and their application to ring-configured diode modulators[1019,1022,1023] (see Figure 4.1) have generally

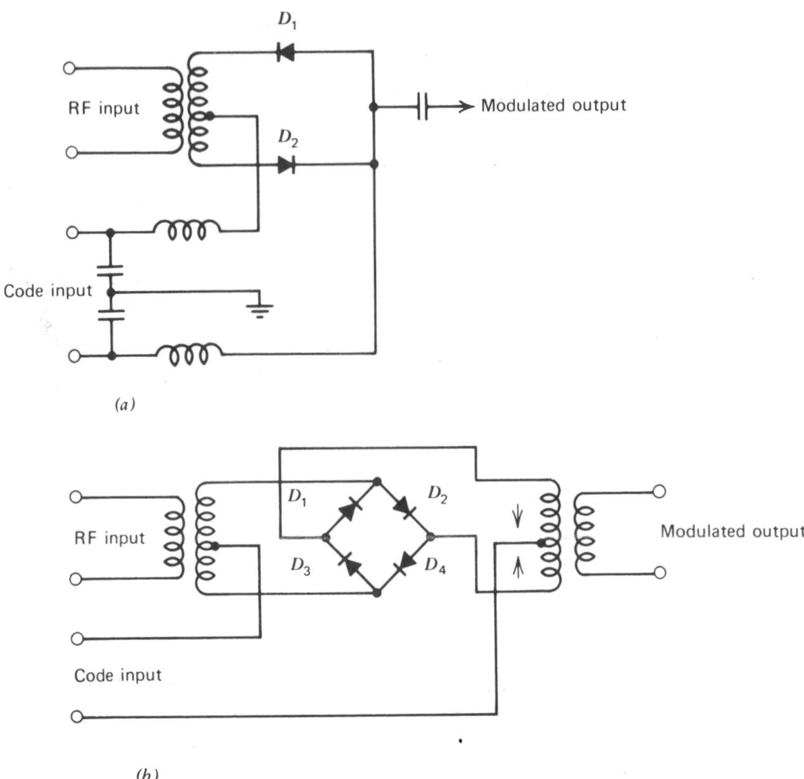

Figure 4.1 Diode-balanced modulator configurations: (*a*) typical single-balanced mixer/modulator; (*b*) typical double-balanced mixer/modulator schematic diagram.

made obsolete all others for most applications. Therefore, our emphasis is on the diode modulator. Commercially available single- and double-balanced diode modulators cover most frequency ranges and offer excellent performance from the standpoint of input signal suppression, low loss, and stable operation.

A simple form of balanced mixer/modulator is a double throw switch, connected so that it switches between two 180°-separated input signals at a modulating signal rate. This configuration would work quite well for digital data modulation and in fact is the same in function as a modulo-2 adder. Figure 4.2 shows a balanced mixer output signal as modulated by a binary switching signal. (The binary switching signal is for our purposes a code sequence, with embedded information. This is true for biphase or quadriphase signals. Note that since the code includes the transmitted information, the single spectrum spreading, balanced-modulation operation satisfied both information transmission and RF signal spreading requirements.)

For code-modulated RF signals the balanced-modulator output is a

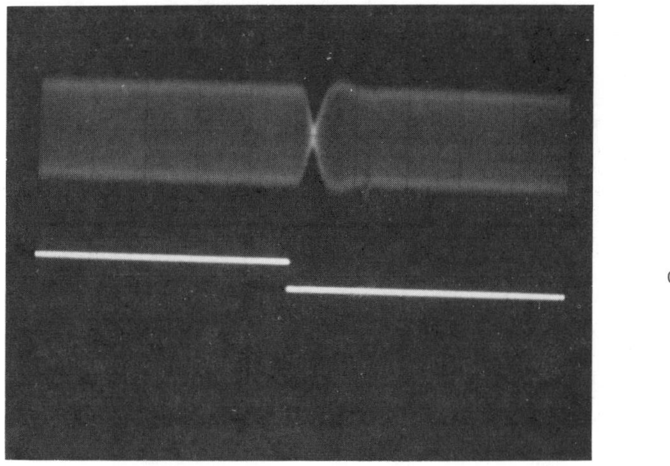

Figure 4.2 An actual balanced mixer output signal.

square wave modulated carrier rather than the familiar sine-wave-modulated balanced signal. For sine wave modulation, the signal may be expressed as

$$A \cos \omega_c t \; B\cos \omega_m t$$

Similarly for data modulation the signal is

$$A \cos \omega_c t \; B\cos \omega_s t$$

where ω_s is a time-varying set of frequencies harmonically related to the data bit rate and run lengths.

Output waveforms from sine-wave- and data-modulated balanced modulators are shown in Figure 4.3.

Spectral outputs of the two modulating signals are $AB\cos(\omega_c + \omega_m)t$ for sine wave modulation in which only two sidebands are represented and $AB\cos(\omega_c + \omega_s)t$ for data. Those familiar with single-sideband systems will recognize the signal in Figure 4.3*a*.

The code-modulated spectrum is not a single pair of sidebands but a set of symmetrical $[(\sin x)/x]^2$-distributed sidebands due to the many code frequency components. In any case, the carrier signal is suppressed an amount that depends on the design of the balanced modulator and on the carrier waveform.

When a code is used to generate a spread spectrum output from a balanced mixer/modulator, a great deal of care must be exercised to provide both carrier and code clock suppression. Figures 4.4 through 4.10 are actual photographs of spread spectrum output signals. The single-balanced mixer of Figure 4.1*a* suppresses only the RF input by inverting

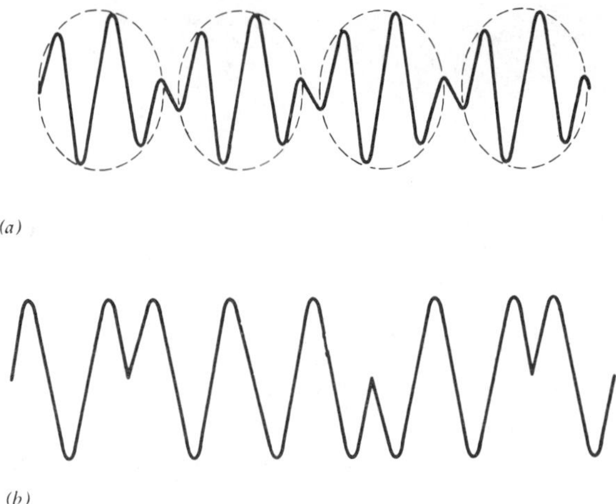

Figure 4.3 Balanced modulator output waveforms: (*a*) sine-wave-modulated balanced mixer output signal; (*b*) data (code)-modulated balanced mixer output signal.

the phase of the RF carrier at a code-determined rate. A code input turns on either diode one (D_1) or diode two (D_2) and turns off the other diode, depending on the relative polarity of the input. Because the code input is complementary, one input is always at a higher DC level than the other, and if the difference between the two exceeds a diode's forward voltage drop, one of the two will conduct. (In practice, the code input voltage should be higher than the carrier input level.) Diode conduction then allows one phase of the RF carrier to pass to the output.

For improved overall input suppression, the double-balanced configuration of Figure 4.1*b* is recommended. Here a diode bridge is used and an output transformer is added. Operation is similar to the single-balanced mixer/modulator in that the polarity of the code input lines determines whether D_1 and D_4 or D_2 and D_3 are conducting and thereby determines which RF carrier phase appears at the output. In addition, the balanced windings on the transformers allow suppression of the code input signal at the output. This suppression occurs because code-related current is in opposition in both input and output transformers and thus tends to cancel.

The spectrum of a properly operating biphase modulated signal is shown in Figure 4.4. The power distribution of this signal is $[(\sin x)/x]^2$, with line separation within this distribution equal to the repetition rate of the code modulating signal; for example, if a 1023 chip code with a 1 Mcps rate were used, the code would repeat $(1 \times 10^6)/1023 = 977$ times per second and the line separation in the modulated signal would be 977 Hz. Because of the pseudorandum character of the code, however, the appearance of any particular spectral component in time is also pseudorandom. Therefore, the direct sequence modulated signal appears to be noise with a $[(\sin x)/x]^2$ distribution.

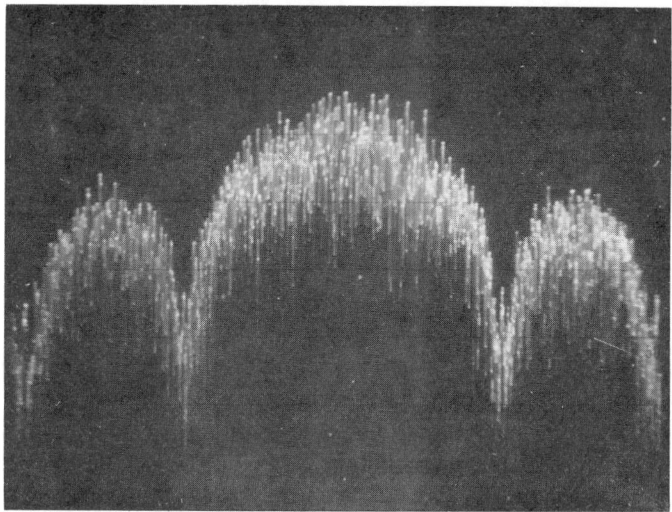

Figure 4.4 Satisfactory balanced mixer/modulator output.

Figure 4.5 illustrates the effect of poor carrier balance, which could be caused by imbalance in the transformers of Figure 4.1, an open diode, or any other cause that might give the effect of unequal 180° carrier inputs. The two phases of the carrier must also be well matched to provide satisfactory balance. (In the diode mixer/modulator of Figure 4.1 the two carrier phases are provided by the input transformer.) Either amplitude inequality or less than 180° phase shift can cause the carrier to be present in the modulated output.

Figure 4.6 is a phasor diagram of the modulated output, which is

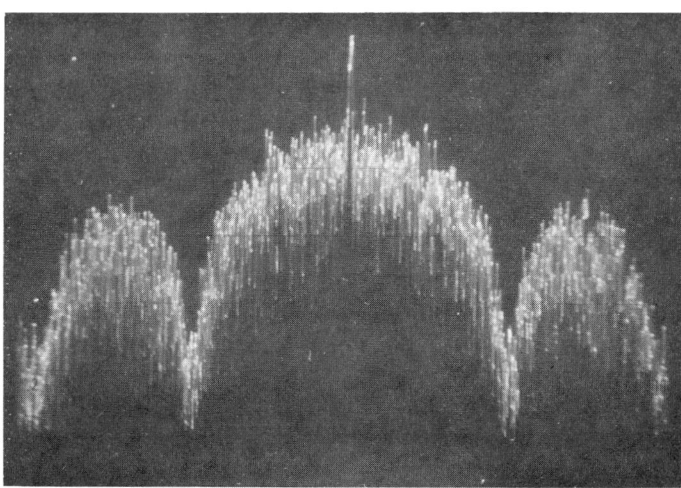

Figure 4.5 Balanced mixer/modulator output showing lack of carrier balance.

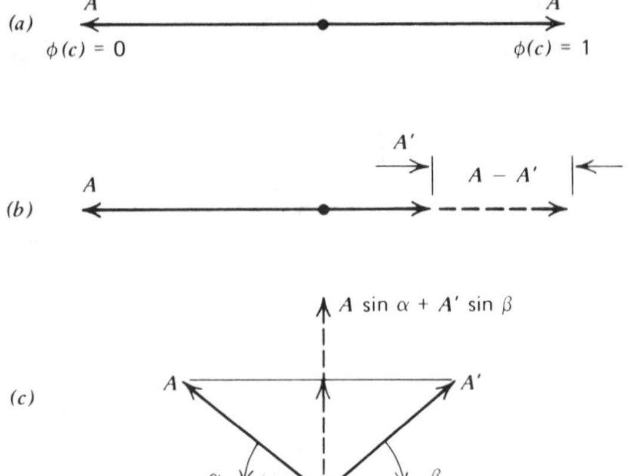

Figure 4.6 Vector diagram of biphase-balanced modulation.

recognized as antipodal. The carrier output that results is zero when the $\phi(c)$ corresponding to the one and zero code inputs are balanced and in phase opposition.

A difference in amplitude of the two output phases results in imbalance $A - A'$ as illustrated in Figure 4.6*b*, and a less than 180° phase relationship is shown by Figure 4.6*c*.

Carrier suppression may be expressed in decibels below the output desired by the expression

$$V = 10 \log \frac{B}{A \sin \alpha + A' \sin \beta} \text{ dB}$$

where B = amplitude of the correct output signal (i.e., $A = A'$ and $\alpha = \beta = 0$),
A = zero degree signal amplitude,
A' = 180° signal amplitude,
α = A phase offset,
β = A' phase offset.

In some cases a controlled amount of carrier signal is deliberately sent for use in carrier tracking without code demodulation as in the Air Force's SGLS space ground link subsystem system. For such an application the system is usually designed so that $\alpha = \beta$, $A = A'$, and $V = (10 \log B)/(2 \sin \alpha)$.

Phase modulation by less than 180° is also used when the modulated signal is multiplied by an even number. Unfortunately, this is not practical for 180° biphase modulation because even-order multiplication of any signal that is symmetrical in the frequency domain removes the symmerical sidebands and produces an unmodulated output at the frequency multiple desired. This problem is overcome by phase modulating by $\pm 90/k$ degrees before multiplication by the factor k. When carrier suppression is critical, however, the technique is not recommended, for modulation at odd phase angles can be quite difficult to maintain and any departure from the desired $90/k$ modulation angle would significantly reduce carrier suppression after multiplication.

Alternatives to less than $\pm 90°$ modulation in which multiplication is involved are the following:

1. Separately multiply the carrier, then modulate. In general, this is a practical solution because balanced mixers and modulators are commercially available to several gigahertz.
2. Multiply by the odd order, because odd-order multiplication preserves symmetrical modulation.

A typical direct sequence output signal in which the carrier signal is poorly suppressed is shown in Figure 4.5. In most instances poor balance might not be so clearly evidenced. Often, however, a constant signal component is detectable in the output spectrum when carrier balance is not as it should be. The best method for checking carrier suppression is to remove the code modulation and substitute a square wave. This permits observation of the carrier component in comparison to the square-wave-modulation sidebands without noiselike code components (see Figure 4.7). Varying the square wave frequency over the range of code frequency components (code rep rate^{-1} to clock rate/2) also allows a check of modulator balance with code modulation.

Figures 4.8 and 4.9 show the effects of (1) nonsymmetry in the code modulating sequence and (2) simultaneous code nonsymmetry and lack of carrier balance. Code symmetry is more difficult to provide in a practical system than carrier balance because of the characteristics of the circuits used to drive the balanced modulator with a code sequence. Many integrated-circuit flip-flips are capable of driving a diode-balanced mixer/modulator directly and offer the best general symmetry; that is, both discrete circuits and separate drivers tend to exhibit more differ-

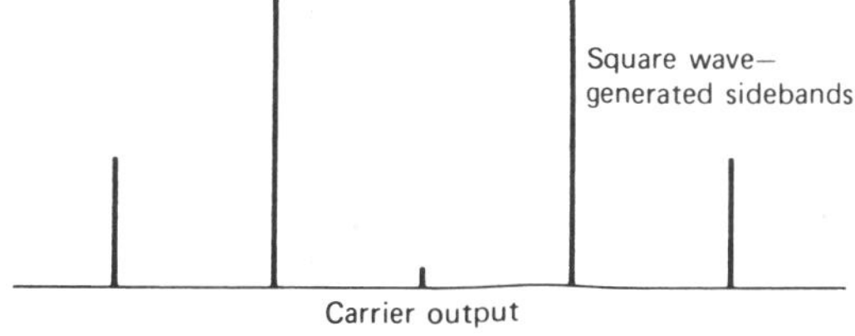

Figure 4.7 Typical square-wave-balance modulation test spectrum for checking carrier suppression.

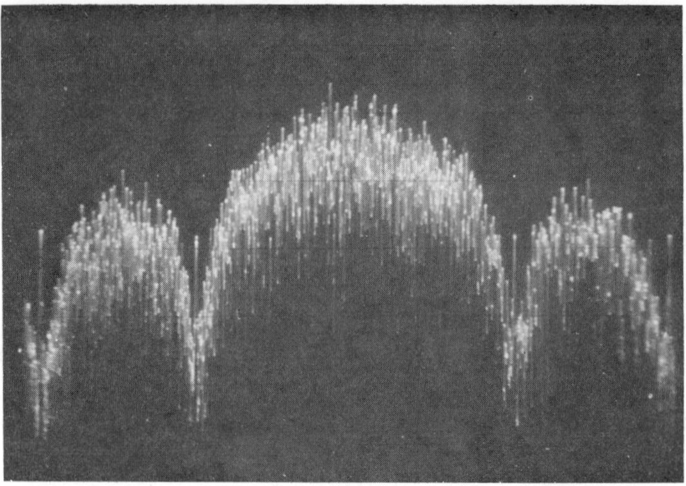

Figure 4.8 Balanced mixer/modulator output showing poor code balance.

ential change in output signal than complementary outputs from a single integrated-circuit flip-flop.

The problem is this: to exhibit good code-rate suppression a balance-modulated output signal must weight the signal corresponding to a one at the same value as a signal corresponding to a zero; that is,

$$A\int_0^{T_1} G_1(c)\,dt$$

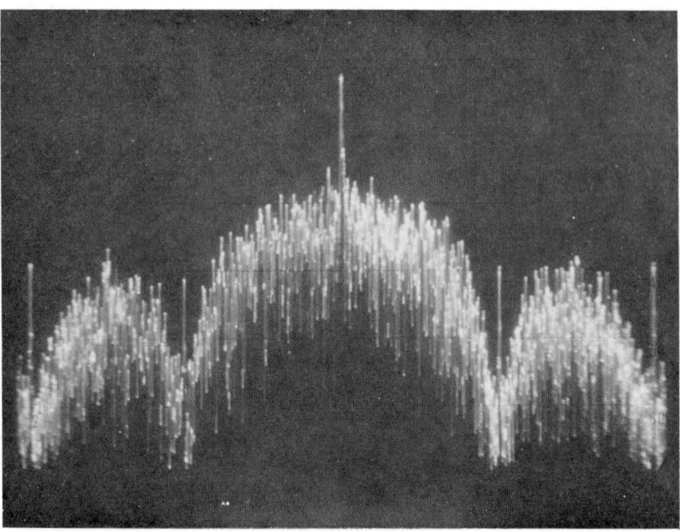

Figure 4.9 Balanced mixer/modulator output showing both carrier and code imbalance.

must equal

$$A' \int_{\tau_1}^{\tau_2} G_0(c)\, dt$$

where $G_1(c)$ is the signal envelop corresponding to a one and $G_0(c)$ is the signal envelope corresponding to a zero. Also, the periods $\{\tau_1 - 0, \tau_2 - \tau_1\}$ and amplitudes $\{A, A'\}$ must equal one another.

Now, in flip-flop output stages the rise and fall times of the output signals are not the same. Furthermore, temperature change causes characteristic changes and the two may not maintain the same relationship. Therefore, we caution the designer that no matter how good the balanced modulator may be some attention must be paid to code symmetry or his or her work will be of little value for code-rate suppression.

Effects of Carrier and Code-Rate Spurious (Lack of Suppression)

What is the effect of having a modulated output like that in Figure 4.9? In the transmitter the most significant effect is that after having gone to the trouble of code modulation our output spread spectrum signal has stationary, easily detected signals in its midst, which defeat one purpose of spread spectrum modulation by causing narrowband interference, transmission of wasted power, and relatively easy detection.

In spread spectrum receivers a narrowband signal due to carrier or code imbalance is even less desirable. These signals can render the receiver unusable because undesired narrowband input signals may be translated directly into the IF (see Section 5.2). Figure 4.10 shows the effect of excessive code clock leakage through or around the balanced mixer/modulator so that the output is not only balanced-modulated by the code but amplitude-modulated by the code clock. This is usually simple to prevent, but the undesired output signals can cause the same kinds of spurious signals already discussed.

Quadriphase Modulation

Quadriphase modulation is an extension of biphase-balanced modulation and is often used in direct sequence systems. The main reasons for quadriphase, despite its extra complexity, are that (1) a quadriphase signal is not so seriously degraded as a biphase signal when passed through a nonlinearity simultaneously with interference and (2) the RF bandwidth for a quadriphase signal is half that required by a biphase signal, given the same data rate. Alternatively, twice as much data might be transmitted in the same bandwidth with a quadriphase signal as with a

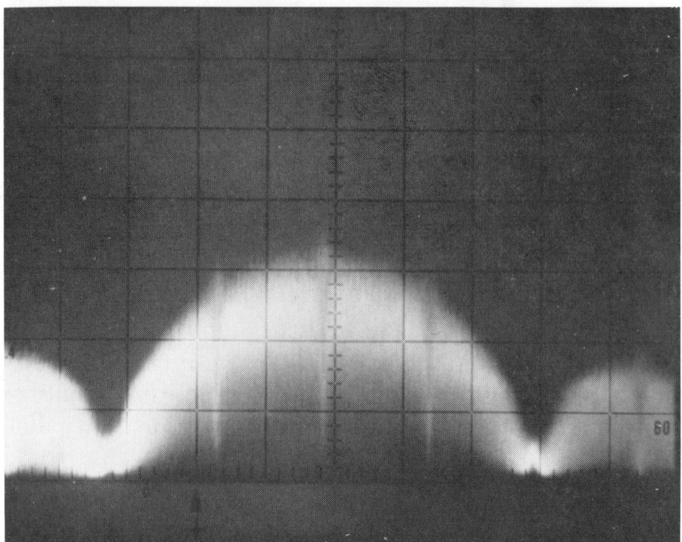

Figure 4.10 Balanced mixer/modulator output showing code clock leakage.

biphase signal. A quadriphase modulator shifts between one of four carrier phases spaced at 90° intervals. Thus 2 bits of binary data are required to select the phase to be transmitted. A typical scheme for such selection might be

Data Bits		
A	*B*	Transmitted Phase
0	0	0°
0	1	90°
1	0	270°
1	1	180°

Of course, any other one-to-one correspondence between the data and the transmitted phase is just as good. The only requirement is that a transmitter and receiver both employ the same data-to-carrier phase relationship.

The block diagram in Figure 4.11 shows a common quadriphase modulator configuration that consists of a combined pair of biphase modulators. An input RF carrier is phase shifted by 90° and applied in the original and quadrature form to a pair of biphase modulators. Separate binary inputs drive the two modulators and their outputs are summed (Figure 4.12). Any combination of carrier phases from the two biphase modulators (a, a' and b, b' are not allowed) results in a summed output that is one of the four desired carrier phases.

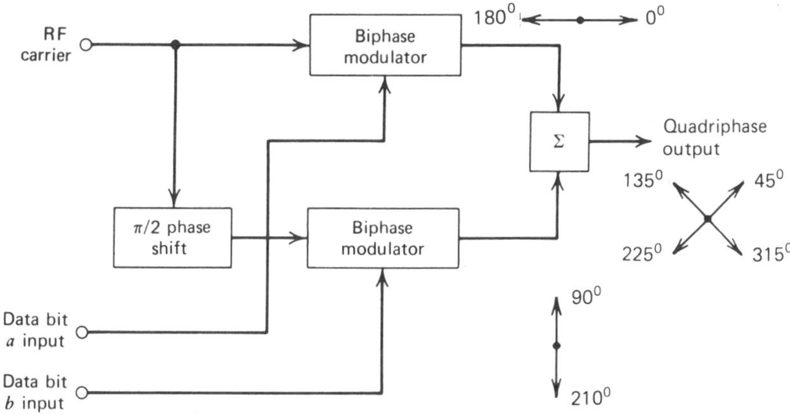

Figure 4.11 Quadriphase modulator block diagram.

The capability of quadriphase signals to pass through limiting with minimal processing loss has a great deal of importance but is left for discussion in Chapter 7.

The power envelope of a quadriphase signal has the same shape as a biphase modulated signal; that is, $(\sin_2 x)/x^2$. The main-lobe bandwidth, however, is only half that of the biphase signal for the same code rate; that is, the first null of the $(\sin^2 x)/x^2$ spectrum occurs at chip rate/2, and successive nulls occur at the same interval. We recall that the biphase-modulated signal has a main lobe whose first null occurs at the chip rate, with succeeding nulls also separated by the chip rate. Figure 4.13 illustrates this relationship.

In typical QPSK-modulated spread spectrum systems the spectrum employed has the same bandwidth as a BPSK-modulated system using

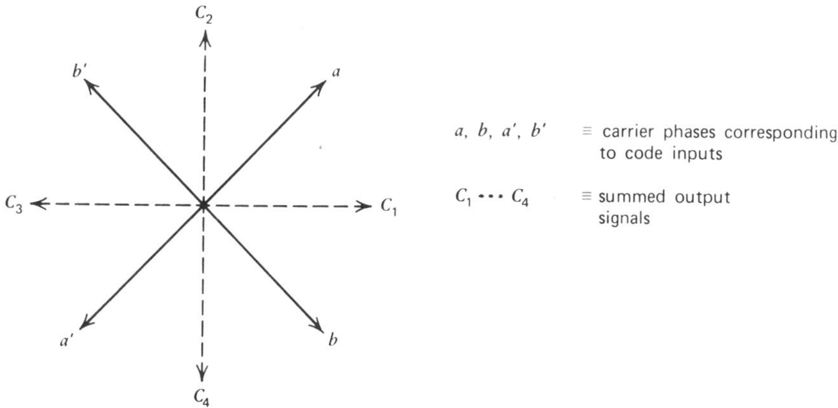

Figure 4.12 Output carrier phase derivation corresponding to data input.

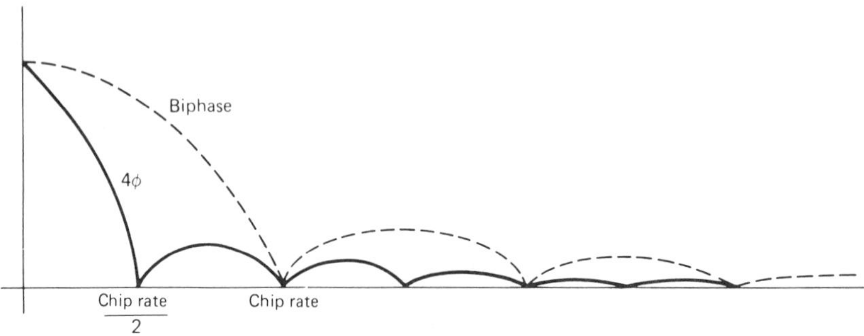

Figure 4.13 Comparison of biphase and quadriphase signal spectra for the same code rate (one code only).

the same code rate. This is accomplished by using two codes in the QPSK systems that have the same rate as the single code used in a BPSK system. This is done to avoid the loss in processing gain that would occur if the RF bandwidth were reduced.

Carrier Modulation Trade Offs

The major concerns in choosing a modulation waveform that must be taken into account in a spread spectrum are:

Effect on synch acquisition technology.

Sidelobe energy output (splatter) and bandwidth.

Complexity of modulators and demodulators.

Effects of jamming on the system.

Impact of the technique on size, weight, power requirements, and reliability.

Effect of Synch Acquisition

The waveform employed is a primary force in determining the synch acquisition capability of a spread spectrum system. Direct sequence modulation, for example, employed high-chip-rate code sequences that force a receiver to synchronize to within a very narrow time window (often as narrow as a few nanoseconds), while frequency hopping with its lower code rates, permits timing uncertainties as great as a few milliseconds. This means that the techniques employed for synchronizing in the various systems are quite varied, depending on the waveform employed. Representative acquisition and tracking techniques are (see also Chapter 6):

1. Direct Sequence:
   (a) Initial acquisition:
      Serial and/or sequential search—slowest
      Sequential estimation—fast but poor AJ performance
      Universal timing—depends on small time uncertainty
      Matched filters/convolvers—fastest, but still has problems
   (b) Tracking:
      Delay lock loops—classic or dithered

2. Frequency Hopping:
   (a) Initial acquisition:
      Serial and/or sequential search—works well because of reduced uncertainty
      Universal timing—works well due to low drift rates
   (b) Tracking:
      Split-bit tracking—similar to delay lock

(Some frequency hopping systems operate without tracking loops, due to their relative stability and long chip time.)

Sidelobe Energy Output

Production of unwanted sidelobe energy is much higher in some common direct sequence modulating approaches than in others, and is typically high in frequency hopping, although frequency hopping sidelobe energy is well contained inside the signal band. (With frequency hopped signals, one must remember that the signal is a pulsed carrier, and that the spectrum of the signal produced by pulsing the carrier is strictly dependent on the pulse envelope.) Typical rectangular-pulse frequency hopped carriers have $[(\sin x)/x]^2$ power spectra, with sidelobes at −13 dB, and that roll off at a 6 dB/ octave rate. Techniques for pulse shaping exist, however, that can reduce the sidelobe energy, and these may be employed in a system design, where they are applicable.

Also of interest are the techniques needed for use in reducing sidelobe energy output in direct sequence modulated systems. Of the modulation techniques available—biphase phase shift keyed (BPSK), quadriphase shift keyed (QPSK), offset quadriphase shift keyed (OQPSK), and minimum shift keyed (MSK)—the MSK-like signals appear to be most promising.

MSK, or minimum shift keying, is actually a family of signal structures that must satisfy two conditions:

1. The modulating pulse $g(t)$ is symmetrical about $t + T/2$ and 0 otherwise.

2. The modulated carrier has a constant envelope. That is,

$$g^2(t) + g^2(t + T/2) = 1$$

Some examples of such waveforms are:

$$\text{(a)}\quad g(t) = \cos\left(\frac{\pi t}{2T} - U\sin\frac{2\pi t}{T}\right) \qquad -T \leqslant t \leqslant T$$
$$= 0 \text{ otherwise}$$

(For classical MSK, $U = 0$ and for SFSK, $U = 0.25$.)

$$\text{(b)}\quad g^{(n)}(t) = \left(\frac{\sin(\pi t/2\mathrm{T})}{\pi t/2T}\right)^n \qquad -2T \leqslant t \leqslant 2T$$

Figure 4.14 shows the (half-pulse) envelope of some signals satisfying equation (a), and Figure 4.15 and 4.16 show the spectral density and out-of-band power, respectively, for three forms of MSK. (OQPSK, though often listed separately, is actually a form of MSK, where the pulse used is square, instead of shaped as in the more usual forms.) Figure 4.17 compares the autocorrelation functions of OQPSK, and two other forms of MSK modulation (classic MSK with $U = 0$, and SFSK with $U = 0.25$).

The OQPSK form does have a more sharply peaked autocorrelation function (this is true only for unfiltered OQPSK), and rolls off linearly. This makes it optimum for ranging systems. MSK and SFSK, on the other hand, have rounded correlation functions, which make them less synchronization-critical.

Complexity of the basic modulators is compared in Figure 4.18. The BPSK and MSK modulators are least complex, and appear identical on a block-diagram basis. The filters employed are very different, however, although the balanced modulators are the same.

BPSK modulation, though simple, is too readily vulnerable, and should not be processed in a limiting channel. Figure 4.19 illustrates the problem seen when passing a BPSK-modulated signal $B \cos \omega_0 t \pm 90°$ through a limiter, together with a strong CW jamming signal $J \cos \omega_0 t$. The result of the process, when the frequency of the jammer is exactly at center frequency, is that the small desired signal amplitude modulates the jamming signal. This results in amplitude modulation that is completely removed by the limiter, so that nothing remains but an unmodulated carrier.

Even in the best case, desired signals are suppressed by larger jamming signals by up to 6 dB and all this must be considered when designing a spread spectrum system. Table 4.1 summarizes the significant characteristics of the various carrier modulation techniques useful for direct-sequence modulation.

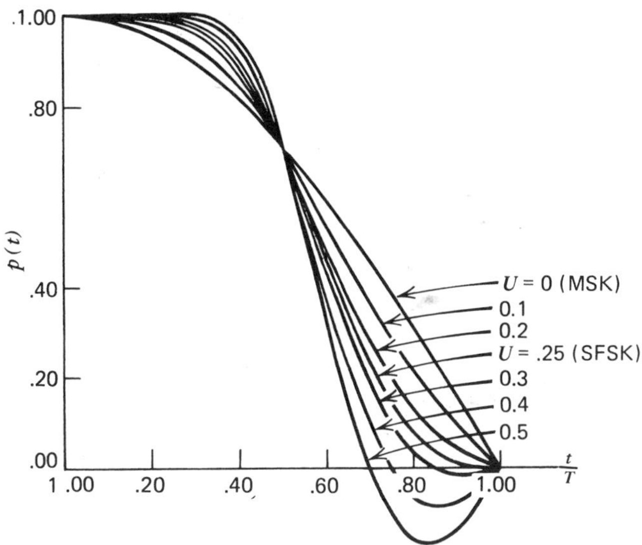

Figure 4.14 Pulses suitable for MSK format.

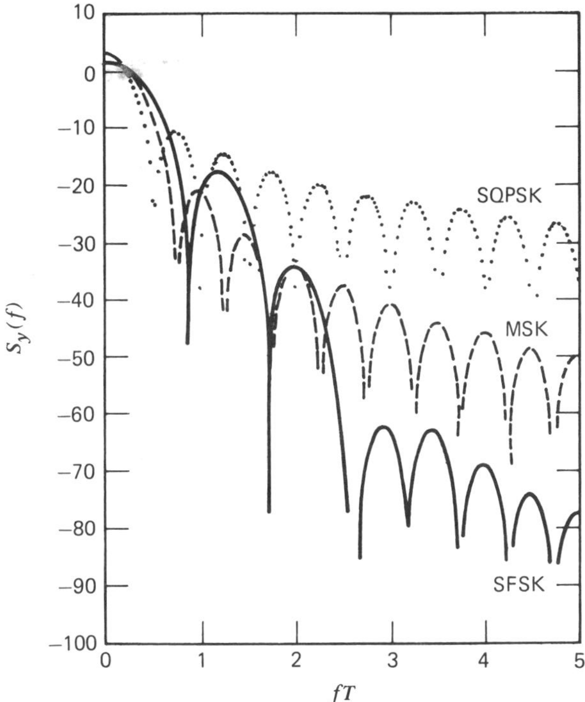

Figure 4.15 Comparison of the equivalent low-pass power spectral densities for OQPSK, MSK, and SFSK.

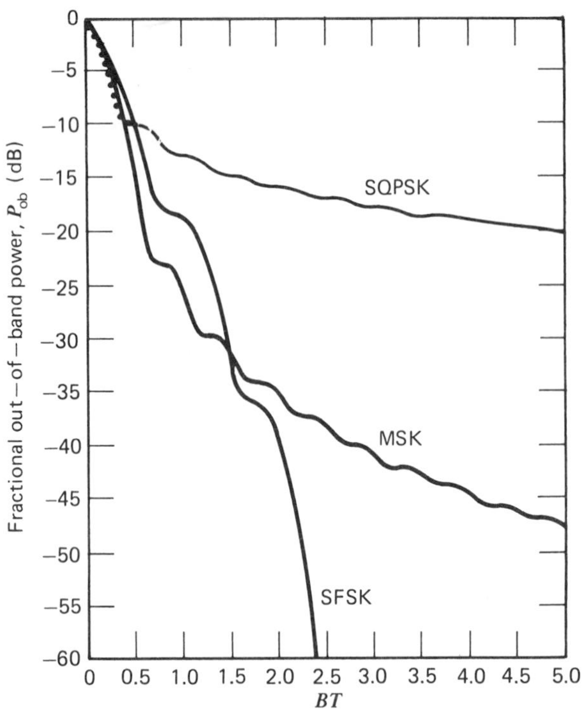

Figure 4.16 Comparison of fractional out-of-band powers for OQPSK, MSK, and SFSK.

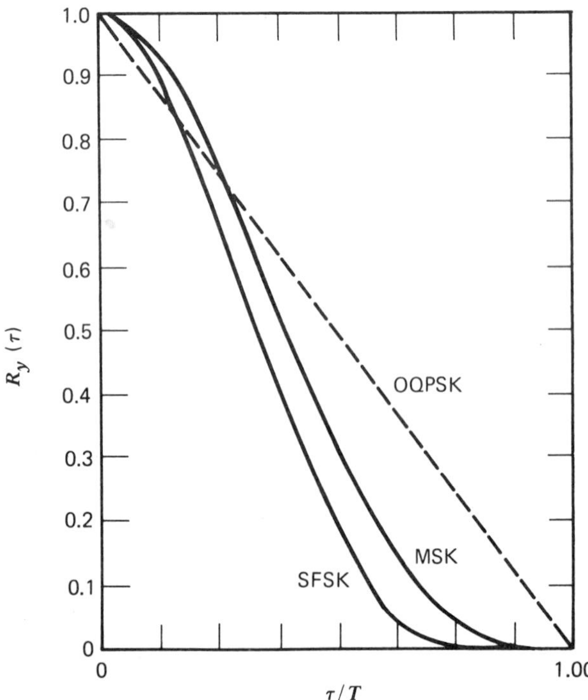

Figure 4.17 Comparison of the equivalent low-pass autocorrelation functions for OQPSK, MSK, and SFSK.

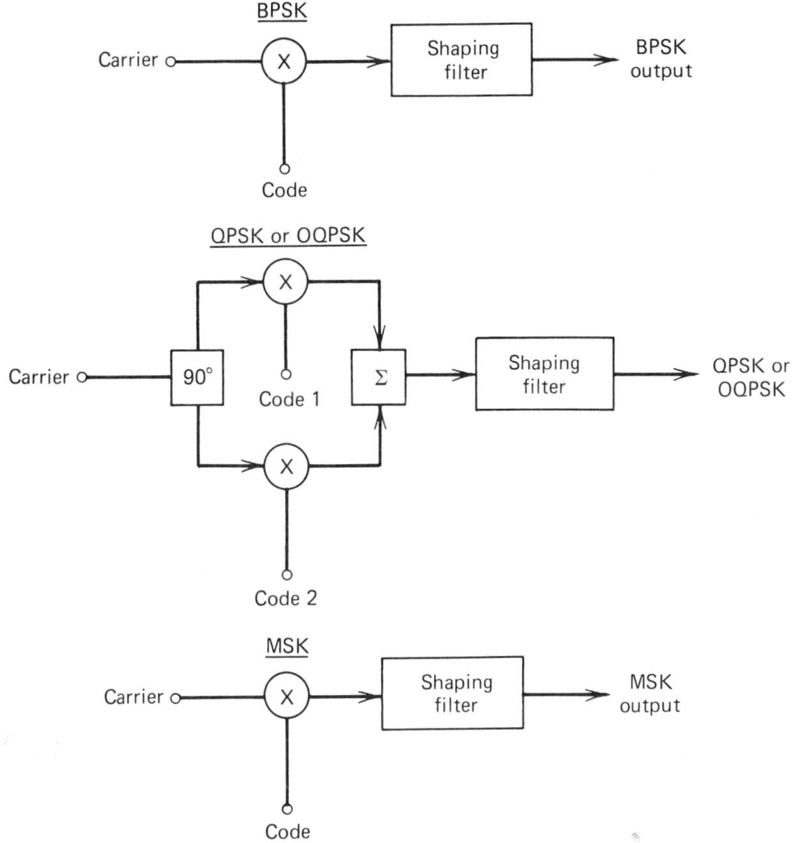

Figure 4.18 Modulator comparison.

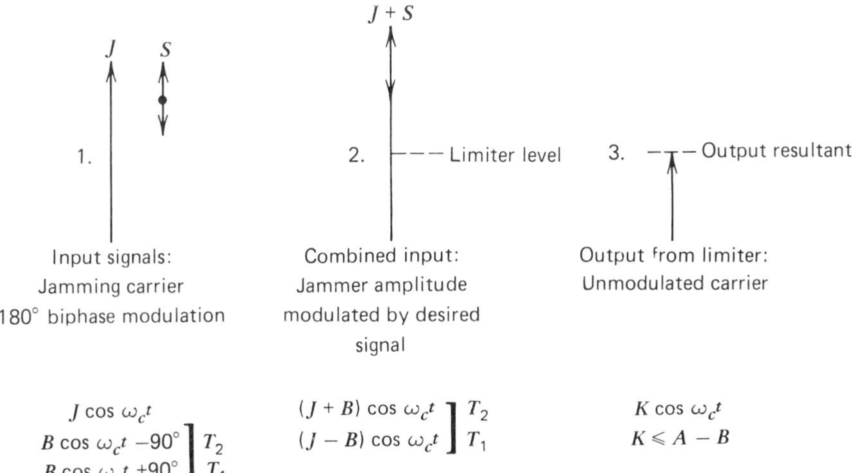

Figure 4.19 CW jammer coherent with BPSK-modulated carrier signal.

Table 4.1 Waveform Trade Off Summary

	Synch Technology	Sidelobe Energy	Nonlinear Tolerance	Moderate Complexity	Autocorrelation
BPSK	M.F. or Convolver	High	Poor	Best	Triangular
QPSK	M.F. or Convolver	High	Poor	2 × BPSK plus two summers	Triangular
OQPSK	Convolver	High, but filterable	Good	Same as QPSK, plus filter	Triangular
MSK	Convolver	Lowest	Best	Same as BPSK	Rounded

4.2 FREQUENCY SYNTHESIS FOR SPREAD SPECTRUM MODULATION

The frequency synthesis section of a frequency hopping system is at once both the heart and the most critical part of the system. The number of frequencies generated and the hopping rate are the quantities that determine system capabilities. They, in turn, depend on the type and configuration of the synthesizer itself. There is no difference, in principle, between the synthesizers used in standard communication transceivers and those in frequency hopping. The frequency hopping synthesizer, however, is designed to switch quickly from one frequency to another, whereas a conventional synthesizer may be required to change frequencies only at long irregular intervals. The rapid-switching requirement does force some difference in the details of synthesizer design. Therefore, some discussion of frequency synthesis for frequency hopping is in order.

Frequency synthesizers are divided into two general types. The first includes all those that generate the desired output frequencies from a source (or set of sources) by mixing, multiplying, and dividing, or any other means not including phase-locked loops. This class of synthesis is called "direct." Any synthesizer employing a phase-locked loop in its frequency-generation elements is called "indirect." Both are of interest here.

The simplest type of direct synthesizer is shown in Figure 4.20. Here a number of frequencies are mixed together in various combinations to give all the sum and difference frequencies. The utility of this simple synthesis approach quickly becomes limited when a large number of frequency choices is desired. The size and weight of the filters required are major factors in themselves.

A great variety of frequency synthesizers has been developed. In general, every new piece of communications equipment has a new synthesizer configuration, and although many commercial frequency synthesizers are available, no two have a great deal of similarity beyond being in the same class (direct or indirect). This is pointed out only to show that there are as many ways to implement a synthesizer as there are

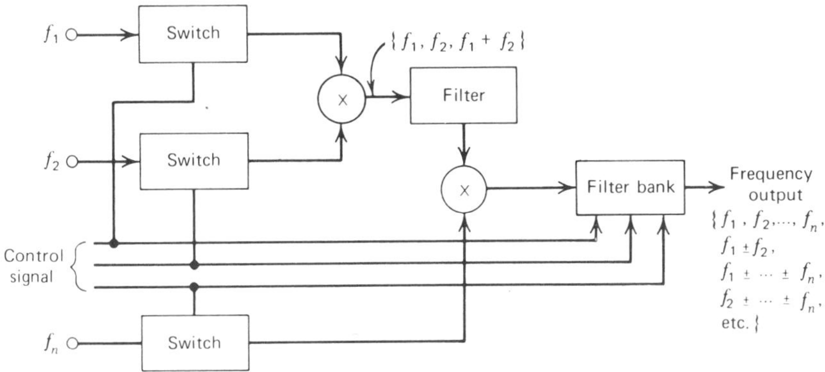

Figure 4.20 Simple direct synthesizer.

synthesizers. One direct synthesis approach has been successfully used in the TATS (tactical transmission system).[111] It is well adapted to generating large numbers of frequencies and can switch from frequency to frequency rapidly. This is the add-and-divide (or mix-and-divide) approach suggested by Hastings and Stone.[814]

Add-and-divide synthesizers are made up of a series of add-and-divide modules, which are most often made to be identical to facilitate construction. Figure 4.21 is a diagram of an add-and-divide model. The number of frequencies that a direct synthesizer of the add-and-divide type can generate is a function of K, the number of mixing frequency choices, and A, the number of times the mixing operation is performed. Frequency separation is determined by the base frequency separation, divided by the number of output frequencies, times the number of base frequencies; that is, the number of frequency outputs is K^A and the output frequency separation is (base separation/ K^A) $\times$ number of bases. As an example, let us postulate a synthesizer with at least 4000 output

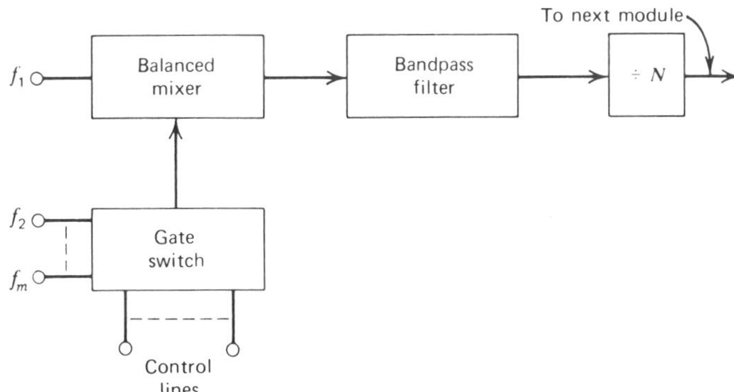

Figure 4.21 Basic add-and-divide frequency synthesizer module.

frequencies and 5-kHz frequency separation. Our output center frequency f_0 is arbitrarily chosen as 50 MHz. Because we need at least 4000 frequencies, we can use $4096 = 2^{12}$ as a goal. Conveniently enough, $2^{12} = (2^2)^6$, so that if we let $K = 2^2 = 4$, then A, the number of mixer stages, is 6; that is $K^A = 4^6 = 4096$. We need $K = 4$ base frequencies separated by K^{A-1} output separation) $= 4^6 \times 5000/4 = 5.12$ MHz. (If 4.88-kHZ spacing is acceptable, then exact 5.0-MHz base separation is satisfactory.) For simplicity let us assume 5.0-MHz base separation.

Now that we know that our base frequencies must be separated by 5.0 MHz, we can define the remaining signals. Because we have decided to use a divider (M) of 4 and an output center frequency of 50 MHz, we see that the first divider output (referring to Figure 4.22) is 12.5 MHz. Preserving the 50-MHz center frequency, we choose the base frequencies f_1 through f_K such that

$$\frac{1}{K}\sum_{n=1}^{K} f_n + 12.5 \text{ MHz} = 50 \text{ MHz}$$

and the f_K are symmetrically distributed about the mean

$$\frac{1}{K}\sum_{n=1}^{K} f_n$$

Therefore, the mean frequency is $50 - 12.5 = 37.5$ MHz and the base frequencies are

$$f_1 = 30 \text{ MHz}$$

$$f_2 = 35 \text{ MHz}$$

$$f_3 = 40 \text{ MHz}$$

$$f_4 = 45 \text{ MHz}$$

which is the set of four frequencies symmetrically distributed about 37.5 MHz at 5.0-MHz spacing.

Our basic design is now complete. Figure 4.22 shows the block diagram of the synthesizer we have postulated. It has 4096 output frequencies spaced at 4.88-kHz intervals across the 40.0–59.9-MHz band. Table 4.2 shows the development of these frequencies.

Summarizing the steps we have the following:

1. Determine the number of output frequencies needed and their spacing.
2. Select the center frequency (f_0) around which they are to be supplied. (This frequency is often bounded by the maximum frequency capability of the dividers available.)

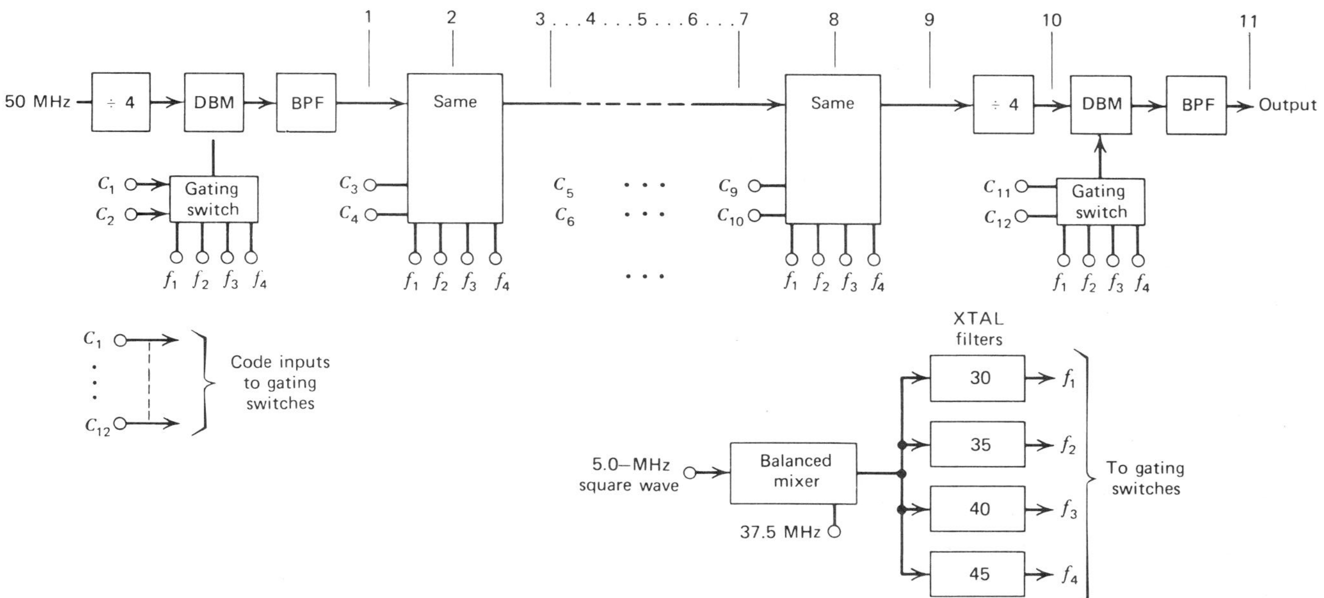

Figure 4.22 4096 frequency direct add-and-divide synthesizer.

Table 4.2 Frequency Table for the Direct Add-and-Divide Synthesizer in Figure 4.22

Point	Band (MHz)	Δf(MHz)	Frequencies
1	42.5–57.5	5	4
2	10.625–19.375	1.25	4
3	40.625–59.375	1.171875	16
4	10.15625–14.59375	0.277343	16
5	40.15625–59.59375	0.303710	64
6	10.034062–14.898437	0.076005	64
7	40.034062–59.898437	0.077595	256
8	10.008515–14.974609	0.019399	256
9	40.008515–59.974609	0.019498	1024
10	10.002128–14.993652	0.004874	1024
11	40.002128–59.993652	0.004880	4096

3. As a trade off between the number of add-and-divide operations (A) and the number of frequencies (K) in the base set ($f_1 \cdots f_K$), choose A and K: $K^A \geqslant$ the number of output frequencies required.
4. Select a divisor (M) that gives enough separation between f_0/M and the nearest of the ($f_1 - f_K$) frequencies to allow a reasonable filter design. [Because f_0 is known and the extent of the set ($f_1 \cdots f_K$) is also known, the separation between f_0/M and $f_0 - (f_1 \cdots f_K)$ is available.]

$$f_m = (f_1 - f_K) = f_0 - \frac{f_0}{M} = \frac{1}{K} \sum_{n=1}^{K} f_n$$

5. Base set ($f_1 \cdots f_K$) may now be determined from f_m and the base frequency spacing (see Figure 4.23).

So far we have not mentioned spurious output signals or switching speed. These must be considered and to some extent traded off against one another because the bandpass filter following the mixer in each section of this synthesizer configuration affects both. Spurious signal output is a function of the level of the unwanted sidebands produced by mixing at the output of an interstage filter. Each unwanted signal is mixed in the succeeding section and produces new unwanted frequencies. Thus the

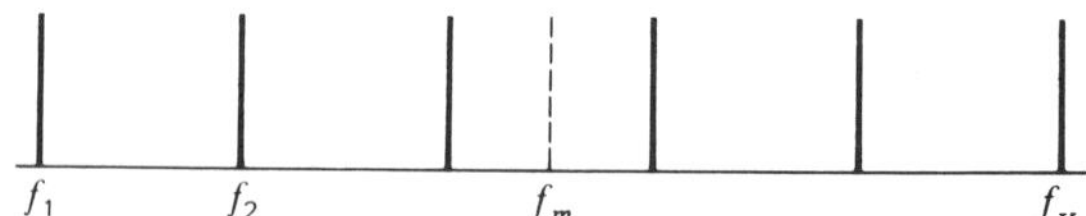

Figure 4.23 Base frequency relationships.

bandwidth and skirt response of the interstage filters is important in a determination of the output spurious signal levels. The filter is also important in determining frequency switching speed for the synthesizer because its response limits the rate of change of phase of any signal passed through it. Noting that several filters are cascaded in the add-and-divide synthesizer, we find it obvious that these filters ultimately limit switching speed. Also, if not well matched between synthesizers, these cascaded filters can cause a loss of frequency-to-frequency phase coherence from synthesizer to synthesizer.

One not-so-obvious source of switching-speed improvement is in the way that a new code command is applied to the frequency gating switches. We have assumed in the discussion so far that all code inputs are applied at the same instant to these selectors. For high-speed frequency hopping, however, it is advantageous to apply new code commands to the synthesizer sequentially. Referring once again to Figure 4.22, we note that the desired output signal is generated in a series of add-and-divide steps and that the propagation through these steps is additive. Thus, when a frequency change is desired, the new signal must travel from the first stage through the last before appearing at the output. Therefore, by applying the new command sequentially, so that the first stage in the chain receives its command with a time delay equal to the signal propagation, all stages switch simultaneously with respect to the output signal. The switch from one frequency to another is faster; that is, less "settling time" is involved and the duration of a signal at one particular frequency may be reduced in proportion to the reduction in settling time.

The usual rule of thumb for determining maximum frequency-hopping rate for a frequency synthesizer is that the period of transmission for one frequency should be 10 times the time required for setting to that new frequency. Thus a reduction in settling time is effective in increasing hopping rate, because a setting-time reduction of 1 μsec would reduce a frequency period by 9 μsec.

Indirect Synthesis

The second type of frequency synthesizer is the indirect (any synthesizer employing phase- or frequency-locked loops is so termed). The simplest type of indirect frequency synthesizer is a phase-locked oscillator whose feedback path is modified to include a variable counter as in Figure 4.24*b*. Addition of this counter enables the phase-locked loop to generate any frequency multiple of its reference signal; that is, given a 1 MHz reference oscillator, a 1-to-1000 counter, and a phase-locked oscillator, theoretically we could generate 1000 frequencies spaced at 1 MHz intervals.

How does it operate? The reference input signal is compared in a phase detector that generates an output signal proportional to the phase difference between the reference and the counted-down VCO. The signal

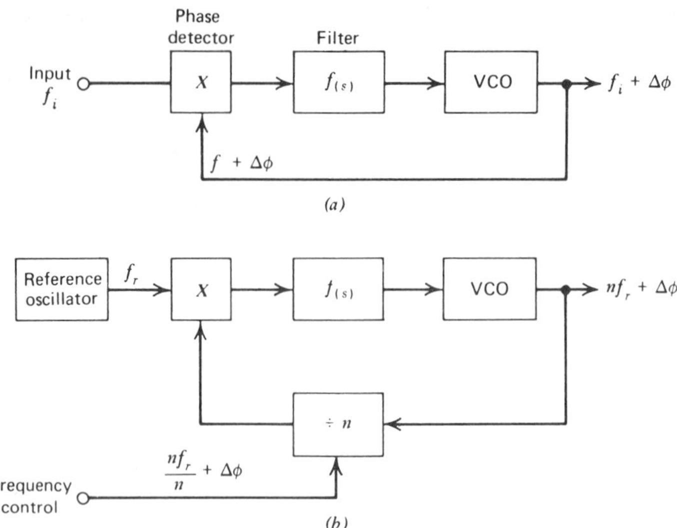

Figure 4.24 Basic indirect frequency synthesizer components: (*a*) simple phase-locked loop; (*b*) simple phase-locked loop modified for multiple frequency output capability.

has a DC component that is applied, after filtering, to the VCO, closing the loop and causing the VCO to be locked to the input signal (or a multiple thereof) in frequency. Thus inserting the variable counter between the VCO and the phase detector causes the VCO to operate at a frequency that is n times the reference frequency (where n is also the division ratio of the counter).

From this point, once the phase-locked loop is built and the variable counter inserted, all that remains is to add a coded input to the variable counter (to command the count) and we have a frequency hopper. In general, however, even the simplest of frequency synthesizers is not quite so easily come by, for a particular design must be specifically tailored to meet the hopping rate and stability requirements of a given communications system. Commonly, short-term stability of the output signal must be balanced against hopping rate because the loop filter has a direct bearing on both; that is, a narrow loop filter reduces VCO jitter due to noise pick up, reference noise, and so on. That same narrow filter can also significantly increase the time required to settle at a new frequency.

Common techniques for reducing frequency switching time while preserving the low-noise characteristics in indirect frequency synthesizers include the following:

1. Use of a sampling loop filter.
2. Digital-to-analog (D-A) conversion of the frequency command word, the analog signal is used to coarse-set the controlled VCO to a frequency near the desired output.
3. Use of more than one loop with alternate outputs taken in sequence.

Sampling loop filter. Although offering little to speed up loop frequency acquisition (in fact, the loop acquisition rate is limited by the sampling rate), as sample-and-hold type of loop filter helps to reduce jitter induced by the phase detector. This, in turn, permits greater freedom in selecting the loop reference frequency and filter parameters. Typical loop filters of the sampling type use field-effect transistors as switches and often employ tandem sampling filter sections to provide for rapid sampling while retaining high effective integrating time constants. Because of its sampling action, the sampling filter may also act as a phase detector.

D-A frequency command conversion. A phase-lock frequency synthesizer can be locked to a new frequency by using the frequency command word (after conversion to a DC control signal) to set the loop VCO close to the desired frequency. Because frequency uncertainty is reduced, lock time is also reduced. D-A command word conversion can also reduce output frequency jitter because it allows open loop gain to be reduced.

When the output frequency range of a phase-lock loop is $|f_n - f_1| = f_0$, open-loop gain is required to be such that AoL $\geqslant f_0$ (Hz). Open-loop gain can be derived from only two sources: (1) phase detector gain (in volts per radian offset) or (2) VCO gain (in radians per volt). Amplification within the loop is equivalent to increasing one of these parameters. Dividing the lock range of the loop by setting the loop VCO in several increments across the desired frequency band therefore allows loop gain to be reduced to cover only the range of a sector (i.e., setting the VCO to 10 intermediate frequencies within a frequency band could allow loop gain to be one-fifth of that otherwise required). The advantage of reduced loop gain is, then, that the VCO can be made less sensitive to noise occurring at its inputs; it appears as a direct reduction in output frequency jitter and noise.

Multiple loop techniques. Two frequency synthesizers, alternating their outputs, can increase the composite hopping rate by as much as 10 times when settling time for a single synthesizer is held to 10% of a chip. Figure 4.25 is a block diagram of a multiple loop synthesizer; waveforms show the possibility of speedup.

Especially well adapted to use in multiple-loop synthesis are the integrated-circuit frequency synthesizers listed in Table 4.3. These devices (depending on the particular unit in question) are capable of generating large numbers of frequencies at output frequencies up to 120 MHz. Their small size permits more than one synthesizer to be included in systems that could employ only one with discrete-circuit techniques. By including two synthesizers in a system, and switching between them to allow for one to output a signal while the other settles to new frequencies, the system's hopping rate can be increased by a factor of 10 over a single synthesizer

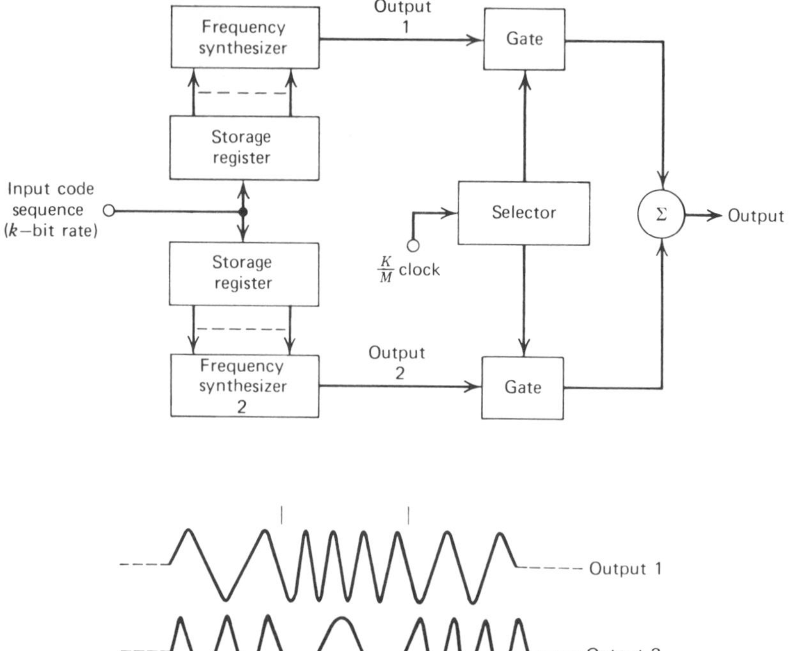

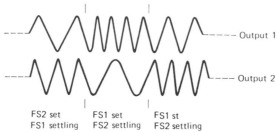

Figure 4.25 Multiple-synthesizer frequency hopper.

that allows for 10% settling time out of a hopping period. That is, when going from one to two synthesizers where the fraction of a hop time allowed for settling is $T_{\text{settle}}/T_{\text{hop}}$, the increase in hop rate for use of two synthesizers is $T_{\text{hop}}/T_{\text{settle}}$. A further increase in the number of synthesizers (beyond two) increases the hopping rate by a further factor of $r - 1$, where r is the total number of synthesizers. Multiple (r) synthesizers, therefore, allow for a hopping rate increase of up to $(r - 1)\,T_{\text{hop}}/T_{\text{settle}}$. For example, a single synthesizer capable of settling in 10 μsec would be useful for frequency hopping at up to 10,000 hops per second by itself. When four such synthesizers are used, however, the hop rate increases to

$$\frac{(4 - 1)\ 100\ \mu\text{sec}}{10\ \mu\text{sec}} \times 10{,}000 = 300{,}000\ \text{hops/sec}$$

Those who wish to explore the phase-lock field further will find that is is exceptionally rich and well documented (see the references in Appendix 2). The bibliography in this book will serve to point those who are interested in the right direction.

Table 4.3 Some Integrated-Circuit Frequency Synthesizers and Their Characteristics

Manufacturer's Type	Number of Frequencies	Reference Frequency	Maximum Divider Input	Control Requirement
Hughes HCTRO347	45/90	50 Hz–500 kHz	10 MHz	8 bits parallel
Nitron 6401,05	82	4.0 MHz	3.5 MHz	8 bits parallel
6410A	100	4.0 MHz	1.6 MHz	8 bits parallel
6403	45/90	3.0 MHz	3.5 MHz	8 bits parallel
RCA TA10336	64	10.24 MHz		6 bits parallel
Motorola 145104 to 12	256/512	10.24 MHz	4.0 MHz	8–9 bits parallel
National MM55104,7,14	256	10.24 MHz	3.0 MHz	8 bits parallel
MM55106,9,16	512	10.24 MHz	3.0 MHz	9 bits parallel
MM55108,10	1024	10.24 MHz	3.0 MHz	10 bits parallel
Fairchild 11C84	128	10.24 MHz	20 MHz	7 bits parallel
Plessey SP8921 to 23	64/128	10.24 MHz	30 MHz	6–7 bits parallel
MPS 7139,49,89	64/128	16 MHz max	30 MHz	6–7 bits parallel
National DS8900	40	10.24 MHz	30 MHz	Up/down only
Signetics 8X08	8192	3.6 MHz	80 MHz	4 bits parallel/serial
Rockwell RS3291	1599		120 MHz	11 bits parallel
National DS8906,07	16384/8192	10.24 MHz	120 MHz	20/18 bits serial

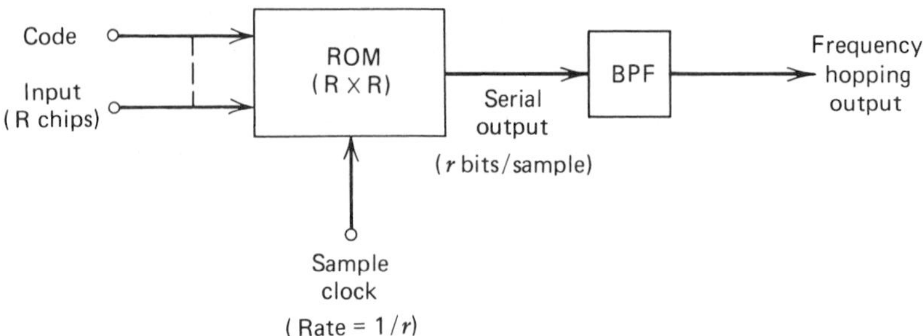

Figure 4.26 Read-only memory as a frequency hopping synthesizer.

Synthesis From Digital Words

One unique method of frequency synthesis that provides fast switching from frequency to frequency is that of digital storage of words corresponding to frequencies. For example, a 01010101 . . . pattern and a 001100110011 . . . pattern each represent a fundamental frequency that corresponds to $1/T(01)$ and $1/T(0011)$, respectively. If a word generator can be caused to output words in increments of bit times, then frequencies separated by $1/T$ can be generated.

One way to accomplish this is to store a set of patterns in a read-only memory, or ROM. Then if a particular output frequency is desired, the digital input (code word in our context here) corresponding to that frequency is repeatedly input to the ROM as long as the output frequency is needed. This is illustrated in Figure 4.26 in block diagram form.

Figure 4.27 shows typical spectra and waveforms for signal generation by a ROM synthesizer.

4.3 SENDING THE INFORMATION

Information modulation in spread spectrum systems is possible in most of the conventional ways; amplitude and angle modulation are both satisfactory. In addition to these more normal methods, information may be sent by various ways of modifying the code sequences used. Of course, if the code sequence is viewed as a subcarrier, which it is, the methods of modulation are revealed as the usual techniques in only a slightly new guise.

So far we have spoken little about information transmission in spread spectrum systems. Any communication system, however, has no *raison d'être* without a method of transferring information. (On–off keying is satisfactory in some cases and could be used. To be exact, on–off keying is a method of transferring information that requires no other operation but conveys only presence or absence of the signal.) In the following pages we

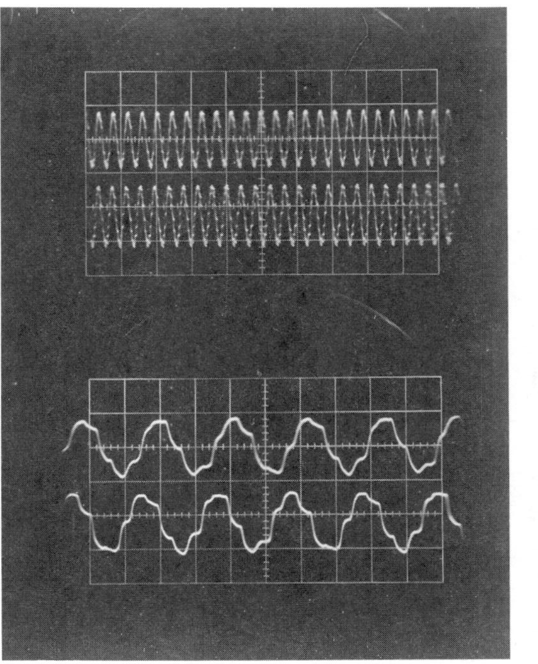

Frequency Spectra

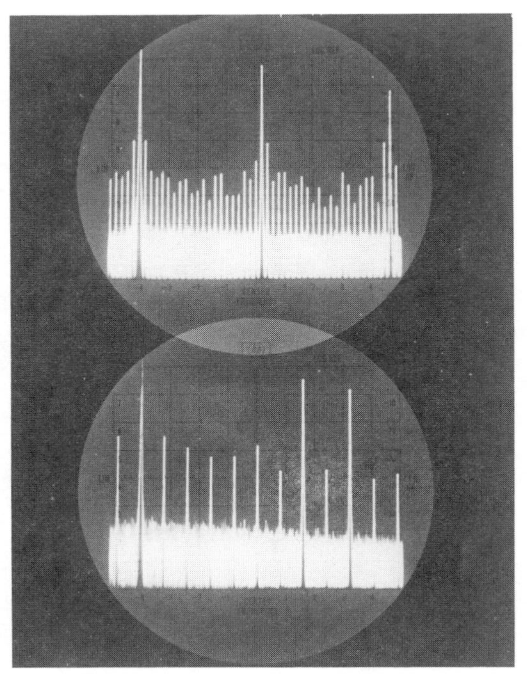

Wave Forms

Figure 4.27 ROM frequency synthesizer spectra and waveforms.

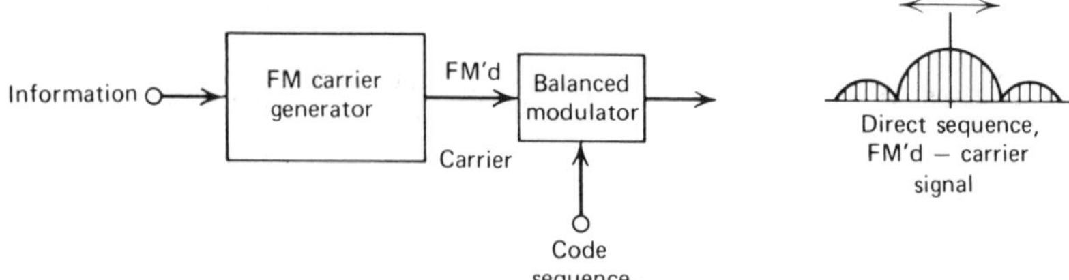

Figure 4.28 Simplified carrier-FM/direct-sequence block diagram.

discuss some of the techniques especially adapted to spread spectrum systems. Before proceeding to specific spread spectrum oriented modulation methods let us discuss the effects of carrier modulation separately.

Carrier Modulation

Amplitude modulation of spread spectrum signals is not normally used because it tends to be at cross purposes with the intent of these systems. Instead of a constant envelope, low detectability signal, one with large peak power and easy demodulation by simple receivers is produced. Carrier-frequency modulation is more useful because it is a constant envelope signal, but the information sent is still readily observed. In both carrier AM and FM no knowledge of the code is required to receive and make use of the transmitted information.

In those examples in which the purpose of employing spread spectrum modulation is only to reject interference, carrier FM in combination with direct sequence modulation is quite useful. Figure 4.28 illustrates such a system. The output spectrum of a carrier-FM/direct-sequence modulator is the covariance of the FM and spread spectrum code modulation. The entire $[(\sin x)/x]^2$ code spectrum is deviated as the information frequency modulates the carrier. If the composite carrier-FM/direct-sequence signal is squared, however, a two-times carrier term is produced which bears all the frequency modulation information. For this reason, when some degree of privacy is required, carrier FM is not recommended.

Using carrier FM combined with frequency hopping (Figure 4.29) is also possible but requires even more care than the carrier-FM/direct-sequence approach because the frequency hopping modulation is itself a form of frequency modulation. Therefore, it is extremely difficult to prevent crosstalk due to frequency hopping the carrier from disturbing the information. Any phase change associated with a hop from one frequency to another is interpreted by an FM demodulator as information and is usually output as pulsed noise.

One aspect of combined-FM/frequency hopping demodulation is that noise pulses produced at frequency transitions are difficult to suppress. Bandpass filtering ahead of the demodulator tends to accentuate the problem. At a phase transition any signal traversing a bandpass filter is

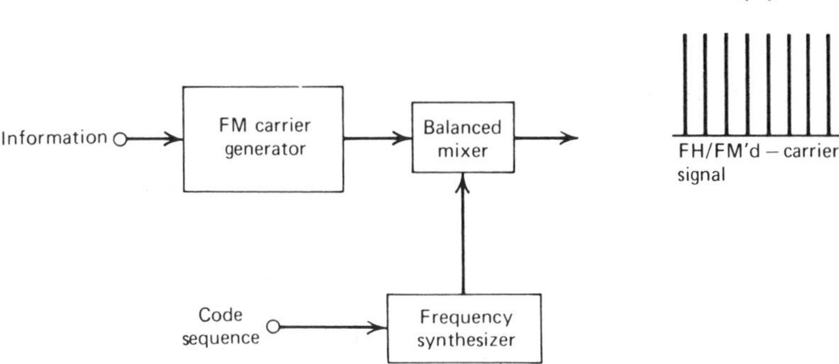

Figure 4.29 Simplified carrier-FM/frequency hopping block diagram.

limited in its rate of change of phase by the phase rate of change of the filter (which is governed by the filter's bandwidth). Therefore, passing a signal with rapid phase shifts through a bandpass filter only stretches the transition time and increases the percentage of time of a disturbance. Postdetection filtering does reduce the effects of frequency hopping on an FM signal but only to the extent that the hopping rate can be removed from the information band.

FM or AM frequency-hop-modulated systems are recommended only when the hop rate is so low that it is unobjectionable. Otherwise transmit-and-receive frequency hopping signals must be highly coherent so that no phase transition will occur in the dehopped carrier signal.

For most purposes we have eliminated the usual forms of modulation from consideration for transmitting information in spread spectrum systems. Two other ways of transmitting the desired information remain: (1) clock rate modulation and (2) code modification.

Clock Rate Modulation

Clock rate modulation is actually frequency modulation of the code clock. Demodulation is accomplished by the clock tracking circuit in the receiver, which must follow the deviation produced by both modulation and frequency offsets. For most purposes clock modulation is not used, because the loss in correlation due to phase slippage* between received

*In a phase-lock demodulator the reference VCO is normally phase shifted by 90° with respect to the incoming signal to which it is locked. This condition occurs when the incoming signal is at the same frequency as the open-loop (rest) frequency of the VCO. As the input frequency deviates away from the VCO rest frequency, more or less phase shift occurs (depending on whether the deviation is higher or lower than the VCO), and this phase shift increases until it is near 180° or zero, at which point the loop will break lock. At any given frequency offset the phase offset in radians is $\Delta f/$loop gain $+ \omega_{\text{mod}}/\omega_n^2$. The second term of this equation contributes little when the loop is the wideband loop required for information demodulation. Therefore, for our purposes $\Delta\phi = \Delta f/K$.

and local clocks could cause degraded performance. Information sent by clock rate FM is not detectable by simple squaring, as is carrier FM, because it produces sidebands that are symmetrical about the carrier frequency and the squaring process cancels all symmetrical signals.

For direct sequence systems an FM demodulator tuned to the RF carrier plus or minus the clock could recover the modulation if code imbalance allowed any clock component to be transmitted. As discussed in an earlier section of this chapter, code imbalance causes a clock frequency component to appear in a direct-sequence spectrum, and this clock component could be tracked to recover any clock rate modulation.

For frequency hopping systems, the code clock is usually at or near the data rate; therefore, clock rate modulation is not useful.

One further difficulty in implementing clock frequency modulation is that of constructing sufficiently stable oscillators that can be deviated over a wide range. Two solutions to this difficulty are given in the block diagram in Figure 4.30. In Figure 4.30*a* a crystal discriminator is used as the stabilizing element—a VCO (not crystal stabilized) is modulated by the information to be sent. Its output is fed to a crystal discriminator whose average output is zero when the VCO signal is at its center frequency. VCO drift causes a correction voltage to be generated.

A similar technique, but one that depends for stability on a reference oscillator, is shown in Figure 4.30*b*. This configuration is more adaptable to high-frequency output signals because the addition of a divider in its feedback loop can cause the VCO to lock to a multiple of the reference. The discriminator stabilized circuit is limited to the frequency range of available crystal discriminators.

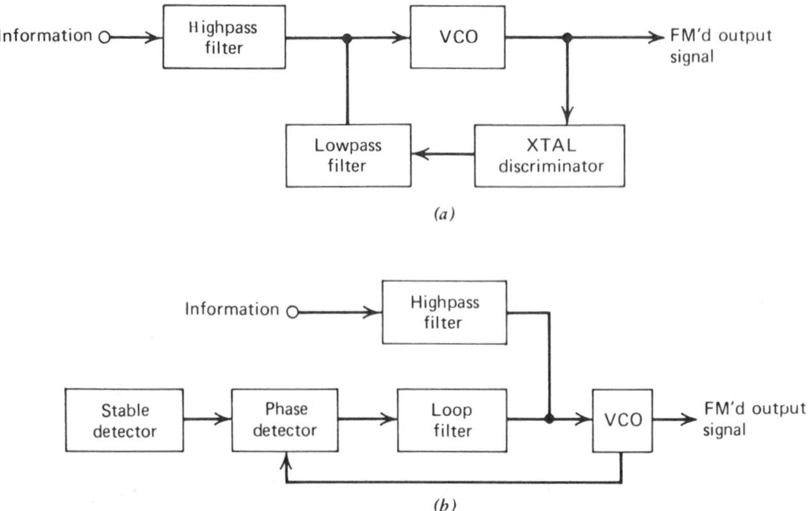

Figure 4.30 Stabilized FM modulator block diagram: (*a*) crystal discriminator stabilized; (*b*) phase locked to stable reference.

Code Modification

Code modification (also called phase inversion modulation or code inversion modulation) as a modulation form is the most useful technique for general use in spread spectrum systems. It works equally well for direct sequence and frequency hopping systems, although the effect on RF output is not exactly the same (as one might expect because the outputs themselves are not the same).

What do we mean by code modification? Here, we refer to the basic code sequence used in the direct sequence system to phase-shift modulate the carrier or the code sequence used in a frequency hopper to choose the pattern of hopping. Modification means that we must change the code in such a way that the information signal is embedded in it and is uniquely decipherable only by someone knowing the original (unmodified) code. In addition, the desired code properties—good autocorrelation and low cross-correlation, which we discussed at some length and are so important in spread spectrum systems—must be preserved.

Satisfying all these conditions at one time would seem to be no mean trick. The simplicity of the circuits used to solve the code modification problem belies its importance. Although many different approaches have been used, all reduce to some technique for digitizing the information to be sent and modulo-2 adding it to the code sequence. This process has the effect of inverting (ones become zeros and zeros become ones) the code each time a transition occurs in the information data stream. Figure 4.31 illustrates the operation in its simplest form.

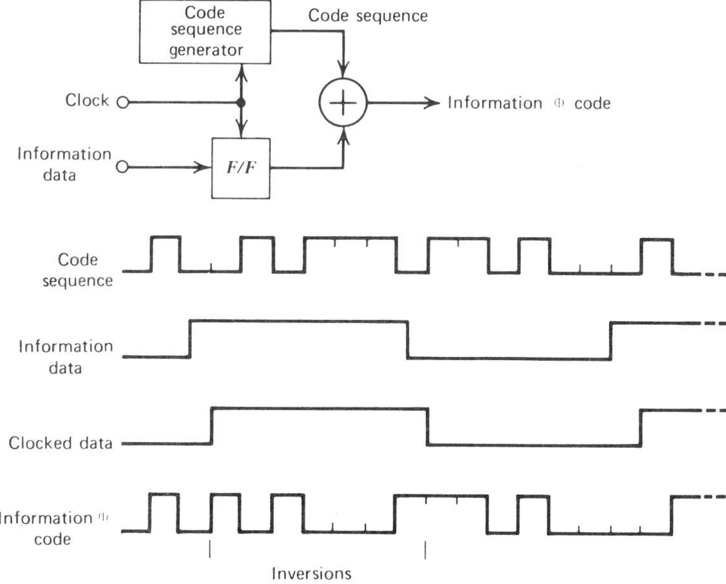

Figure 4.31 Basic code modification (modulation) using modulo-2 addition.

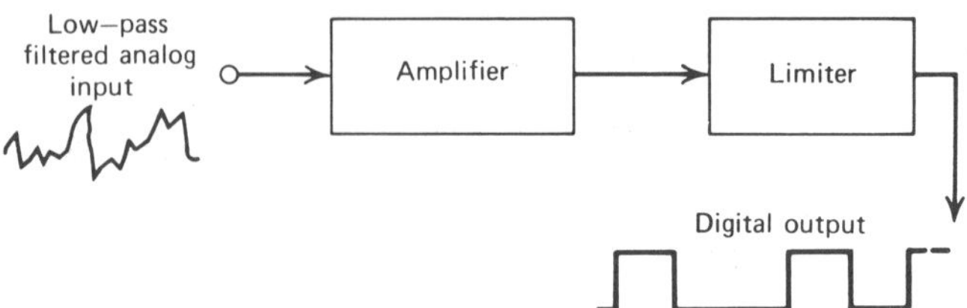

Figure 4.32 Simple limiting digitizer.

Note that the input data are in digital format and reclocked by the code clock before being added to the code. This is done to ensure that inversions of the code due to information are present only at times when a code transition would normally occur. To an observer who does not know the original code sequence the new product of information and the original code is of no more use than the original. A receiver who does know the original code needs only once again to modulo-2 add the composite (information $\oplus$ code) sequence* to its reference code and, then, the information is recovered.

Because the modified code is not actually transmitted in an RF communications system as binary data, but is sent as phase transitions on an RF carrier, the receiver uses another RF carrier which is also phase-shift keyed but with the unmodified code. These two phase-shift keyed carriers are multiplied together and the effect is the same as that described for binary signals. Here we consider the transmission process as completely binary for purposes of simplicity.

Digitizing Methods

Several ways to digitize analog signals (principally voice) for transmission in spread spectrum systems have evolved. The simplest is equivalent to clipped speech. It consists of applying the analog input signal to an amplifier and limiter whose output is a squared version of the analog signal. Voice processing of this kind has been used in both AM and FM to increase average modulation.

Figure 4.32 is a block diagram of this digitizer. After clocking, the binary signal generated by the limiter is modulo-2 added with a code sequence just as any other binary data stream would be.

In practice, digitizing in this way produces usable, though distorted, signals. When voice recognition is valuable, this technique is not recommended, although from the standpoint of getting a message across, word intelligibility is actually increased.[911] When simplicity is important and voice quality must be sacrificed, the limiting digitizer is worthy of consideration.

*Here we assume synchronization between the transmitter's and receiver's code sequences. Also because $A \oplus B \oplus A = B$, then data $\oplus$ code $\oplus$ code = data.

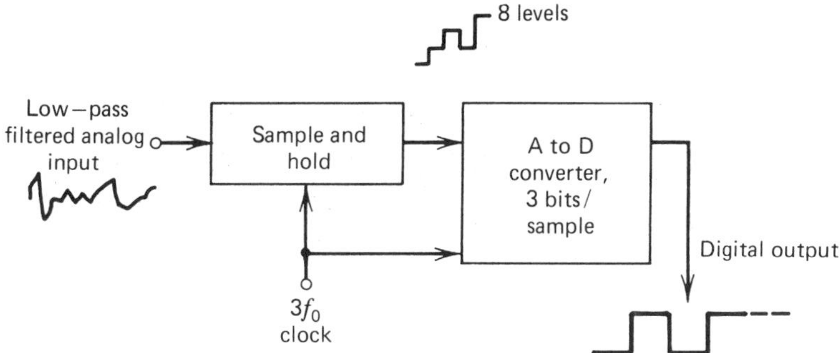

Figure 4.33 Three-bit PCM digitizer.

Pulse code modulation (PCM) is also of interest for signal digitization if only because of its wide use. Here we consider a simple form of PCM although in principle it is the same are more complex forms. Figure 4.33 shows a three-bit PCM generator capable of quantizing an analog signal into eight levels. For most applications in voice transmission systems three-bit PCM offers satisfactory quality.

Higher-fidelity requirements may be met by increasing the number of quantization levels and thus the number of bits transmitted per level. Remembering that for a spread spectrum system, the process gain available is a function of the ratio of the RF bandwidth to the information rate, we see that increasing the number of bits transmitted to represent a given signal is not a highly thought of practice. Therefore, PCM is not often used in spread spectrum systems for information transfer, except as in the following.

A technique that could readily be considered to be a form of PCM is "delta" modulation, except that only a single information bit is sent for each sample. This is accomplished by comparing a given sample with the previous sample and transmitting a one if the current sample is larger and a zero if the current sample is smaller than the previous sample. Thus, only representations of changes are sent as information, and the transmission rate required is greatly reduced. (For example, the transmission rate is equal to the sample rate, with delta modulation, where the transmission rate would be three times the sample rate for three-bit, or eight-level, PCM.)

A basic delta modulator is shown in Figure 4.34. This simple digitizer consists of a sampled comparator (i.e., whose output changes only at sample times) and an integrator. The integrator is used to integrate the comparator's output, which output is the digitized representation of the signal prior to the present input. The integrated signal then is compared to the input signal, and the result is a one or zero output at each sample time.

An advantage of the delta modulator's design is that, in addition to a minimum transmission rate, the same circuit is useful for demodulating the data stream. That is, the delta-modulating analog-to-digital converter

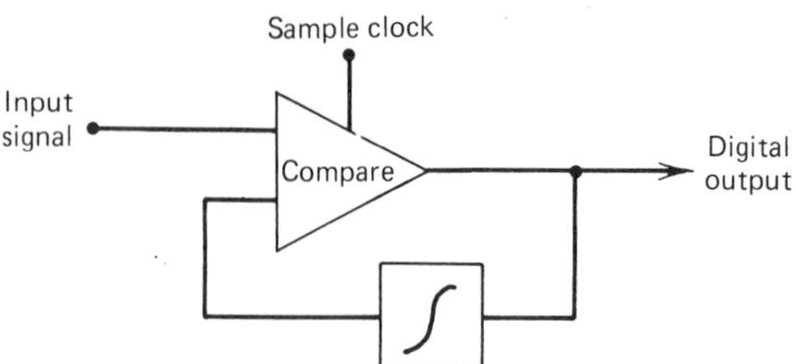

Figure 4.34 Delta modulator block diagram.

also serves as the digital-to-analog converter that regenerates a representation of the original analog information, in the receiver.

In the receiver mode, a received delta-modulated data stream is input to the same delta modulator, and the analog output is taken from its integrator (refer to Figure 4.34). In a simple delta modulator such as is shown, the fixed integration time is a disadvantage, however, if the signal to be transmitted has significant dynamic range. That is, if a larger and/or rapid change occurs, then the maximum step* taken does not effectively follow the intended waveform, and distortion is introduced. On the other hand, for small and slow changes, a smaller step size would be better.

Figure 4.35 illustrates the integrator's response to an analog signal.

The solution to providing for varying signal rates of change is to sense the variation and modify the integration time to compensate for the changes. Such a technique has been elegantly implemented in the "companded variable slope" or "continuously variable slope" delta modulators, both of which are typically called "CVSD" modulators. These modulators adapt to signal-level changes by counting the number of ones and zeros that have occurred, and when some specific number (usually three or four) have occurred in a row, by adjusting the integrator's time constant. Thus, if the requisite number of ones occurs in a row, which signifies that the signal level is tending to increase, then the integration time decreases and this enlarges step size. Conversely, if the same number of zeros occurs in a row, step size decreases.

*The response of an integrator to a pulse is to charge to a given level and remain at that level. Since the integrators in use here usually consist of a single RC integrator, the input pulses, together with the integrator's resistor from a current source that charges or discharges the capacitor by a voltage step

$$\Delta V_{oc} = \frac{1}{C} \int \frac{E_p}{R} \, dt$$

where E_p is the pulse level. If a series of ones is sent, then the integrator's output is a positive-going stepped (staircase) function, for zeros a negative-going stepped function, and for a steady-state signal, an alternating one–zero pattern is sent (which integrates to zero charge).

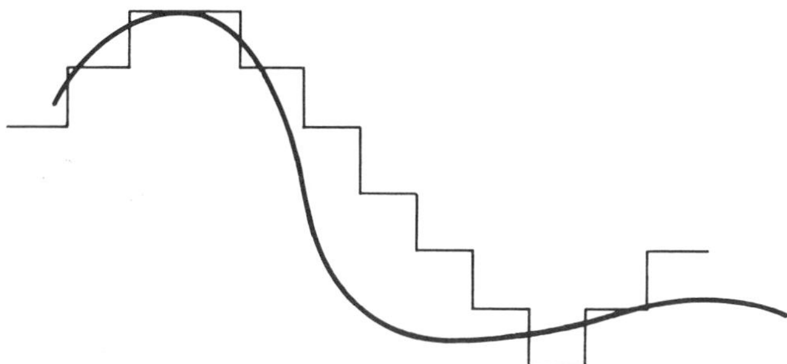

Figure 4.35 Delta-modulated representation of an analog signal showing distortion due to integrator step size.

Diagrams showing typical CVSD modulator/demodulators are shown in Figure 4.36. These IC units are simple to use and inexpensive, and dominate most applications for contemporary digital voice, including spread spectrum systems.

Performance of CVSD modulator/demodulators is good, with capability to operate at error rates as low as 1×10^{-2}. They also provide good intelligibility (see Figure 4.37). The prime drawback to their use for spread spectrum applications is that they tend to require operation at 16 Kbps or higher for voice applications, and sound noisy even though their intelligibility scores are good. Figure 4.38 shows the signal to noise performance that may be expected for CVSD modulation at 16 Kbps and 32 Kbps as input signal level varies.

As an aside, we note that vocoding techniques are also used to convert analog signals to digital (usually PCM) format. In general, these techniques break the voice band down into subbands. Information in regard to which subbands contain the power in the analog signal is then transmitted. Periodic sampling of the subbands and transmission of a digital representation of the significant subband is an analog-to-digital conversion method. Vocoding as a technique has been known for many years, but has not come into wide use largely because of its implementation cost.

Another method is suppressed-clock PDM (pulse-duration modulation). This approach is shown in Figure 4.39. An input analog signal is input to a sample-and-hold circuit operating at p times the highest frequency in the input signal. A sawtooth generator, synchronous with the data samples, is compared with each sample-and-hold output, and when the sawtooth ramp exceeds the stored level, the comparator output changes state. If the comparator output is set to the "one" state at the beginning of each sample and it goes to zero when the ramp exceeds the stored sample level, then a series of variable length ones is generated. The length of these ones depends on the stored level because the time required for the ramp to reach the stored level depends on that level. Noting that each PDM one varies between zero length and $1/R_s$ (the clock period)

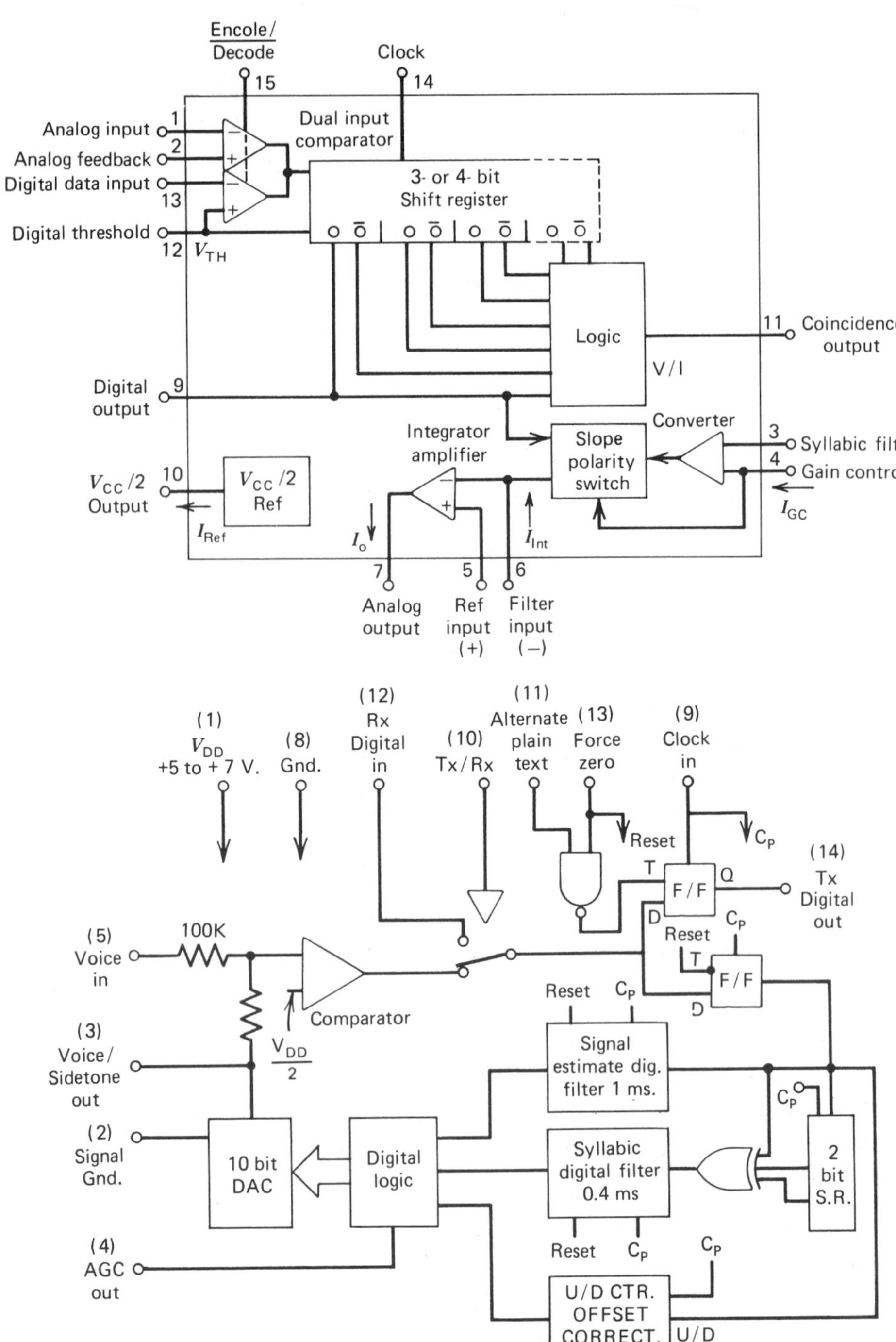

Figure 4.36 Typical CVSD integrated circuits: (*a*) Motorola MC 3417/18; (*b*) Harris Corp. HC-55516/55532

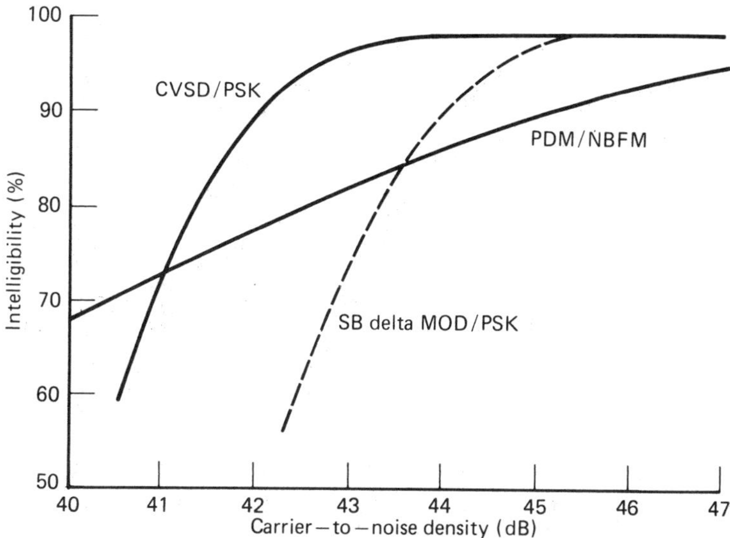

Figure 4.37 Comparison of PDM and CVSD voice.

and is synchronous with the clock, we modulo-2 add a square wave generated by dividing the clock by two. This modulo-2 addition subtracts the clock from the data stream and gives the same information but with an effective $p/2$ sample rate and therefore smaller information bandwidth. Replacing the clock information at the receiver is as simple as modulo-2 adding the half-clock rate square wave to the suppressed clock signal. This restores the original data stream.

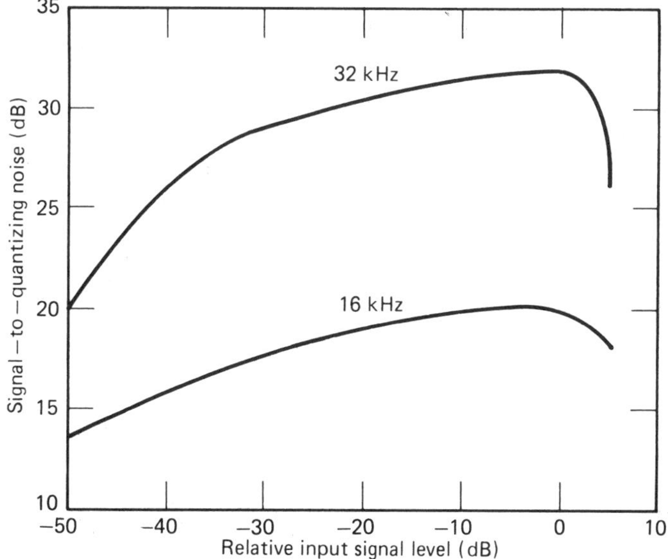

Figure 4.38 SNR using 1-kHz sine wave for CVSD.

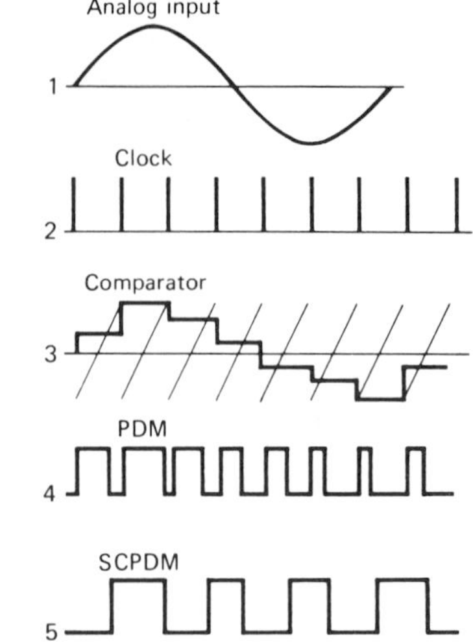

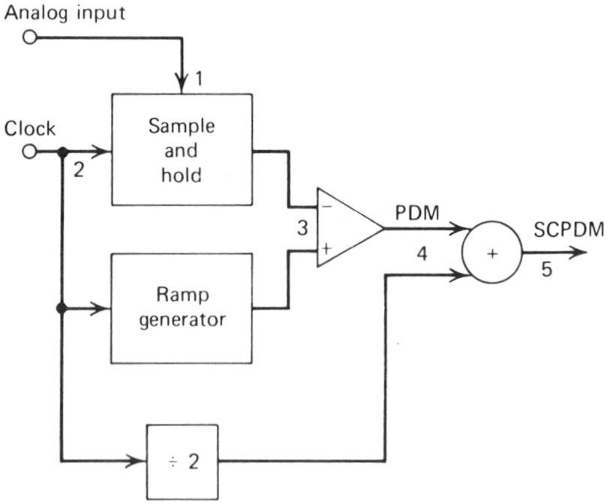

Figure 4.39 Suppressed-carrier PDM generator.

Suppressed-clock PDM works extremely well in spread spectrum systems, for it meets the requirement of being quantizable at the code clock rate. Curves that measure performance of SCPDM and FM for various FM deviation ratios, as carrier-to-noise ratio varies, are shown in Figure 5.25. The improvement derived from SCPDM over FM is in the range of 4–7 dB, which must be considered a significant gain when we note that to get a 4 dB improvement in process gain by increasing code rate the RF transmission bandwidth must be increased by a factor of 2.5.

We now come to the subcarrier-FM code-modification method of modulation. In this technique, as in the others we have discussed, code modification actually consists of sequence inversion keying by information in digital form. The subcarrier-FM method employs an oscillator, frequency modulated by the information to be transmitted, whose output is shaped, reclocked, and modulo-2 added to the code sequence. Subcarrier frequencies well below the code clock rate are chosen but must be high enough to be above the highest modulating frequency. Subcarriers of 50–500 kHz with code chip rates in the 1-to-100 Mbps range, are typical. Keeping the separation large between information, subcarrier, and code rate helps to reduce noise introduced in the digitizing process.

A voltage-controlled astable multivibrator is an excellent subcarrier oscillator for this use. Not only is its modulation linearity good but it also has a square wave output. Thus only reclocking at the code rate is necessary before addition to a code sequence.

Several of the information-transfer techniques discussed are more useful in direct sequence approaches than in frequency hopping. Subcarrier FM, for instance, or any other code modifier that increases the required hopping rate, is undesirable in a frequency-hopping system.

Remembering that data rate reduction produces as much processing gain improvement as does a bandspread increase, then we see that a lower rate voice digitization approach would be desirable. One such technique that is promising is linear predictive coding (LPC). Although not yet developed to the point of being implemented on a single IC chip as the CVSD units have, the LPC approach offers a process gain (and jamming margin) increase of 10 log (16,000/2400) = 8 dB over CVSD since LPC can provide a 2400 bps digitized voice rate. This may not prove to be quite as great as improvement as it initially appears, however, since the CVSD approach works to bit error rates as poor as 1×10^{-2}, while it is doubtful that LPC techniques can tolerate such high error rates and therefore require higher signal to noise ratios.

Frequency Hopping Code Modifiers

Figure 4.40 illustrates a code modifier adapted to frequency hopping. It uses only a modulo-2 addition of the code and input data. Reclocking at the code rate is not shown; however, this would be accomplished by transferring frequency-select information to the synthesizer only at clock times.

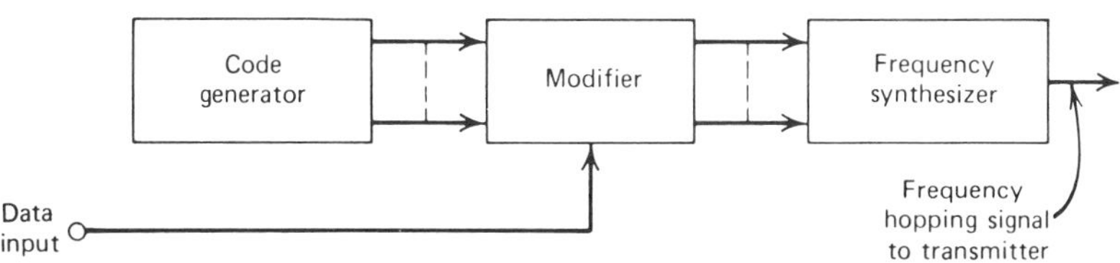

Figure 4.40 Frequency hopping code modifier (modulator).

This simple frequency hopping modulator is made up of three parts: (a) a code generator, (b) a code sequence modifier (information modulator), and (c) frequency synthesizer. In this scheme, in which P frequency choices are required from the synthesizer, the number of parallel bits of information from the code generator would be $\log_2 P$. The code generator itself would be made up of n stages, each feeding an output line. Of course, more than n stages could be used, but n outputs would still be satisfactory.

Each code generator input line connects to a data modifier whose other input is the data to be transmitted. Figure 4.41 is a diagram of a modifier for a frequency synthesizer with 2^5 frequencies to choose from.

The effect of the code modifier is to pass on a word that is the complement of a one in the code generator output, whenever the data input is a one. When the input data is a zero, however, the code generator output word is uncomplemented. Thus one frequency command is sent for a data one and another is sent for a data zero; for instance, if

$$\{Q_1, Q_2, Q_3, Q_4, Q_5\} = \{1, 0, 1, 1, 0\}$$

then

$$\{Q_1, Q_2, Q_3, Q_4, Q_5\} = \{0, 1, 0, 0, 1\}$$

Data-modified n-tuple $\}Q_1, Q_2, \cdots, Q_n\}$ is then used to command the frequency synthesizer to give a corresponding output frequency.

There are, of course, many other ways to modify the frequency command code. Simple inversion of the command word as illustrated

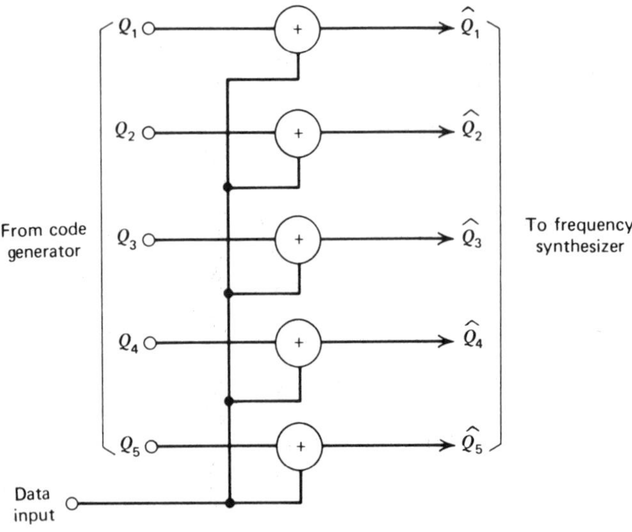

Figure 4.41 Simple code modifier.

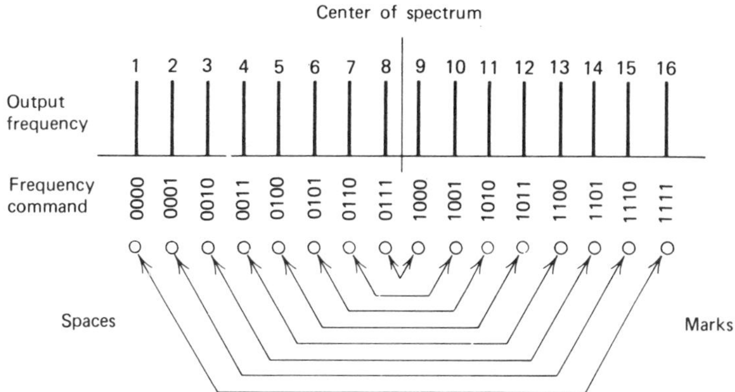

Figure 4.42 Illustration of space/mark symmetry for simple code modifier.

could cause the mark-space frequency choices to be symmetrically distributed, which is not desirable in most systems; that is, if the code word were 1 0 1 0, then a mark would be sent at frequency 1 0 1 0 ≐ 1 1 and a space would be sent at frequency 0 1 0 1 ≐ 6 (see Figure 4.42). The correspondence between the frequency command word and the output frequency may be modified to alter this pattern, but any fixed command-output correspondence would still give a fixed mark-space relationship; that is, if f_m were chosen for a mark, then f_s would always signify a space (or vice versa). A would-be interferer, knowing the correspondence between f_m and f_s, could render the system inoperative by transmitting the complementary frequency whenever any in the set were received. The only protection from such action by an interferer, then, is to modify the code-frequency correspondence periodically or change the code modifier so that a nontrivial relationship will exist between frequency complements. One simple method, useful for changing the one–zero relationship, is illustrated in Figure 4.43 (only one line is shown).

Here, instead of simply inverting the coded frequency to go from a zero command to a one command, a different code input is taken for each of the possibilities. Each of these coded frequency commands is taken from a different source (i.e., from two separate code streams). The two code inputs could come from separate code generators or from different taps

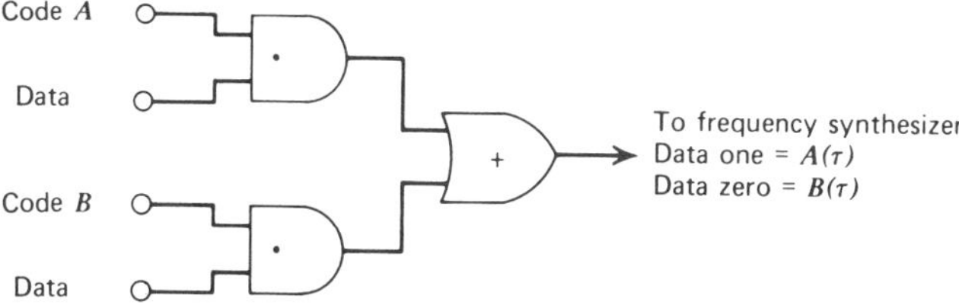

Figure 4.43 Gating for modifying code/frequency correspondence (showing one line).

on the same generator, depending on system requirements. In either case the one–zero relationship would be nonconstant and difficult to determine for an outside observer.

It is worthy of note that a frequency hopping system is inherently capable of protecting its internal code sequences to a higher degree than can the direct-sequence systems. This is because the direct sequence systems modulate their carriers with their code, and it is therefore directly observable. In a frequency hopping system, the code itself is not transmitted. Instead, a frequency representing a code vector is sent. Therefore, an observer must know the translation between frequency and code vector in order to determine the code that is in use.

PROBLEMS

1. For a given data rate, compare the RF bandwidth of quadriphase modulation with biphase modulation.
2. How many mix-and-divide modules would be required to develop 8000 or more frequencies from a set of 16 base frequencies in a direct frequency synthesizer?
3. If the synthesizer is the simple phase-lock (indirect) type, draw a block diagram to produce 8000 frequencies separated by 100-Hz spacing.
4. What minimum data rate would be usable for transmitting voice by a three-bit PCM digitizing method? (Let voice bandwidth be 3 kHz.) How much RF bandwidth would be consumed in providing 40-dB process gain for this digitized voice signal?
5. Compare the frequency-hop-rate limitations of direct and indirect synthesizers.
6. Determine the mean and base frequencies for a direct, 64-frequency add-and-divide synthesizer with output to 70 MHz and 10 kHz output spacing.
7. Show that carrier modulation is not effective for information protection in a direct sequence system.
8. How many chips of code are required to provide unambiguous frequency selection for a synthesizer with 2000 frequencies available?
9. Compare the relationships of base frequencies and divisors possible for a 256-frequency direct synthesizer. What is the optimum number of bases to minimize the synthesizer's complexity?
10. What is the effect of dividing the PDM signal, before clock subtraction, by 2 (toggling a flip-flop)?

5
CORRELATION AND DEMODULATION

Spread spectrum signals are usually demodulated in two steps: first, the spectrum spreading (direct sequence, frequency hopping) modulation is removed, and second, the remaining information-bearing signal is demodulated. In either the direct sequence or frequency hopping systems removal or demodulation of the spectrum spreading modulation is accomplished by multiplication with a local reference identical in structure and synchronized in time with the received signal. (Again we reserve the problem of synchronization for Chapter 6 and assume that received and local reference signals are synchronous.) Baseband recovery is a matter of demodulating the conventional kinds of signals—FM, FSK, PSK—in the face of noise. The noise of most interest seen by the baseband demodulator, however, is generated by cross-correlation of a local coded reference with an interfering signal.

This chapter describes the correlation (despreading or bandwidth collapsing) and baseband demodulation processes. In the area of spread spectrum communications there is no more important pair of operations, for it is here that process gain and code discrimination are effective. The other circuits used and other functions performed are peripheral to them; that is, the special properties of spread spectrum systems are realized in the correlation and baseband demodulation functions. It can be said that all other operations are performed just to make them work properly.

5.1 REMAPPING THE SPREAD SPECTRUM

The first of the operations performed by a spread spectrum receiver (neglecting amplification, downconversion) is that of despreading or removing the spread spectrum modulation. Any received spread spectrum signal is mapped from a large set of signal components, spread over a wide band, into a small or even single-valued set occupying ideally just enough bandwidth to pass any baseband modulation components. In this operation, called correlation, a received signal is multiplied with a locally stored model of the signal, and the resulting output is integrated. Demodulation then follows.

Quoting R. L. Rex and G. T. Roberts,[226] "As it applies to waveforms, correlation is a method of time domain analysis that is particularly useful for detecting periodic signals buried in noise, for establishing coherence between random signals, and for establishing the sources of signals and their transmission times." Here we are most concerned with establishing coherence between (pseudo-) random signals. Detection of the periodic information signal buried in (pseudo-) noise is left to the baseband demodulator.

For frequency hopping and direct sequence systems the correlation process is the same. The main difference is that the local reference is a frequency hopping or direct sequence signal, depending on which is to be received; that is, the local reference is matched to the desired signal. In Chapter 2 we discussed the way in which frequency hopping and direct sequence receivers realize their processing gains by transforming desired wideband signals into narrowband and undesired signals into wideband. The remainder of the process, as we also stated in Chapter 2, consists of separating out the desired signal.

Our purpose is to describe the implementation of typical correlators, which have much in common with the modulators discussed in the preceding chapter. Two types of spread spectrum correlators are discussed here. We have labeled them "in-line" and "heterodyne" correlators for obvious reasons.

In-Line Correlation

All correlators for direct sequence systems shown up to this point have been the in-line type, not because they are used more often or are superior to others but because their operation is exactly the same as a direct sequence modulator (the circuits can be identical) and can perhaps be a little more readily understood on that basis.

In the in-line correlator in Figure 5.1*b*, in which a phase-shift modulated received signal is one input, the signal is multiplied by a code identical to that used in generating the phase-shift keyed input signal. The effect of this is complementary to the code modulation in the transmitter: the input signal is phase inverted each time the local correlating code sequence has a one–zero or zero–one transition. If the transmitter's code is identical to the local code and the two codes are time synchronous,*

*Actually it is important to realize that the receiver must synchronize its code sequence to the code modulation on the signal received at its input. In practice the code sequence, as seen at the receiver, is quite different from that being sent by the transmitter at any given instant. This difference in code phase is, of course, due to the propagation time between transmitter and receiver and is the basis of spread spectrum ranging.

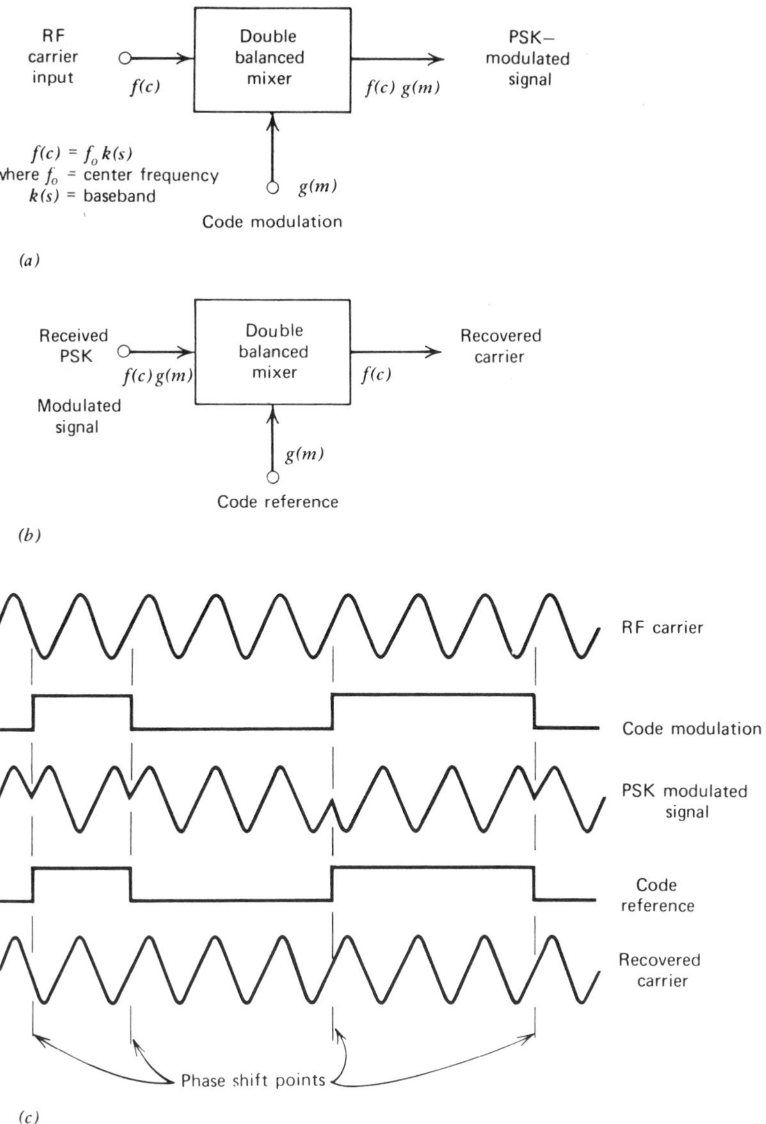

Figure 5.1 In line correlator carrier recovery: (*a*) modulating multiplier (PSK modulation); (*b*) in-line correlator (complementary PSK modulator); (*c*) waveforms in PSK (biphase) carrier recovery.

then at each phase shift of the transmitted signal the receiver phase shifts it again. These two complementary phase shifts combine to cancel one another and to restore the original carrier. The spectrum spreading–despreading operation is a "transparent" process in that carrier modulation, such as AM or FM, is passed through without change.

In-line correlation, though extremely simple, is not used in the majority of spread spectrum systems. The reason for this lies in the area of practical implementation rather than being a theoretical block. First note that the input and output frequencies are the same (see Figure 5.1); that is, if the correlator's PSK-modulated input signal is centered at f_o, the recovered carrier will also be at f_o. This, in itself, would be of no consequence were it not for one of the main purposes of the correlator—to be specific, rejection of interfering signals.

The correlator in a spread spectrum receiver is that point beyond which interfering signals cannot be permitted to pass, at least in their original form. Otherwise a narrowband interfering signal could enter the post-correlation circuits and masquerade effectively as a desired, properly despread signal. Remembering, then, that interfering signal inputs are likely to be extremely large compared with the desired signal, we must consider the likelihood of direct leakage of an interfering signal around the correlator. When leakage occurs, the carrier suppression capability of the correlator's balanced mixer is of no avail because the interfering carrier does not pass through the correlator but around it. For this reason in-line correlation is not often used.

Heterodyne Correlation

A heterodyne correlator, as implied by its name and the preceding discussion, is one in which the correlated signal output is at a different center frequency from the input signal. In the process of despreading or removing the code modulation the information-bearing signal is translated to a new center frequency. This avoids the possibility of direct feedthrough and in some instances simplifies receiver design, because the circuitry following a heterodyne correlator can be at a lower frequency. Frequency hopping correlators are almost always of the heterodyne type (to be more exact, I do not know of any frequency hopping correlator that is not). Direct sequence correlators, in general, also employ heterodyne correlation.

Heterodyne correlators are shown in block diagram form in Figure 5.2 for direct sequence and frequency hopping receivers so that a direct comparison between the two can be made. (Filters that would normally be employed between the local reference modulation and the correlating mixer are not shown.) The direct sequence correlator in Figure 5.2*a* generates a local reference signal that is a frequency-offset replica of the expected direct sequence signal (with the exception that the local reference is not information-modulated). If we were to observe the signals generated in a direct sequence heterodyne correlator, we would see the spectra in Figure 5.3.

The local reference signal in Figure 5.3 is generated in exactly the same way as the transmitted signal. In fact, when we have a receiver and transmitter than do not operate simultaneously, the same circuit could be

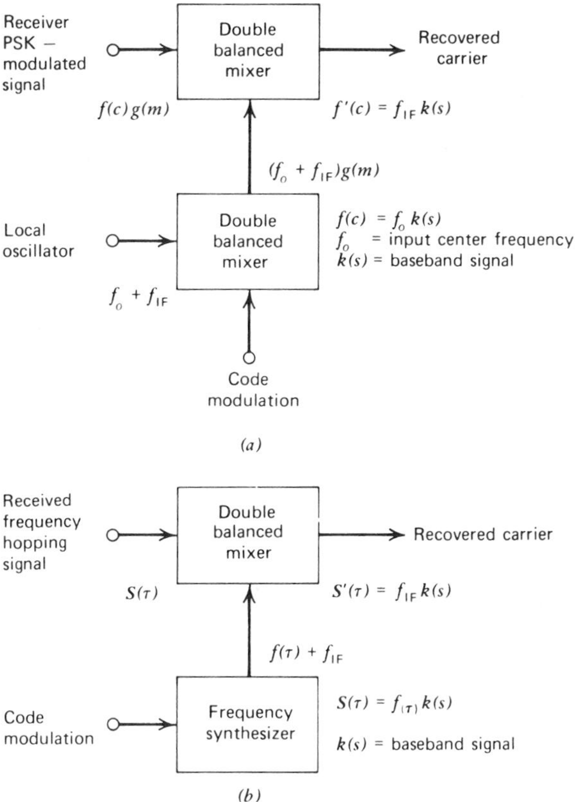

Figure 5.2 Comparison of heterodyne correlators for direct sequence and frequency hopping: (*a*) direct sequence heterodyne correlator; (*b*) frequency hopping heterodyne correlator.

shared between the transmit modulator and receive local reference generator.

In both direct sequence and frequency hopping systems the correlation process is a frequency-by-frequency multiplication (corresponding one-to-one with a chip-by-chip code comparison in the time domain) of the local reference with incoming signals.

The frequency-by-frequency correlation process is obvious when considering frequency hopping, but is not quite so obvious in direct sequence systems. If, however, one considers that each run of ones or zeros is a half period of a square wave related to the code chip rate (i.e., $1/RC, 2/RC, 3/RC, \ldots, n/RC$), then we see that the modulated carrier has sidebands that are related to these runs. Thus, when the local reference code is synchronized with the incoming signal code, the local reference sidebands correspond frequency by frequency to the input signal sidebands. This correspondence in turn allows the two signals

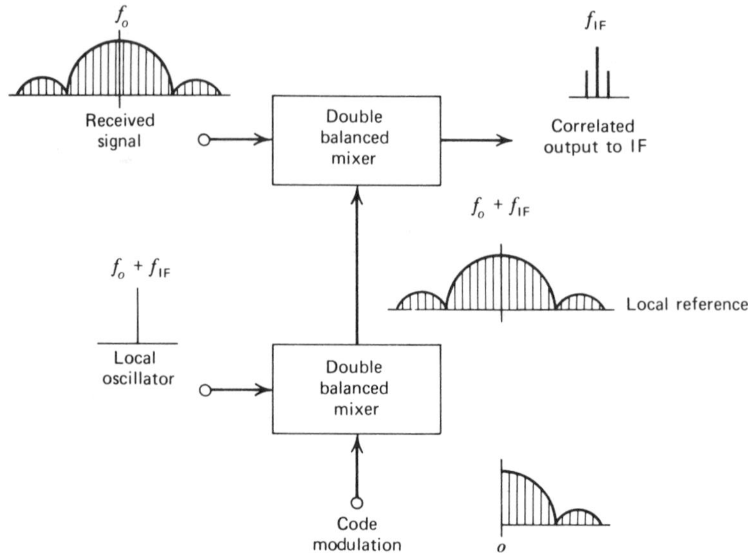

Figure 5.3 Signal spectra in direct sequence heterodyne correlator.

(local reference and received signal) to correlate and produce a despread output signal.

Typical correlator output signals are shown in Figure 5.4*a* and 5.4*b*. Figure 5.4*a* shows the unmodulated despread carrier, and Figure 5.4*b* shows the suppressed carrier spectrum typical of PSK modulation by a square wave. In this case the modulation is a 1200-Hz square wave, simulating a one–zero series at 2400 bps. No interference is present in either case. Note that in Figure 5.4*b*, the modulation sidebands are symmetrical around the suppressed carrier, and that only odd harmonics are present. (It is also noted that the two primary sidebands are not equal in amplitude, as they ideally should be. This was found to be due to amplitude nonlinearity in the receiver, but did not result in a detectable loss in system performance.)

Code Timing Offset Effects On Correlation

The prime purpose of the correlator, whether direct sequence or frequency hopping, is to match a local reference signal to a desired incoming signal and thereby reproduce the embedded information-bearing carrier as an output. So far we have assumed that a perfect match exists between our receiver's local code sequence and the received signal code modulation, both in the pattern of the codes and in their timing. Such a happy circumstance, however, seldom exists in a real world system; therefore, it is of some interest to examine the effect of less-than-perfect code synchronization on correlator output.

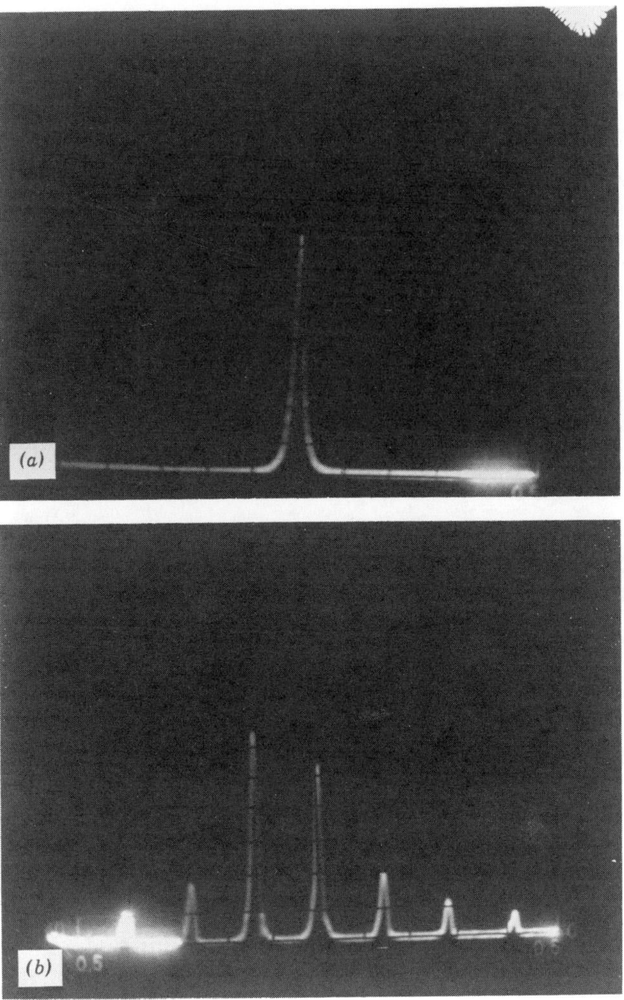

Figure 5.4 Typical correlator output signals: (*a*) no modulation, no interference; (*b*) PSK modulation, no interference.

In Chapter 3 we showed that the autocorrelation function for a maximal linear sequence is the triangle in Figure 5.5. For signals encoded by such a sequence the maximum signal output from the correlator occurs when the two codes are in the zero-shift position. (For that matter maximum autocorrelation output occurs for any code type in the zero-shift position. The only reason for speaking specifically about maximal linear sequences here is that their autocorrelation is well behaved in all shift positions; this, in turn, saves a great deal of qualifying explanation—such as this statement.) The optimum signal-to-noise ratio occurs when the local and received codes are exactly aligned, because this is the point

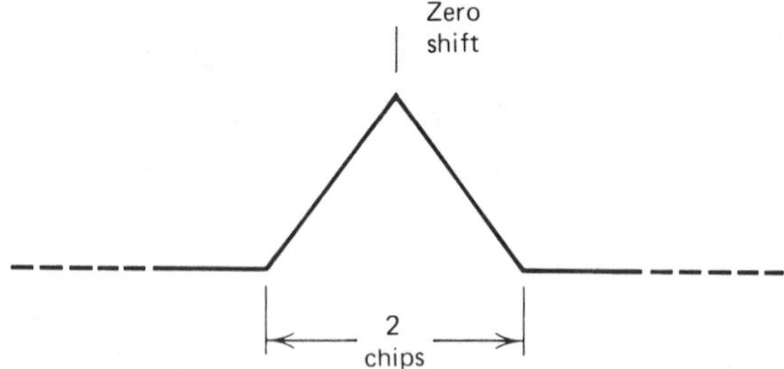

Figure 5.5 Autocorrelation for binary code sequency in a region of perfect synchronization.

of maximum signal output. (Where a receiver employs AGC to control the signal level, the autocorrelation curve is not one of signal level, but instead is a plot of signal to noise ratio.)

Noise seen at a correlator's output derives from several sources:

Atmospheric and circuit noise × code modulation.

Undesired signals × code modulation.

Desired signals × code modulation.

Noise output due to atmospheric and circuit noise is overshadowed in those circumstances in which undesired signals (jamming or other interference) are present. This is due to the ability of spread spectrum signals to operate in environments in which undesired signal power is much larger than desired signal power. Atmospheric and circuit noise are of real concern only in maximum-range, interference-free conditions.

When the incoming desired signal is exactly matched with the receiver's reference, there is no noise generated by the desired signal. Before synchronization, or during imperfect synchronization, a part of the desired signal is output as code-related (pseudo-) noise. The amount of output noise depends on the degree of synchronization; when there is no synchronization (the local and received signals are more than one code chip apart), the output produced is all noise.

Considering the diagram in Figure 5.6 in which a pair of identical codes is not quite synchronous, we see that the corresponding frequencies produced by code modulation for direct sequence,* or the frequency

*Here we recognize that direct sequence modulation and frequency hopping are actually similar. A frequency hopper generates a new frequency for each new code (generator) state by commanding a frequency synthesizer to output a particular frequency corresponding to that state. A direct sequence modulator does nearly the same thing by directly modulating with a code that has varying one and zero durations. These variable-duration pulses then produce variable-frequency sideband spectra in the modulator output, which

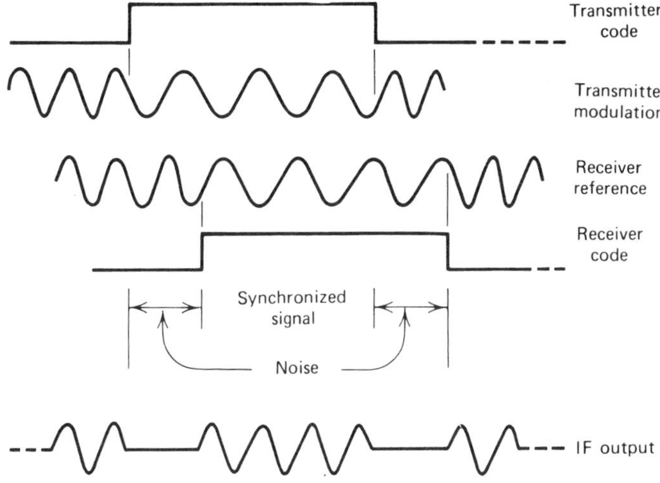

Figure 5.6 Effect of code timing offset on correlator output.

transmitted for a frequency hopper, overlap and that for a part of each code run the signal is properly translated into the IF. During the nonoverlapping chip time, however, the product of the incoming and reference signals lies somewhere in a region bounded by the covariance of the two signals. This part of the output is noise, some of which will fall within the IF bandwidth.

It is interesting to note that as the two codes slip from exact synchronization the correlator output code noise increases an amount corresponding to three times the offset. Observing Figure 5.6, we set that for each r reduction in the synchronized signal, due to code timing offset, there is a $2r$ increase in the noise contribution which then results in a signal-to-noise function

$$f(\tau) = A\,\frac{T-\tau}{2\tau}$$

where A = maximum output,
T = period of one bit,
τ = timing offset.

---

charge each time the modulating code has a one/zero transition. Both frequency hopping and direct sequence signals are then seen to be frequencies or sets of frequencies which change at a code determined rate, and frequency hopping can be said to be a single sideband version of direct sequence modulation. In the heterodyne correlator the two work in the same way, translating the desired signal, frequency-by-frequency, to a single output frequency while translating everything else to a (pseudo-) randomly determined frequency.

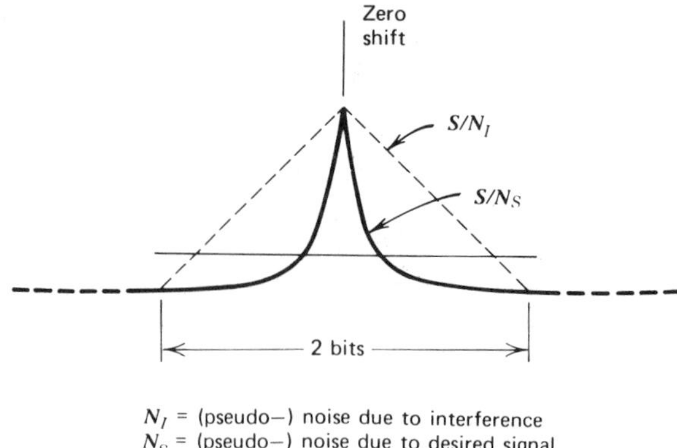

Figure 5.7 Comparison of signal-to-interference-noise and signal-to-self-noise for code turning offset of ±1 chip.

This function is plotted in Figure 5.7 when no interfering signal is present and the desired signal is above threshold. If a significant amount of interference occurs, however, the signal-to-noise ratio increases as a linear function of synchronization accuracy. Why? Because the code × noise interference produced is not so much a function of code timing accuracy and is large enough in any case of interest to an antijamming system to swamp out any self-noise effects completely. Figure 5.7 compares signal-to-noise functions for jammed and unjammed input signals. (Note that the signal-to-noise function for the high noise condition is the same as the code correlation function.)

The difference in the two curves in Figure 5.7 from the standpoint of their width would seem to imply that more accurate range measurements are possible under high signal-to-noise (i.e., no interference, high-signal) conditions. This is certainly true and should be expected. After all, any signal-to-noise sensitive device (and what device is not) used to detect synchronization would be forced to recognize the synchronized signal within, for example, approximately ±0.2 chip of the zero shift under high-signal conditions in which it would resolve to only ±0.8 chip for the same threshold point when being jammed.

5.2 EFFECT OF NONSYNCHRONOUS INPUT SIGNALS

The performance of a spread spectrum system when faced with nonsynchronous input signals is as important as its handling of desired signals. Remember that we said that the correlator is the point beyond which an undesired signal must not pass? The problems illustrated in the preceding chapter for balanced modulators also apply to the balanced modulators that are part of a correlator—and apply more stringently. Referring to

Chapter 4 and the various possible balanced-mixer problems illustrated there (see Figure 4.4 through 4.10), let us examine their effect on a spread spectrum receiver. First, what effect has carrier suppression (or lack of it) on a correlator and its output?

For the in-line type of correlator it should be obvious that insufficient carrier suppression would allow a jamming signal to leak directly through and into the detection circuits. When an interfering signal is multiplied by a coded signal, or directly by a code as in an in-line correlator, the signal is suppressed by the multiplier's suppression factor, whereas the modulation products output are equivalent to any other multiplier. Therefore, an interfering input signal is (or should be) reduced to a level well below that of the pseudonoise produced by multiplication with a code.

The spectra shown in Figure 5.8 and 5.9 illustrate the effect of jamming signal leakage in a direct sequence system, due to lack of carrier suppression. Note that the leakage-related signal appears as a narrow band carrier. Proper manipulation of the jammer's frequency can in turn cause this unsuppressed carrier to capture the receiver's carrier tracking loop, making the jammer very effective.

Figure 5.10 shows a direct sequence spectrum in which both carrier and code-balance suppression are unsatisfactory. Such a signal, if applied to a narrowband detector, such as a phase-lock loop, would cause the phase-lock loop to lock on the unsuppressed carrier. Proper carrier suppression, however, would depress the signal in the noise and prevent the detector from locking to this spurious signal.

The signal outputs seen at the nulls in Figure 5.10, caused by poor code

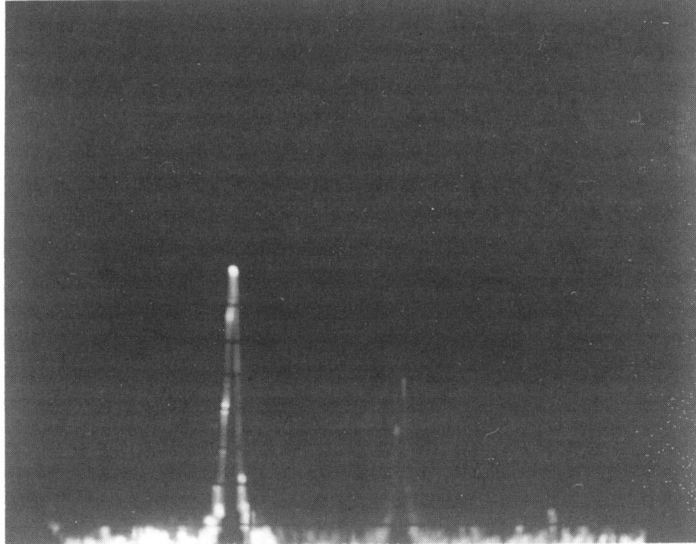

Figure 5.8 Balanced mixer leakage at correlator output. 8191 code. No modulation.

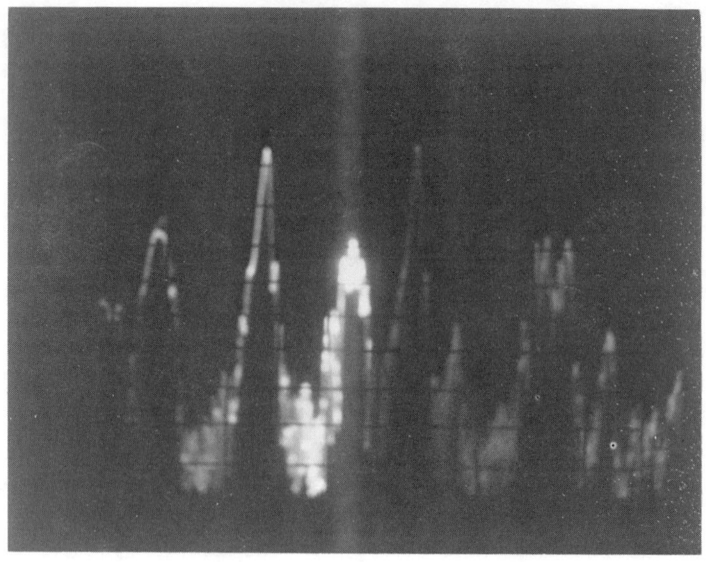

Figure 5.9 Balanced mixer leakage at correlator output, 8191 code. No modulation.

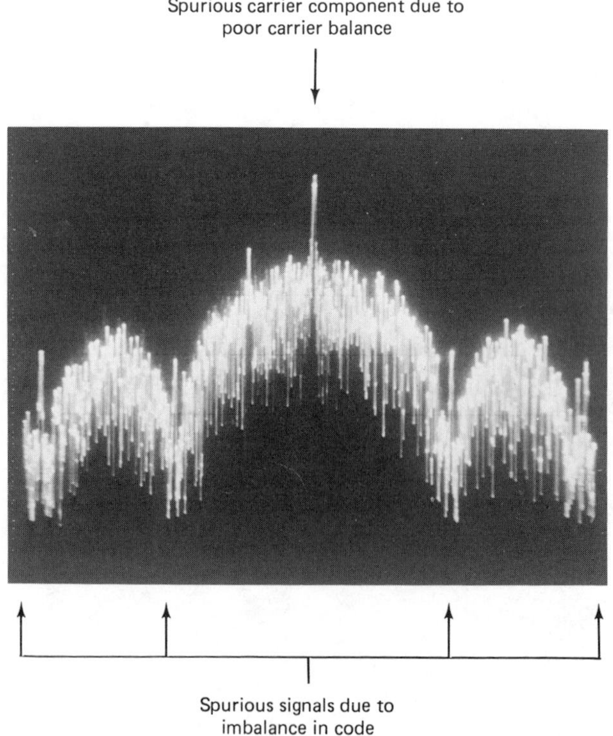

Figure 5.10 Typical direct sequence correlator output with poor suppression of carrier and code-unbalance components.

balance, can also cause the detector to lock up and falsely register signal detection. Worse yet, these narrowband signals lie at a point where they are not well masked by code-related noise. Even when good carrier suppression exists poor code balance can allow false detector lockup due to interference. (The interference, rather than occurring at center frequency, would necessarily be offset by an amount equal to the clock rate.)

Does the same thing happen in a heterodyne correlator? Yet, it does, but not quite in the same way. In the heterodyne correlator the local reference is the one that must show good suppression of spurious signals. Consider that the spectrum of Figure 5.10 is a local reference signal to a correlator in which the received signal is at f_o, the local reference center frequency is $f_o + f_{\text{IF}}$, and the signal nulls are at $f_o + f_{\text{IF}} + f_{\text{clock}}$. Again, interference translated into the receiver IF causes the same problem (i.e., recognition of interference as a desired signal) as in a poorly operating in-line correlator.

Several methods (and some empirically generated rules) have been employed successfully in efforts to overcome these problems. One is the intuitively reasonable rule of thumb which states that the carrier suppression of a balance mixer used in a direct sequence system should be at least equal to the process gain desired in the overall system; that is, if the system RF bandwidth is 20 MHz and its detection bandwidth is 10 kHz, the process gain expected is 33 dB. The correlator then should have 33 dB or more carrier suppression.

When very high processing gain is required, beyond the capability of a simple balanced modulator, tandem arrangements of more than one have been used successfully in correlators to attain the required carrier suppression. Otherwise, many practical correlators are limited to about 30-dB carrier suppression if environmental conditions are considered.

A tandem correlator consists of a pair of mixers arranged as shown in Figure 5.11. An auxiliary code (code 2) is modulo-2 added to the primary code (code 1, which matches the received signal's code). When code 1 and

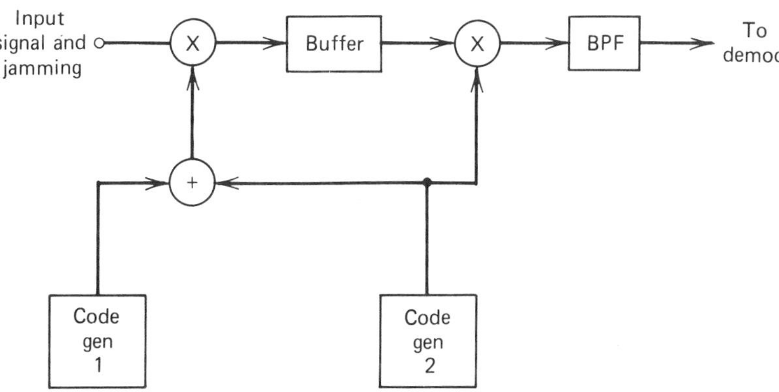

Figure 5.11 Tandem in-line correlator.

the received code are synchronous, the output from the first multiplier is still modulated by the auxiliary code. Thus any leakage through the first multiplier (mixer) is masked by a broadband signal and does not appear as a coherent input to the second multiplier. At the second multiplier, the auxiliary code is multiplied with the output of the first multiplier, and since it is synchronous with itself, remaps the desired signal into the desired narrow bandwidth. At the same time, it spreads the residue of the first multiplication due to the jamming signal. Thus, the effect to the demodulator is that of a correlator whose balanced mixer has the carrier suppression capability of the first multiplier plus that of the second. In this way, two balanced mixers with nominal 30 dB carrier suppression can produce an overall suppression of 60 dB—a significant result.

To overcome the possibility of false receiver-lock on code-unbalance-related signals (which can occur at all multiples of the code clock around the signal carrier), some form of filter is required ahead of the receiver's correlator. This filter may take the form of a simple fixed-frequency bandpass filter in an IF preceding the correlator or it may be a more complex tuned filter in a receiver preselector. At any rate, the bandpass characteristic of the filter must be such that spectrum roll off beyond the main lobe prevents locking due to code imbalance from signal input to the correlator at center frequency plus or minus the code clock and its multiples. Filter requirements vary from system to system because of code rates, gain and AGC values, and other parameters. In general, however, the code-unbalance lockup-prevention filter should be approximately as wide in bandwidth as the code rate (BW $\approx$ code rate) and roll off sharply thereafter.

It is also important that a filter having bandwidth characteristics similar to those of the IF filter be inserted at the output of the local reference. This has two effects:

1. It reduces the level of any stationary signals at the nulls of the local reference signal.
2. It provides a match between the level of the various components in the local reference spectrum and the input signal spectrum. (This is necessary to prevent generation of undesired crossproducts in the cancellation process.)

Figure 5.12 illustrates the correlator with these two filters in place.

We have spoken about the effects of carrier suppression and other effects of improper direct sequence correlator operation. There is much less in a frequency hopping correlator to concern us in that neither carrier suppression nor code imbalance have any meaning to the frequency hopper. The nearest equivalent is a stationary spurious signal in a frequency synthesizer output, fed to the receiver correlator as a local reference. This signal could cause undesired signals to be translated into the receiver IF. Frequency hopping synthesizers, however, do not

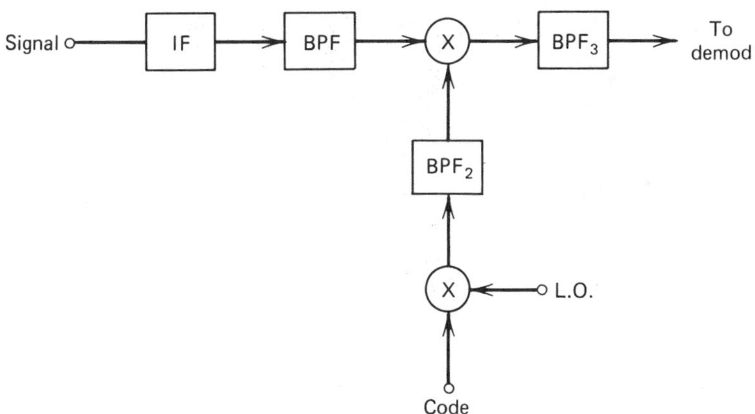

Figure 5.12 Correlator showing positioning of bandpass filters. (BPF, passband same as BPF_2 passband, but offset in frequency. BPS may inegral to IF.)

typically exhibit stationary spurious outputs.* Spurious output signals from such synthesizers normally vary from frequency to frequency at the same rate as the desired output because they are generated as a function of the component signals and synthesizer configurations necessary to generate the desired signal output, and those change as the output frequency changes. For this reason frequency hopping systems are not so susceptible to "ticklish" spots, in which an interfering signal is likely to cause reception problems, as direct sequence systems.

It should be noted that the very process of frequency hopping inherently produces spurious signal sidebands. The frequency hopping signal is, of course, a series of pulses at the hopping rate, and these have spectra that are a function of the pulse shape. Typical frequency hoppers with rectangular pulse output can be expected to have (sin x/x) distributed sidebands. Work has been carried out in the area of sideband reduction through pulse shaping, but such systems are currently the exception. The key is to remember that a frequency synthesizer may have extremely low spurious output when at a fixed frequency but high spurious levels as soon as it begins to hop, because of the pulsed signal characteristic when hopping.

Correlator Output With Interference

The output of a spread spectrum receiver's correlator (before filtering) is interesting to observe. When both interfering and desired signals are

*Here the word "stationary" must be emphasized. Frequency hopping synthesizers have spurious outputs. It would be miraculous if they had none. The spurious outputs change as rapidly as the desired outputs, however (and it is hoped are at lower levels). Therefore, it is at least as hard for a would-be interferer to find them as the intended output.

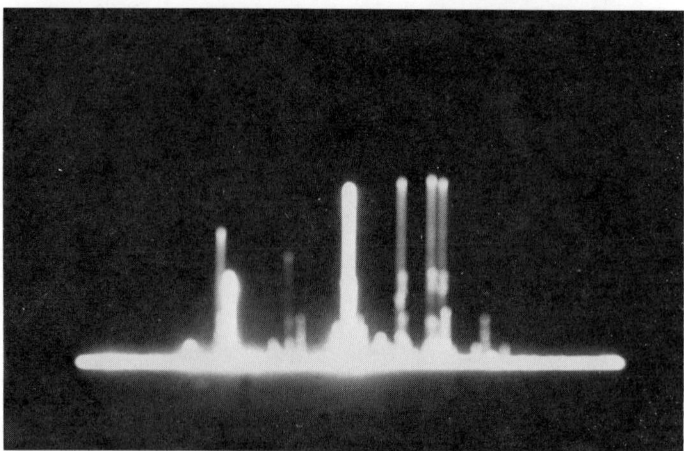

Figure 5.13 Output of frequency hopping correlator with both desired and interfering signal inputs (exposure for approximately five chip intervals).

present, a great deal of insight into system operation is available to the viewer. Figures 5.13, 5.14, and 5.15 show actual correlator outputs under interference conditions for frequency hopping and direct sequence receivers.

The first of these photographs shows a frequency hopping correlator's output for an exposure time covering a few frequency chips. Inputs to the correlator are a synchronous desired signal and an equal power CW interfering signal. The recovered carrier, which has been remapped to a single frequency, is at the center; the CW interfering signal (a single frequency input to the correlator) appears to be at several different frequencies—one for each new local reference frequency. An increase in the interference level by adding more jamming signals only causes an increase in the number of signals seen at each chip interval. (Figure 5.13 would also be identical to a correlator with five CW jamming inputs and a desired signal but with exposure for the duration of a single chip. For this reason only one photograph is shown.)

Figure 5.14 shows the output of a direct sequence receiver's correlator with various levels of CW jamming. This series of photographs is more interesting in that it shows the increase in the level of the (pseudo-) noise output with respect to a constant-level desired signal. As jamming level is increased, we can see that the signal to (pseudo-) noise ratio out of the correlator decreases. The demodulator which follows must pick the desired signal out of the noise generated by the correlator in the same way that any other demodulator picks its desired input from noise. The only difference is that the noise from a correlator is not strictly random but is structured by the particular code sequence that generates it.

At the 15-dB interference-to-signal level, the desired signal apparently

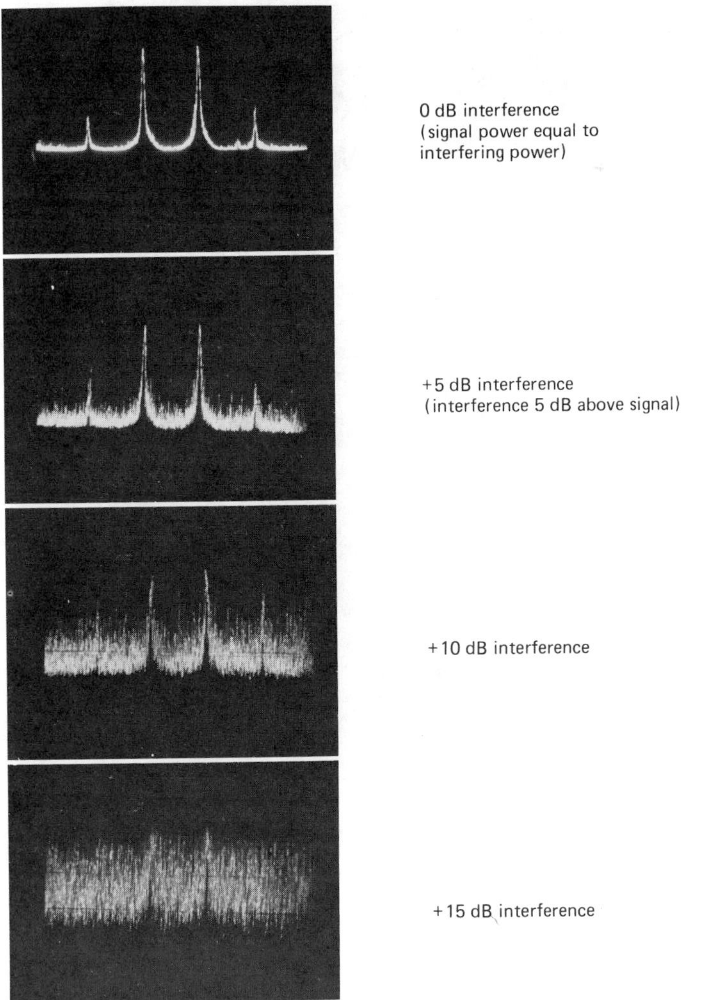

Figure 5.14 Typical direct sequence correlator output with various interference levels.

has almost disappeared into the noise. Operation is still satisfactory, however, and the demodulator still produces a usuable output signal. With equal signal and interference (O-dB) levels, the correlator output appears to be completely free of interference.

Figure 5.14 shows correlator output where the carrier is modulated, using the subcarrier-FM technique. Figure 5.15, on the other hand, shows correlator output where the carrier is unmodulated.

Two characteristics of the spectra shown in Figure 5.14 and 5.15 should be noted. One is that the noise shown is that produced by cross-correlation of the CW jamming signal with a long ($2^{31} - 1$) chip code. The

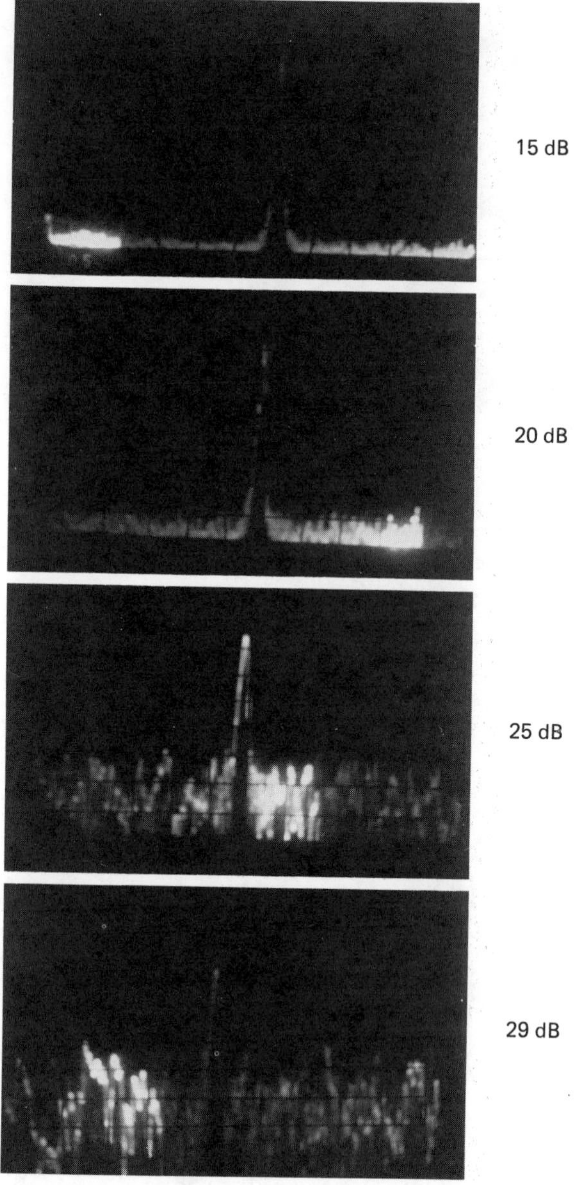

Figure 5.15 Long code, CW interference. Demodulator input signal at various J/S ratios. Unmodulated signal input.

other characteristics is that the desired signal level decreases as the jamming signal becomes large.

The first characteristic is important to us in that the noise produced is approximately "white." The second characteristic is due to the receiver's AGC action. We will refer to both characteristics in our later discussions.

Direct-Sequence Reaction to Interference

One of the major properties advertised as a feature of all spread spectrum systems is their ability to reject interference. Difference in signal structures and system implementations, however, lead to variations in reaction to the same kind of interference. Single-frequency CW interference is theoretically most effective in disrupting a direct-sequence receiver, for instance, although another type of receiver may well be unaffected.

Let us inspect the action of a direct sequence receiver when operating in a hostile environment. Figure 5.16 is a model we can use to examine this operation. Correlator input in Figure 5.16 is $S + I + n_{sys}$, and in the case of interest to us here $I >> N_{sys}$. Ignoring the system noise contribution, we see that correlator output as related to its input is

$$\left(\frac{S+I}{I}\right)_{out} = G_p\left(\frac{S+I}{I}\right)_{in}$$

except that the two inputs are transformed as in Figure 5.17. The bandpass filter at the correlator output completes the picture by accepting the narrowbanded desired signal and rejecting all of the interference lying outside the filter band. Because the interference is transformed into a wideband signal by multiplication with the local reference, most of its power lies outside the information band. The ratio of the bandwidth of the local reference to the bandwidth of the information filter is then the receiver's process gain; that is,

$$G_p = \frac{\text{BW}_{\text{RF}}}{R_{\text{info}}} \approx \frac{\text{BW}_{\text{local reference}}}{\text{BW}_{\text{info filter}}}$$

and the filter output due to interference is I/G_p.

Relating the information filter output to correlator input, we then see that

$$\frac{S + I/G_p}{I/G_p} = \frac{SG_p + I}{I}$$

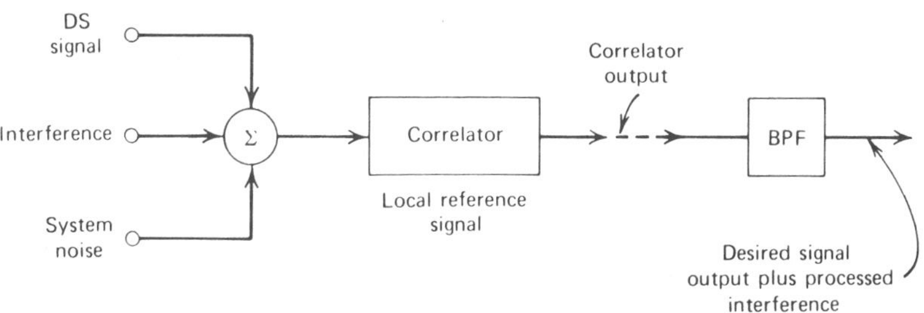

Figure 5.16 Direct sequence model.

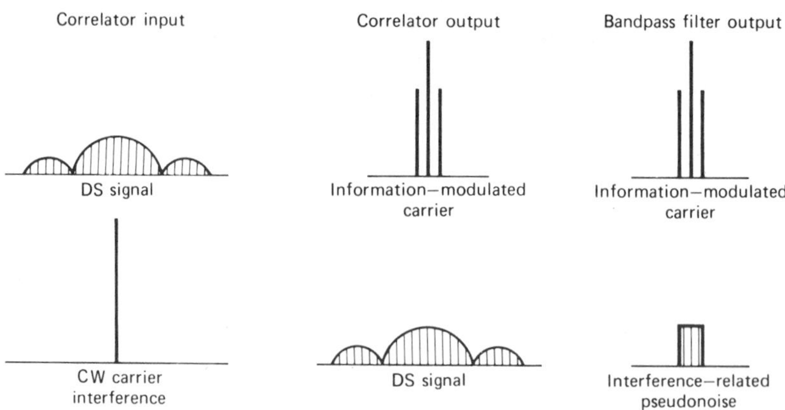

Figure 5.17 Signal transformations due to multiplication with local reference in direct sequence correlator.

where S = signal power,
I = interference power,
G_p = process gain.

Thus, if process gain is defined, filter output S/N is approximately

$$\frac{SG_p + I}{I}$$

which is the signal-to-noise ratio delivered out of the correlator output filter to the receiver demodulator and synchronization detector. The demodulator and synch detector following the correlation process operate as if they were not included in a spread spectrum system at all. In fact, the signals presented to them are not wideband but consist of the reconstituted, modulated carrier signal and of interference decomposed as pseudonoise. (We continue to ignore the small contribution of system internal noise compared with interference. System noise effects are considered in Chapter 7.) Differences in the effect of various kinds of interference depend on the type of interference and the specific demodulator and synchronization circuits used, plus the cross-correlation of the code used with the interfering signal.

At first glance we might consider that a spread spectrum system would be most affected by spread spectrum interference. Except under very restricted conditions, however, in which the interfering signal has high correlation with the desired signal (implying high code cross-correlation), this is not true. Any signal (including the desired signal when not in exact code synchronism) is spread in bandwidth by multiplication with the local receiver reference. Furthermore, the bandwidth of a signal at the correlator output is equal to its own bandwidth before multiplication plus the bandwidth of the local reference signal. An interfering signal with 2 MHz bandwidth being input to a direct sequence receiver whose expected signal (and local reference) is 10 MHz wide would produce a correlator output signal into its bandpass filter that is 12 MHz wide. The wider the bandwidth of the input interference, the wider the output signal. Thus the wider the input signal, the less is its effect on the system because the power density of the signal after processing is lower and less power falls in the bandpass filter.

Conversely, then, we might guess that the most effective interference to a direct sequence receiver is that which has the narrowest bandwidth—a CW carrier. In fact, CW interference is the most effective generally against a direct sequence system. This has been proved in tests of numerous direct sequence systems because the power density in the correlator output due to a narrowband signal is higher than that due to wider band signals.

How big is the difference in signal output due to wide and narrowband interference? If we consider that a CW carrier is the narrowest interference to be seen and that the widest interference bandwidth is equal to the local reference bandwidth (a reasonable assumption because a direct sequence receiver is likely to filter all input signals to equal its reference bandwidth), then the correlator's output varies between one and two times the reference bandwidth. Noise in the correlator output filter bandwidth then varies over a range of two (3 dB) for equal power but varying bandwidth interference. Figure 5.18 illustrates correlator input and output signals and bandpass filter output signals for the cases of CW and wideband interference.

It would be derelict to continue without mentioning the role that the code sequence plays in determining interference effects. We have

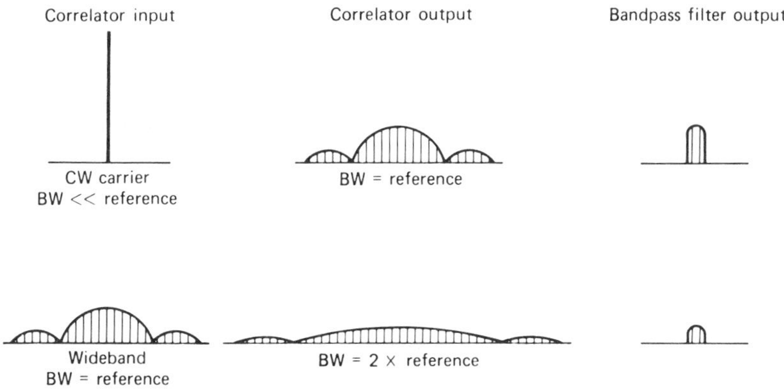

Figure 5.18 Correlator input and output and bandpass filter output for CW and wideband interference.

mentioned code cross-correlation and discuss it in Chapter 3 in some detail. Here, we must point out that any significant degree of cross-correlation between a spread spectrum receiver's local reference and an interfering signal can detract from performance. A common point of failure in the design of direct sequence systems lies in using codes that are too short (too few chips between repetitions) for the projected interference levels. Worse yet, the shorter codes when multiplied with interference (especially narrowband interference) tend to produce correlations that are not at all noiselike. Therefore, a synchronization detector or demodulator in such a system is likely to give surprisingly poor performance when it has been assumed (as is the usual case) that correlator output products due to interference are characteristically Gaussian.

Noting again that cross-correlation is expressed by

$$\int_0^\infty f(t)\, g(t - r)\, dt = \psi_c$$

we see that the implementation is a multiplier with two signal inputs and an integrator, which is usually a bandpass or lowpass filter. For $f(t)$, and interfering signal, and $g(t - \tau)$, a coded local reference, ideally the resultant ψ_c should be zero, where product is pure noise. A demodulator or synch detector would then register no difference between the condition of having interference and having none. Unfortunately this happy circumstance is not even approached unless the codes used are satisfactorily long.

For short sequences code/interference cross-correlations often produce a repetitious correlation envelope like that shown in Figure 5.19. This signal obviously does not integrate to zero and can therefore cause false synch recognition or offsets in signal demodulation.

It is of some interest to note that when observing cross-correlation patterns like those in Figure 5.19, variation of the interference frequency causes a variation in the cross-correlation pattern. With a CW interfering signal, for example, each time the interference frequency is changed by an amount approximately equal to $BW_{IF}/2$ a whole new cross-correlation pattern appears. Because some of these patterns have higher peaks than others, it is normal for a direct sequence receiver to be more sensitive to CW interference at some frequencies than at others. The shorter the code sequence (in general), the more pronounced the frequency-sensitivity property in a direct sequence receiver.

A good rule of thumb for selecting minimum code length in a direct-sequence system is to choose the sequence used: code chip rate/code length $< F_{low}$, where F_{low} is below the lowest frequency of interest in the information being demodulated. What this really says is that the code

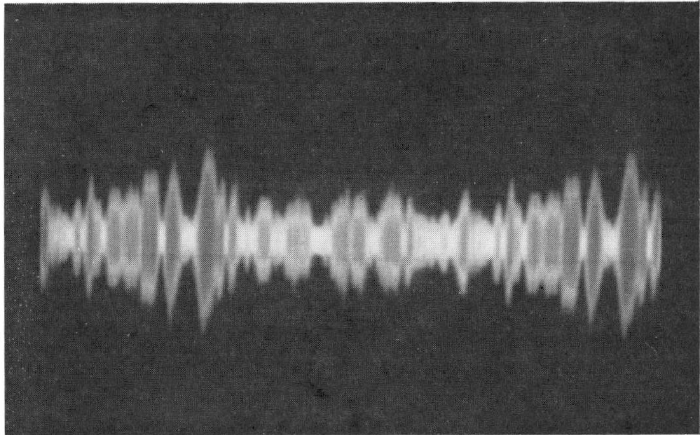

Figure 5.19 Typical short sequence/CW interference cross-correlation pattern (note repetition).

repetition rate should not fall in the information passband.) Otherwise, code/interference crossproducts may fall into the demodulated signal band and thereby reduce receiver output S/N. Such crossproducts can also greatly increase the incidence of false synch recognition. For this reason the response of the initial synch detector in a direct-sequence receiver is often rolled off at low frequencies—usually by simple AC coupling (a discussion of synch detection will be found in Chapter 6).

Figure 5.20 shows the frequency spectrum of the signal shown in Figure 5.19. This figure features a line spectrum, with lines separated at the code repetition rate. It shows beyond question that the cross-correlation of the short code used here (8191 chips long) with a CW jamming signal is *not* Gaussian.

At any rate, it is simple to say that a direct sequence system has a given amount of process gain. We must exercise much more case when specifying useful jamming margins, because it is necessary to know the type of interference to be seen, the interference frequency, and the amount of cross-correlation between that particular interference and the code sequence being used.

A series of photographs showing spectra internal to a direct sequence receiver are shown in Figure 5.21 through 5.26. These figures show the signal at the input to the demodulator in a typical system. Figure 5.21 is a typical noise spectrum produced by cross-correlation of jamming signal with the receiver's local (long code) reference. It is noteworthy that there is little difference in this spectrum, whether there is no signal present, the desired signal is present, or a jamming signal is present. The demodulation sees a noiselike input (depending on the code) under all conditions.

When synchronization occurs, the demodulator's input changes from a noiselike signal to a coherent one, as illustrated in Figure 5.22. In this

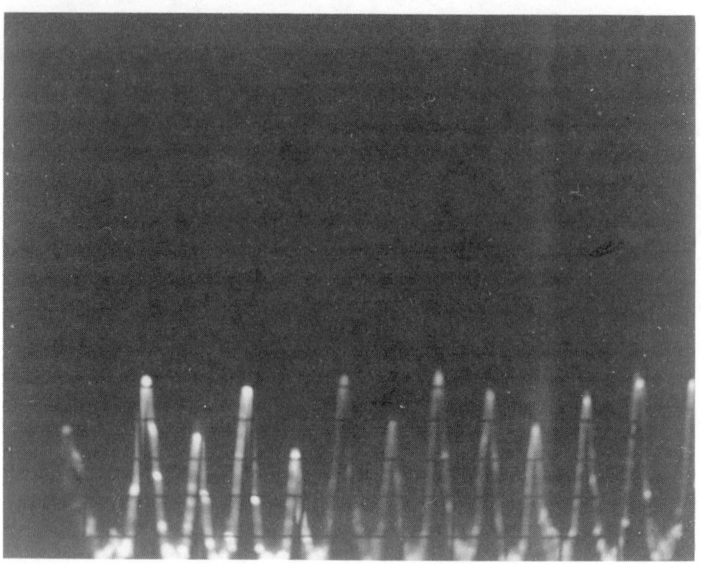

Figure 5.20 Spectrum of the signal produced by cross-correlation of an 8191-chip linear maximal code with a CW jamming signal.

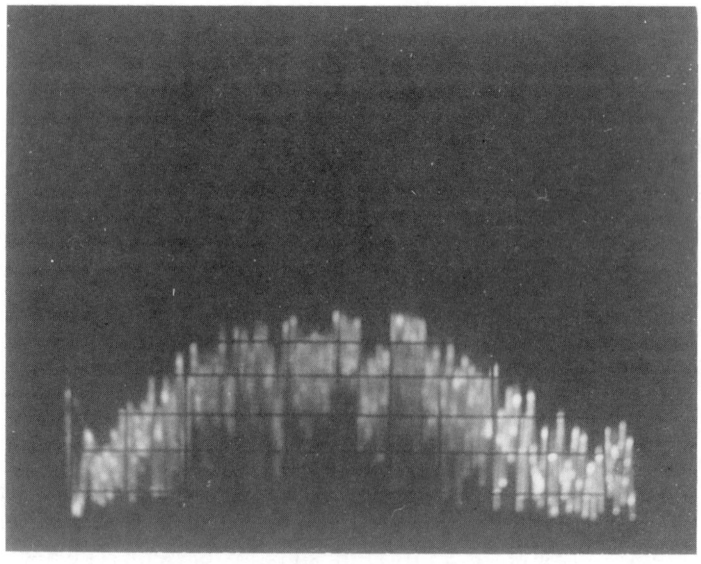

Figure 5.21 Demodulation input (correlator output, band limited by IF) prior to code synch.

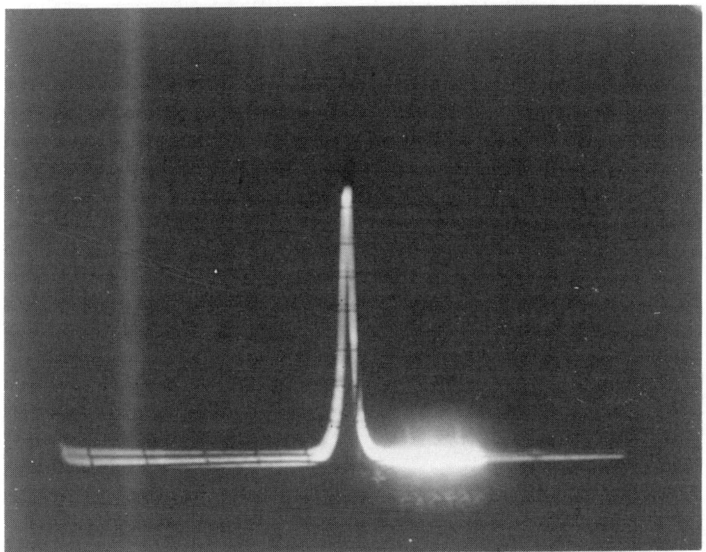

Figure 5.22 Demodulation input (band-limited correlator output) at code synch, 0-dB interference. No data modulation.

figure, the spread spectrum signal has been collapsed to a coherent carrier, which is unmodulated by data.

In the next photographs, Figures 5.23*a* and 5.23*b*, the carrier is phase shift keyed at a 2400 bit per second rate, with a one–zero–one–zero repeating pattern (a 1200-Hz square wave). Figure 5.22*a* shows the two-sided spectrum that results, with sidebands at ±2400 Hz, ±7200, ±12,000, etc. (all odd harmonics). With interference, in Figure 5.23*b*, only the major sidebands are visible above the noise. Figure 5.23*a* shows this receiver with no jamming and 5.23*b* shows the same receiver with a CW jamming signal 27 dB above the desired signal at the RF input. In both photographs in Figure 5.23, the code being used is 8191-chip linear maximal. Figures 5.24 and 5.25 show the same system at slightly lower jamming levels.

For comparison, Figures 5.26 and 5.27 show the same system, with similar jamming-to-signal ratios, but with a long ($2^{31} - 1$) chip code and the 2400 bps modulation removed. Note the striking difference in the level of the noise with respect to the desired signal level in these photographs. Also, note that as in previous photographs, the absolute signal level decreases as the receiver approaches its jamming threshold. This is due to AGC action, and is a desirable characteristic of the design.

More-detailed spectra are shown in Figure 5.28, which is an expansion of the desired, correlated signal spectrum under two conditions, with 2400 bps modulation present. In Figure 5.28*a*, with 0 dB jamming, only the modulation-produced sidebands are apparent, although a jammer-

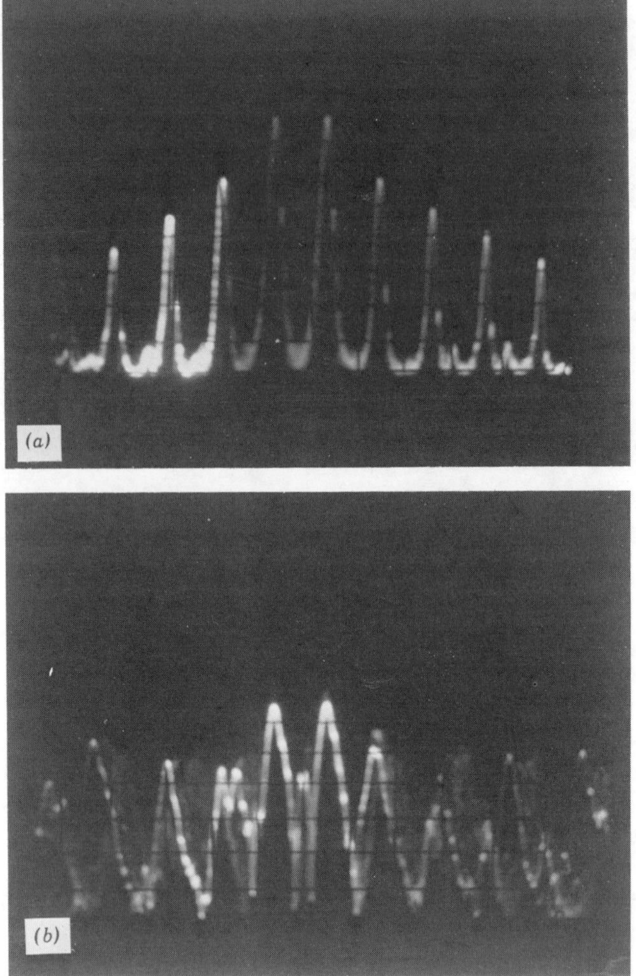

Figure 5.23 Correlator output signals. PSK data 8191-chip code: (*a*) no interference; (*b*) 29-dB interference.

produced carrier is just visible at the center of the spectrum. With the CW jamming signal increased to 15 dB above the desired signal, jammer-produced lines are visible between the desired signal sidebands. This illustrates a serious possible design flaw: In this system, the data rate is 2400 bps, while the 8191-chip code repetition is 2500 Hz. Thus, the lines produced by cross-correlating a CW jamming signal with the 8191-chip code are almost indistinguishable from the desired signal. The demodulator, under the correct circumstances, cannot tell the difference, and may synchronize to the jamming-produced spectrum.

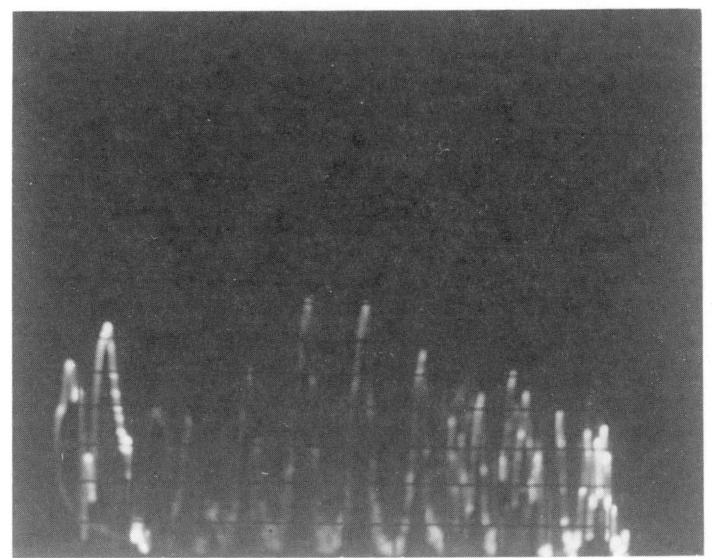

Figure 5.24 8191-chip code. PSK modulation. 25-dB interference.

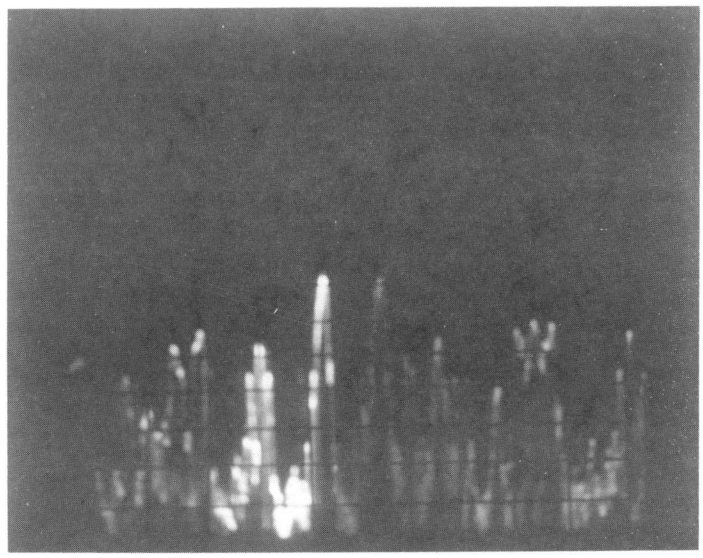

Figure 5.25 8191-chip code. PSK modulation. 27-dB interference.

Figure 5.26 Long Code. No modulation. 25-dB interference.

Figure 5.27 Long code. No modulation. 29-dB interference.

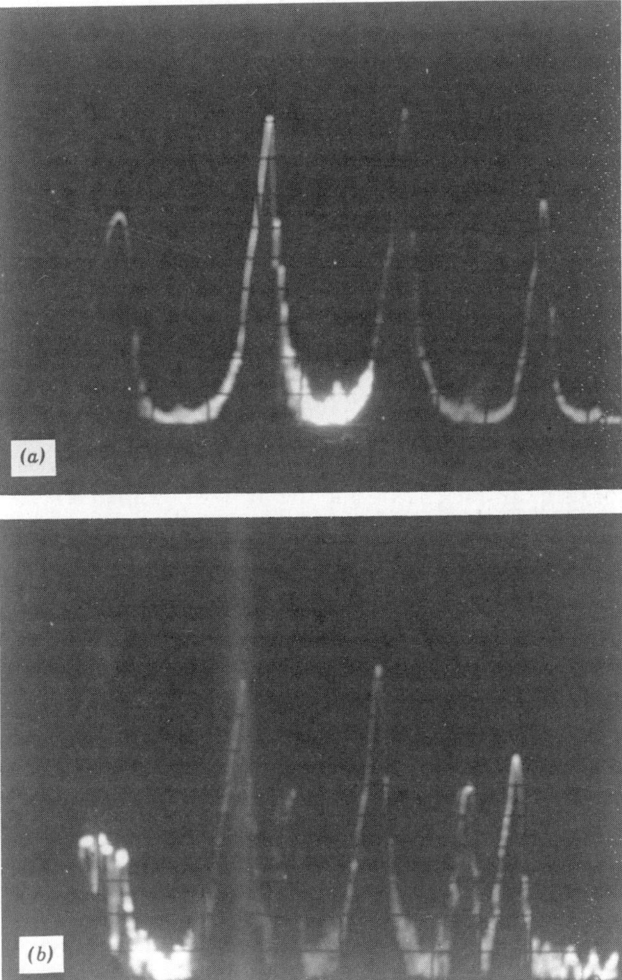

Figure 5.28 Correlator output signals showing more detail: (*a*) 2400-bps modulator with 0-dB interference (8191-chip code); (*b*) 2400-bps modulation with 15-dB interference (8191-chip code).

These photographs show typical signals seen by a direct sequence system's demodulator. They illustrate possible difficult tasks of differentiating between desired and undesired signals that are left to the demodulator under high jamming conditions.

Frequency Hopping Reaction to Interference

Frequency hopping systems are often termed "avoidance" systems from the standpoint of rejecting interference or jamming; direct sequence systems by comparison employ "averaging." The direct sequence approach spreads an interfering signal's power over a wide band of frequencies, and this power is thereby said to average over the band of frequencies even though the power is not evenly distributed.

A frequency hopping receiver also spreads any received interfering signals over a broad band,* but the process is different. While code modulation causes an incoming CW signal to a direct sequence receiver to occupy many frequencies, the frequency hopping receiver simply translates it (a CW signal) to a new frequency, and each time the frequency hopper goes to a different frequency the CW signal is translated differently. From a modern algebraic standpoint the operation of correlation in a direct sequence system is homomorphic,† *whereas in a frequency-hopping system it is isomorphic.*‡

We have already shown the direct sequence approach in which the local reference phase modulates incoming signals at a code-determined rate. Now let us examine the frequency hopping approach.

The frequency hopping receiver's local reference signal, which it multiplies with all incoming signals, is a replica (neglecting modulation) of the signal it expects to receive from a desired transmitter; that is, the local reference hops from frequency to frequency on command from a code generator (again identical to that contained in the transmitter). There is one difference—the local reference is offset by an amount equal to the center frequency of the correlation filter following the local/received signal multiplier. When the received signal is f_1, the local receiver must generate $f_1 + f_{IF}$, etc., and the local reference must continue to generate new frequencies at the same rate and frequency (plus IF) as the transmitter to remain synchronized. The difference frequency (IF) is output through the correlation filter.

What, then, is the effect of a single frequency (CW) interfering signal? First, we point out that it is the least effective interference for a frequency hopping receiver—exactly the opposite of its effect on a direct sequence receiver. Next we define a CW interferer as one whose power is contained completely within one frequency hopping channel and need be only ϵ greater than that of the desired signal to cause an error. Here ϵ is considered to be the decision threshold of the receiver's bit-decision device; for example, if the desired signal transmitted is in a channel corresponding to a one, the decision device could be forced to make an error (decide that a zero was sent) by an interfering signal with $S + \epsilon$ power in the zero channel. For a single interfering signal, however, the

*In both direct sequence and frequency hopping receivers an interfering signal is spread over a band equal to the covariance of the interference and the local reference; i.e., $BW_{spread} = BW_{interference} + BW_{reference}$.

†A one-to-many mapping from a set G_1 to a set G_2.

‡A one-to-one mapping from G_1 to G_2.

frequency hopping receiver could be expected to average one error in every n transmissions (n is the number of channels available).

If only one frequency (or chip) is sent per bit, a single CW interferer would cause an error rate of $1/n$. For this reason frequency hopping systems are forced to use some form of redundancy such as sending multiple chips per bit of information. The particular frequency or power (once the $S + \epsilon$ criteria is exceeded) level of a single CW interferer is of no consequence to a properly designed frequency hopper, for the probability of using one channel is the same as for any other channel.

To be effective against a frequency hopping receiver, a would-be jammer must meet at least one of the following criteria:

1. Place $S + \epsilon$ power in a large number of the possible channels.
2. Be able to predict the next frequency the hopper will use (involves breaking the code and the command pattern).
3. Be able to receive the hopped signal, determine its frequency, and retransmit a signal calculated to change the effect of the desired signal at the receiver (a "look-through" jammer). All this must be done in significantly less than the hop time to be effective.

Here we are speaking of binary frequency hopping systems in which the choice of a one or zero frequency is made on a chip-by-chip basis; that is, each chip transmitted has only two possible frequencies for transmission; f_1 to signify a one and f_2 to signify a zero.

Effective methods are available for a frequency hopping system to combat both the second and third approaches: the codes used can be made so difficult to break that it is not worth the effort, and by hopping at high rates and/or avoiding disadvantageous geographical situations the look-through threat can also be circumvented.

It is also possible to negate the look-through jamming threat by employing simple, noncoherent pulsed transmission techniques. Let us suppose that we employ a frequency hopped carrier that is turned on to send a one, and off when a zero is intended. (In other words, we employ on–off keyed frequency hopping.) At the receiver we make the bit decision as a function of whether a signal is present at the frequency of interest. Therefore, if we intend to send a one, we transmit a carrier at fone, and if we intend a zero, we do not transmit. The look-through (or frequency-following) jammer then can:

1. When we send a one, transmit at the frequency we are using, which only increases the probability of our making the correct decision, since any energy at fone signifies a one.
2. When we "send" a zero, transmit randomly, as he has no transmission to follow.

This means that the look-through frequency-following jammer is no more effective against such a system than the simple jammer who spreads his or her power over as many frequencies as possible. In the case of a one being intended, the probability of a jammer-related error is the probability that the jamming signal will be equal to and opposite in phase with the desired carrier (to cancel it completely). Where a zero is intended, the desired transmitter does not transmit, so the jammer cannot follow, and his probability of causing an error is just

$$\frac{J}{n} = \frac{\text{the number of frequencies jammed}}{\text{the number of frequencies available}}$$

The most effective overall threat to any frequency hopping system, then, is that of placing $S + \epsilon$ power in a large number of the possible channels.

So far, the effect of varying the relative phase of an interfering signal with respect to the phase of the desired signal in a channel has been ignored. Unfortunately, this does not cause it to go away. We have concerned ourselves with jammers that transmit $S + \epsilon$ in J/n percent of the channels available to a frequency-hopping receiver (J = number of channels jammed; n = number of channels available) and said somewhat naively that the probability of chip error (remember that a chip is a single-frequency transmission signifying a one or zero) is simply J/n. This is not quite true because the probability of interference hitting an intended channel is just as good as that of hitting a complementary channel. Thus interference can cause errors in several ways:

1. By hitting the complementary channel.
2. By hitting both the complementary channel and the intended channel.
3. By hitting the intended channel with a signal that cancels the intended signal.

We have already considered the effect of interference hitting the complementary channel only. The $p(e)$ for this is just J/n.

The probability of hitting both complementary and intended channels is $(J/n)^2$, and the probability that the intended signal and a jammer hitting in the intended channel will add in phase is 0.66* (see Figure 5.15). Therefore, the probability that an error will occur when both complementary and intended channels are hit is $(1 - 0.66)(J/n)^2 = 0.34\,(J/n)^2$.

Finally, we must consider the possibility that interference will hit in the intended channel without hitting the complementary channel. Again the probability of hitting the intended channel is J/n. The probability that the interference is in phase with the intended signal so that they add is shown

*optomistically, from the jammer's viewpoint

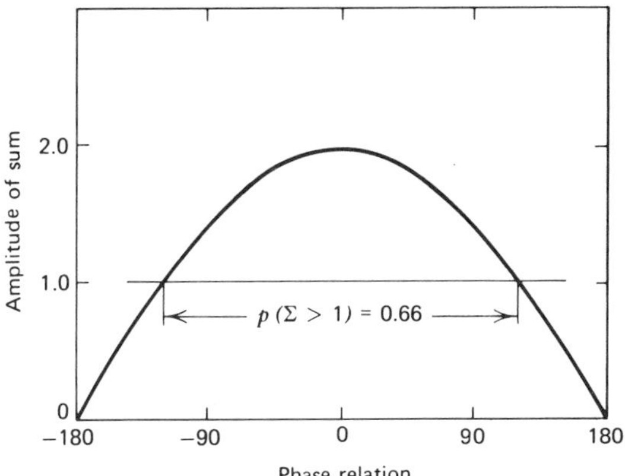

Figure 5.29 Plot of A sin $\omega\tau$ + A sin ($\omega\tau + \phi$) showing pdf for phase addition of random sine waves.

in Figure 5.29. This figure is the probability distribution function (pdf) for all phase positions of two signals $A \sin \omega\tau$ and $A \sin(\omega\tau + \phi)$ and is such that $p(\Sigma > 1) = 0.66$.

Considering that the effect of phase addition must be to give a sum not greater than ϵ more than zero, we see that the probability of an interfering signal causing complete loss of an intended signal due to phase cancellation is

$$1 - p[A \sin \omega\tau + A \sin(\omega\tau + \phi)] > \epsilon$$

For small epsilon,* which is the desired case, this probability should be negligible when compared with the other two sources of error.

To give an example, let us assume that we have a frequency hopping receiver and an interfering signal that is effective in 10% of the receiver's channels; that is, J/N is 0.1. Total chip errors caused by the interference would then be

$$\begin{aligned} p(\mathrm{e}) &< \frac{J}{n} + 0.34\left(\frac{J}{n}\right)^2 + f(\epsilon)\left(\frac{J}{n}\right) \\ &= 0.1 + 0.0051 + 0.001 \text{ [giving } f(\epsilon) \text{ a value of } 0.01] \\ &= 0.1061 \end{aligned}$$

(With nothing else considered, this in itself is a horrible error rate. A five-chip-per-bit redundancy, however, with simple majority coding, reduces this error rate to 0.008, which is much more reasonable and could

*Remember that ϵ is the bit-decition threshold in the receiver and that the designer's goal is (or should be) to make ϵ extremely small to avoid bias errors.

be usable.) At any rate, it has been shown that the contribution of phase effects is small but must be considered with designing low-error-rate frequency hopping systems for data transmission under jamming.

We must also reiterate here that frequency hopping systems should not be designed without some form of redundancy and/or error correction coding.

5.3 BASEBAND RECOVERY

The baseband recovery process in a spread spectrum receiver is identical to the baseband recovery process in any other receiver employing the same type of baseband modulation.* Frequency-modulation, frequency-shift-keying, phase-shift-keying, and even amplitude-modulation forms are quite compatible with direct sequence or frequency hopping (admittedly some combinations display more compatibility than others.) Whenever they are employed, the conventional modulation forms work well with their usual types of demodulators, but space has not been taken here to give such demodulators more than honorable mention.

In the remainder of this chapter we discuss some techniques that have proved to be well adapted to demodulating the recovered carrier in a spread spectrum receiver. (These same techniques are just as useful in nonspread spectrum applications.) In general, all types of threshold extension demodulators are useful: phase-lock loops, FM-feedback demodulators, Costas demodulators; others are best suited to the task at hand, operating well at the lowest possible signal-to-noise ratio.

Frequency hopping systems, unlike direct sequence systems, are difficult to implement with coherent transmit/receive functions. Coherence to a frequency hopper means that every new frequency generated must be not only precisely defined in frequency but must also bear an exact phase relation to the preceding frequency. Because coherence in the signal itself is difficult to achieve or maintain, few frequency hopping systems employ coherent detectors such as phase-lock loops. The reason is that a phase-lock loop would be forced to reacquire a lock each time the signal changes frequency. This could take a significant amount of the overall chip time, especially when fast frequency hopping or high process gain is a goal.

Frequency hopping demodulators often employ simple envelope detection in conjunction with integrate-and-dump matched filtering to make the most of the chip timing available.

Baseband recovery in a direct sequence receiver is inherently a coherent process in that at least the receiver's local code reference must be

*Remember that we said that spread spectrum systems are transparent—that removal of the spectrum spreading modulation restores whatever signal existed before it was overlaid with a coded signal.

an accurate estimate of received code modulation. Also, because of the superior threshold performance of coherent detectors over alternative types, they are almost always used. Phase-lock loops, feedback-FM loops, Costas demodulators, and their variations are often found in direct sequence receivers. Another type of demodulator that takes advantage of the fact that squaring any double-sidebanded signal produces a carrier (see Appendix 4) at twice the signal frequency may also be used. It is called a squaring loop[1220] and is a phase-lock loop used to track the twice-frequency carrier; the loop VCO is divided by two and used as a coherent reference for demodulating the signal. Performance of the squaring loop has been shown to be equivalent to that of the Costas loop, although the phase ambiguity introduced by having a twice-frequency (though divided down) coherent reference must be resolved.

Phase-Lock Loops

A phase-lock loop is a complex device to which literally thousands of pages of scientific material[1201-1223] have been dedicated and deservedly so. Yet its function can be described in terms so deceptively simple that we tend to lack appreciation of its power. A word of caution: do not let the simple terms of the simplified design process lull you. Optimal design often requires intimate knowledge of a phase-lock-loop's internal workings, including third-order reactions to step-function frequency changes. Because the simple phase-lock loop contains at least one nonlinearity, an exact prediction of its reactions can be a rather difficult proposition. Fortunately phase-lock loops have been submitted to thorough examination by both analytical and empirical media, as an examination of the references will reveal.

The integrated-circuit art, for example, has reached a point such that a complete (but simple) phase-lock loop is available in a single small package to handle many applications.

A phase-lock loop is a feedback loop that contains an oscillator, a phase detector, and a filter, connected in such a way that the oscillator "locks on" to an input signal. (Figure 5.30 shows a basic phase-lock loop for reference.) Some of its outstanding characteristics are:

1. A phase-lock loop tracks the frequency of an input signal and not its phase. This makes it a unique frequency-modulation detector or, with some extension, a coherent AM detector. Demodulated-FM information is taken from the line that controls the loop's (voltage-controlled) tracking oscillator, keeping it at the same frequency as the input signal.
2. The loop oscillator tends to track in phase quadrature with the input signal to which it is locked. This phase angle is not maintained, however, the input frequency departs from the rest frequency of the

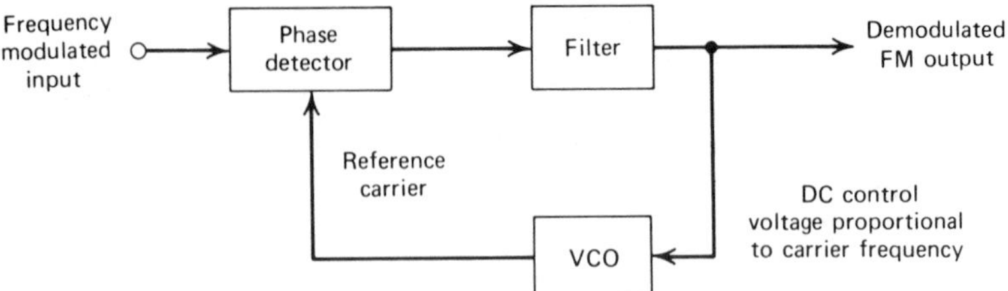

Figure 5.30 Simple phase-locked loop.

loop oscillator. At any given input frequency the phase angle between loop oscillator and input carrier is approximately

$$\frac{2\pi|f_0 - f_i|}{k}$$

where f_0 = loop oscillator rest frequency (open-loop frequency),
f_i = input carrier frequency,
k = open-loop gain for the phase-lock loop.

3. The loop acts as a tracking filter such that its demodulated output has a loop-filter determined bandwidth over the range in which the oscillator can be caused to track the input signal. A phase-lock loop is the equivalent of a tunable bandpass filter which centers itself at the frequency of a desired signal, accepting only the information (and noise) contained in a two-sided (mirror image) replica of its low-pass response.
4. Best of all a phase-lock loop is a low-threshold demodulator. Because it offers a coherent reference carrier (the loop VCO input to the phase detector), a phase-lock demodulator can operate at a signal-to-noise ratio approaching 0 dB.

Why does a phase-lock demodulator offer so much in a spread spectrum receiver? One reason is its low threshold. A second reason is that the loop VCO, in tracking the incoming signal, generates a local reference carrier that can be used in the receiver for code clock tracking, Doppler compensation, or any other receiver-frequency optimization requirement. Third, the tracking-filter characteristic allows a single base-bandwidth filter to pass the baseband modulation, even though the carrier signal frequency varies over a wide range.

From the black box point of view, a phase-lock loop does exactly the same thing as any other FM demodulator. Its input signal is a frequency-modulated carrier and its output is the carrier-modulating waveform. The magnitude of this output signal depends on two things—the amount of carrier-FM deviation and the loop VCO gain K_2; for example, a phase-lock loop containing a VCO whose sensitivity, or gain, is 10 kHz/V would output a 1-V (peak-to-peak) signal when its input signal deviates ±5 kHz.

Output frequency response is determined by the loop filter, which limits the VCO tracking rate. The proper loop filter is a lag-*RC* type, and typical loop frequency response due to such a filter is shown in Figure 5.31. Often it is advantageous to modify the filter by adding a second capacitor in parallel with C and R_2 (see Figure 5.32) to give a third breakpoint, beyond which the loop response again rolls off at a 6-dB/octave rate. This reduces the high-frequency response of the loop without significantly affecting the loop operation in the bandpass region. (The high-frequency rolloff capacitor C_2 is usually chosen to be no greater than one-fifth to one-tenth of the main filter capacitor C_1.) Output signal-to-noise ratio can be significantly improved by this simple addition, which also decreases loop jitter with noisy input signals.

A phase-lock loop can easily be augmented to act as a coherent amplitude demodulator, and this makes it possible to generate coherent AGC or to generate an "in-lock" or "signal present" output. Figure 5.33 is a block diagram of a phase-lock loop, which is augmented to include both AM and FM demodulation. The only addition to the simple loop that is necessary is a phase shifter, a second phase detector, and a low-pass filter, with the phase-lock portion of the overall demodulator itself unmodified.

The phase-lock loop of an augmented loop demodulator such as that in Figure 5.33 continues to operate in its usual manner so that an input signal, when locked to the loop VCO, is in phase quadrature with the VCO. In the additional circuitry a 90° phase shifter supplies a VCO signal to a second phase detector, which is in phase with the incoming signal. The result of multiplication (in the phase detector) of the two in-phase

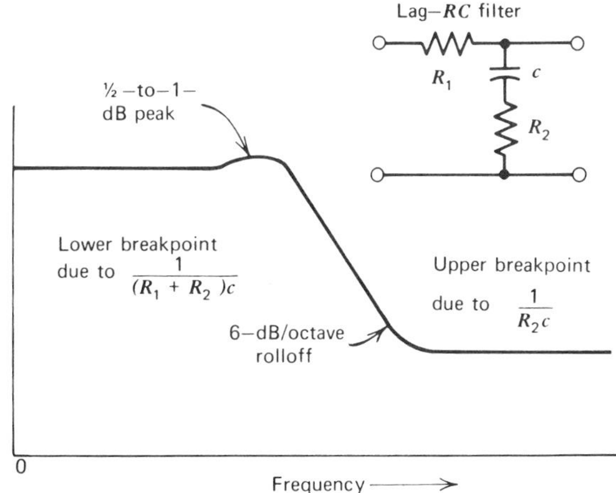

Figure 5.31 Typical low-pass frequency response for phase-lock loop with lag-RC loop filter.

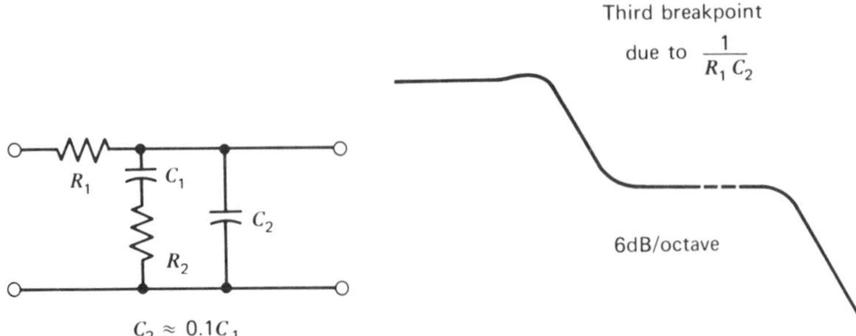

Figure 5.32 Modified loop filter for additional high-frequency rolloff.

signals is a signal that, when filtered, is the amplitude information we desire. The signal waveforms of Figure 5.34 illustrate why this is true. The output signal from the quadrature detector, after integration, is zero when the signal is symmetrical around the zero axis; that is,

$$A \int_{\pi/2}^{0} \sin \omega_c t \, dt + A \int_{0}^{-\pi/2} \sin \omega_c t \, dt = 0$$

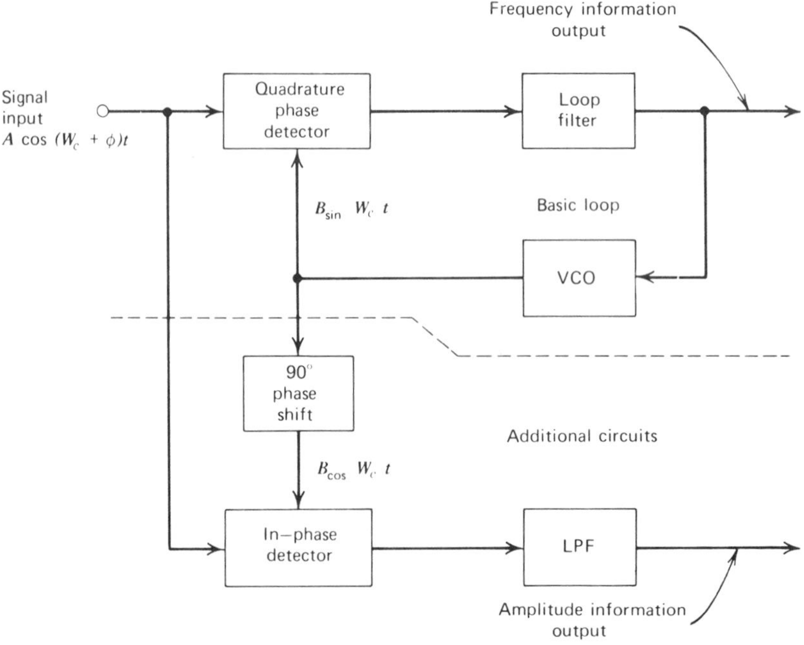

Figure 5.33 Phase-lock loop augmented to furnish amplitude demodulation.

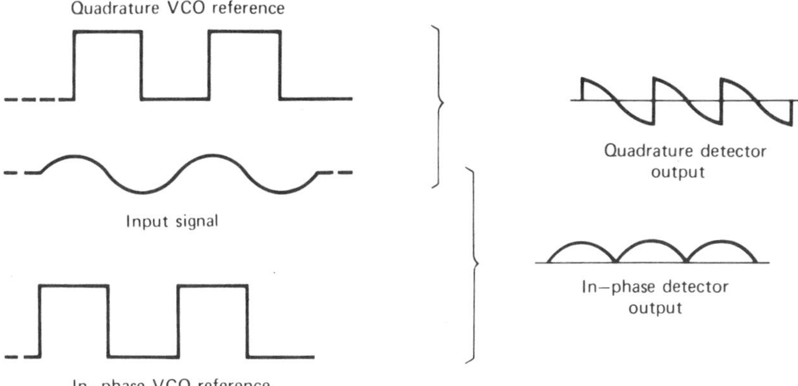

Figure 5.34 Phase-detector output waveforms.

if the phase-lock loop is locked and the VCO rest frequency is the same as the input signal frequency.

The in-phase detector output, on the other hand, is similar to the output of a rectifier and when integrated gives a value

$$\frac{2}{T} A \int_0^{T/2} \sin \omega_c t \, dt = 0.63 \text{ A}$$

In one case, then, we have an output that varies as the phase of the input signal varies. In the other case the output varies as A, the signal amplitude varies.

Phase-lock loops are used primarily in direct sequence systems, for unless a frequency hopper is highly coherent a phase-lock loop must reacquire each time the frequency changes.

Coherent frequency hopping systems have not been demonstrated successfully at this time. It is possible to construct hardware that provides coherent frequency hopping signals, and transmitters and receivers can be built that are coherent when operating side by side. When these same systems are separated, however, and the signal must pass through the atmosphere, it is no longer coherent when it reaches the receiver. This is due to differences in delay from one frequency to another.

Squaring Loops

We have not specifically discussed squaring-loop demodulators here, for they differ primarily from phase-lock loop in that a square-law device is placed ahead of the phase-lock loop that tracks the twice-frequency carrier. The tracked carrier is then divided by two, phase-shifted by 90°,

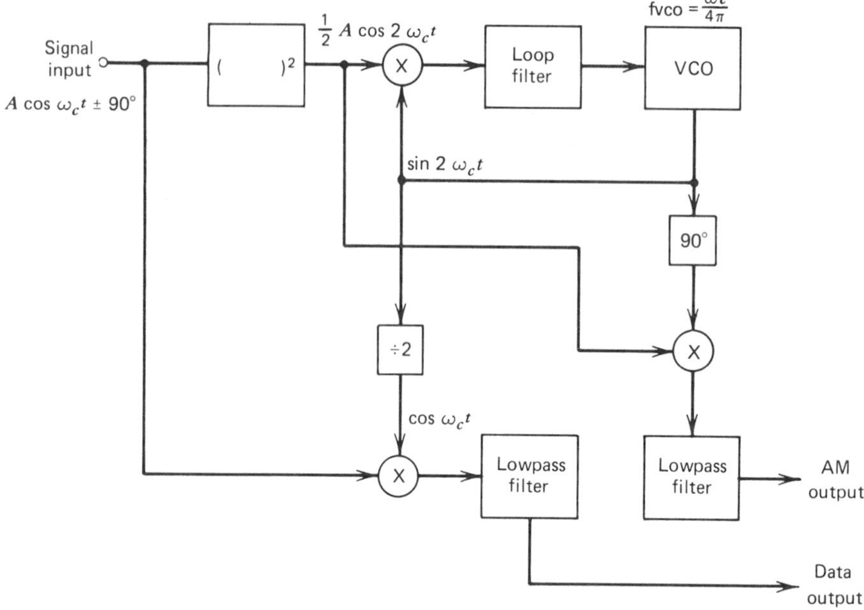

Figure 5.35 Squaring loop demodulator

and multipled with the input signal to demodulate it. Squaring loops, of course, are used for demodulating double-sideband suppressed-carrier signals because the phase-lock loop could be used directly at the carrier frequency without the necessity of squaring if the carrier were not suppressed.

Performance of a squaring loop is a function of the loop bandwidth, the squaring device's performance, and any presquaring filter bandwidth. A squaring loop has performance exactly equivalent to that of a Costas loop.

Figure 5.35 is a block diagram of a squaring loop demodulator. The input biphase signal (PSK) is squared,* which produces a twice-frequency unmodulated carrier. A phase-lock loop is then locked at twice the input frequency to the squared signal. After dividing the VCO output by two, a carrier is produced that may be multiplied with the input PSK modulated signal to produce the desired data. (Multipliers shown can be double-balanced mixers.) Coherent-AM demodulation for AGC, signal detection, and other purposes is also easily accomplished with a squaring loop. All that is necessary is to phase shift the VCO signal by 90° and

*In practice, a frequency doubler can be used instead of the squaring device. These are readily available from a number of sources.

multiply it with the phase-lock-loop input signal, as in the simple phase-lock-loop approach.

Costas Loop Demodulators

A special type of compound phase-lock loop, called an *I-Q* or Costas[109] loop (after J. P. Costas), is often used for demodulating double-sideband suppressed-carrier signals, and this, of course, is exactly the type of signal produced by biphase or quadriphase phase-shift keying. (See phase-lock references, Appendix 3). Also, if code-inversion modulation is used, Costas demodulation is a good choice. Like the other phase-lock loops, Costas loops have their primary application in direct sequence reception.

The basic configuration of a Costas loop is shown in Figure 5.36. At first glance it appears to be the same as the augmented phase-lock loop in Figure 5.33, and it is the same in some ways. Again a VCO is used as the carrier reference, which is multiplied in-phase and in quadrature with the input signal, and the multiplier outputs pass through low-pass filters. The difference is in the addition of a third multiplier to which the filter outputs are applied and whose output is used to control the loop VCO.

How does it work? An input signal $\pm A\cos(\omega\tau + \phi)$ (biphase

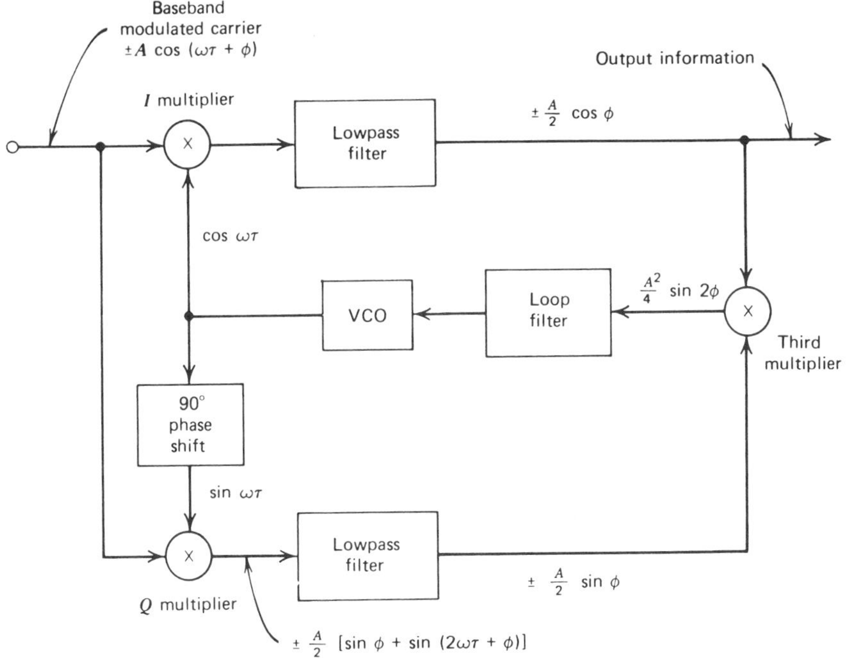

Figure 5.36 Costas loop demodulator for double-sideband suppressed-carrier signals.

modulated) is applied to the I and Q multipliers, where it is multipled by signals B cos $\omega\tau$ and by B sin $\omega\tau$, respectively. Outputs from the multipliers are then

$$\text{for the } I \text{ multiplier,} \quad \pm\frac{A}{2}[\cos \pi + \cos(2\omega\tau + \phi)$$

$$\text{for the } Q \text{ multiplier,} \quad \pm\frac{A}{2}[\sin \phi + \sin(2\omega\tau + \phi)]$$

which, when passed through the I and Q low-pass filters, become $\pm(A/2)\cos\phi$ and $\pm(A/2)\sin\phi$.

These two signals, which contain both the phase-shift-keyed information and carrier phase, are then multiplied together to obtain $(A^2/4)\sin 2\phi$. After filtering, this signal is used to correct the loop VCO and cause it to track the incoming (virtual) carrier. (Of course, we remember that there actually is no incoming carrier signal. That is the purpose of a Costa demodulator—to demodulate double-sideband suppressed-carrier signals.)

Information is available from two points, depending on the modulation method used. Phase-shift-keyed information that results from the data transmission is available at the output of the I multiplier output filter. Frequency-modulation information is available as in the simple phase-lock loop—at the loop filter output. The Costas loop has no amplitude output and cannot in this form furnish either "signal present," "in-lock," or coherent AGC signals. Its main advantage over a simple phase-lock loop is its ability to demodulate phase-shift-keyed* and/or suppressed-carrier signals.

The I filter output $\pm(A/2)\cos\pi$ is just $\pm A/2$ for small π and $\pm A/2$ is the desired binary data. cos π is approximately one when the loop in lock ϕ approaches 0°. It should be noted that the loop does not know, and has no way of knowing, which is one and which is zero. Therefore, we must use a form of data transmission that in itself is unambiguous. Such techniques as differential phase-shift keying or transmission of polarity-determining bits are well enough known not to be discussed further here.

When a signal-presence, in-lock, or coherent AGC signal is required, the Costas loop must be augmented. Three more multipliers, connected as in Figure 5.37, may be used. The signal input applied to the Costas loop, $\pm A\cos(\omega\tau + \phi)$, is also applied to the multipliers. Their other inputs are ±45° an −45° shifts of the VCO.

*Phase-lock loops cannot demodulate PSK information. The abrupt phase shift causes only a transient change in a phase-lock-loop's VCO control voltage, after which it settles to its previous level. Furthermore the transient change may be positive or negative, depending on whether the loop VCO is advanced or retarded to bet back into proper phase lock with the incoming signal.

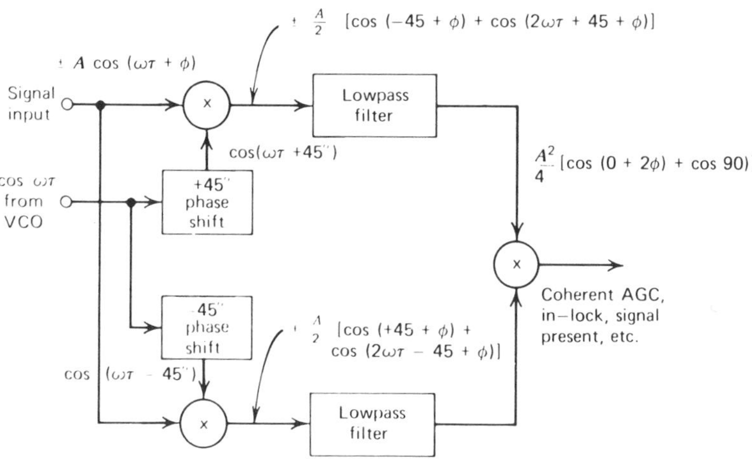

Figure 5.37 Multiplier configuration for coherent amplitude output (added to Costas demodulator).

The signals that result are shown in Figure 5.33. After filtering, which removes the twice-frequency components ($\cos 2\omega\tau + \cdots$), a third multiplier multipies the resultants and gives an output

$$\frac{A^2}{4}[\cos(0^\circ + 2\phi) + \cos 90^\circ] = \frac{A^2}{4}\cos 2\phi.$$

For ϕ small, the in-lock condition, $\cos 2\phi$, is 1 and the output is $A^2/4$. If the Costas loop is not locked to the incoming signal, however, $\cos 2\phi$ rapidly approaches zero. Therefore, the configuration in Figure 5.37 is a true coherent amplitude demodulator and can be employed to give coherent AGC, etc.

Filters used in the amplitude detector would usually be the same as those employed in the Costas demodulator itself, although they would not necessarily be the same. The low-pass filters used in the arms of the Costas demodulator would be designed to pass the information band, with cutoff at the highest desired frequency or at some other frequency chosen to maximize the information signal-to-noise ratio while offering minimum effect to the desired signal. The loop filter used in the Costas loop is the same type used in phase-lock loops but may be quite narrow because it need pass only carrier frequency changes, and unless information is sent on the carrier only Doppler shifts need to be tracked.

Of course, the exact filter bandwidths used depend on the use of the Costas detector or the coherent amplitude detector. Gain constants have the same effect on a Costas loop as on a simple phase-lock loop. It should be noted, however, the imbalance in the I and Q paths can cause offset in the third multiplier output and possible tracking problems as a consequence.

Other approaches to providing a coherent-AM output from the Costas loop involve squaring the inputs to the loop's third multiplier, which gives I^2 and Q^2 signals (that is, the in phase and quadrature multiplier output signals, after low-pass filtering). These two signals are then added or subtracted to give either $I^2 - Q^2$ or $I^2 - Q^2$ signals, which can be used as representations of the carrier amplitude.

Both Costas and squaring-loop demodulators have degraded performance with respect to the simple phase-lock loop. That is, the phase-lock loop's threshold is 3-dB better than either the Costas or squaring loop.

Frequency-Modulation Feedback Demodulators

Frequency-modulation feedback (FMFB) demodulators are akin to phase-lock-loop demodulators in that a controlled oscillator is caused to vary as a function of the input signal frequency, and the control input to the oscillator is taken as the demodulated signal. A typical demodulator is illustrated in Figure 5.38. The components making up this FMFB loop are nearly the same as might be expected in a common FM receiver, especially one containing an AFC (automatic frequency control) loop. The demodulated output signal in a conventional receiver would come from the discriminator, however, and not from the VCO control line as it does in an FMFB loop.

The loop filter in an FMFB demodulator is different from an AFC loop filter in that the AFC loop is designed to track only carrier drift; the discriminator outputs the desired signal. For the FMFB loop the filter is designed to pass the information modulation, which causes the VCO to be varied at the information rate. The VCO signal, mixing with the incoming

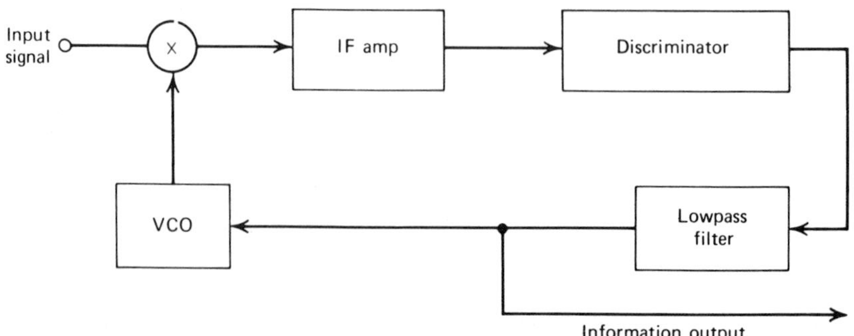

Figure 5.38 FM feedback demodulator block diagram.

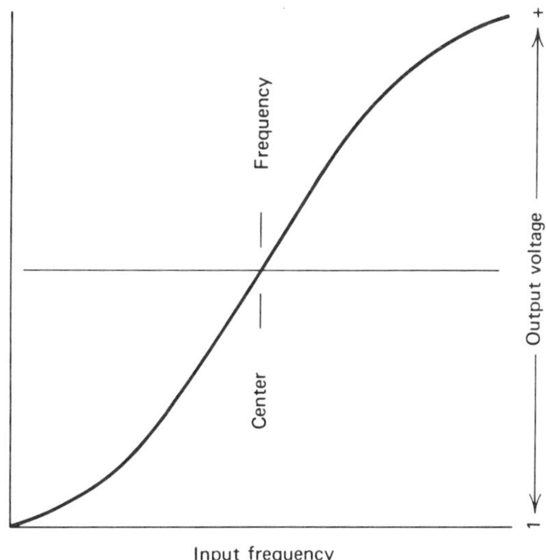

Figure 5.39 Typical discriminator block diagram.

signal, then tends to track the incoming signal and hold the frequency input to the discriminator at the discriminator's center frequency. Any change in the input frequency causes the discriminator to put out a positive or negative DC voltage (depending on the direction of frequency offset) which in turn corrects the loop VCO to zero the discriminator output. (A typical discriminator frequency-to-voltage curve is shown in Figure 5.39.) We will not discuss the design of discriminators for FMFB loops here because excellent, highly stable crystal discriminators that can perform far better than any conventionally constructed unit for this type of system are available.

FMFB loops are not used in spread spectrum systems as often as phase-lock loops, though their performance from the standpoint of threshold is nearly the same for deviation ratios greater than 3, as shown by J. T. Frankle.[I013] The phase-lock loop does offer better performance for low-deviation ratios, but another reason exists for the prominence of phase-lock usage; that is, a phase-lock-loop's internal oscillator is locked in frequency to its input frequency, with only a phase shift between the two which is never greater than $\pm 90°$. The FMFB loop's oscillator does exhibit frequency error, which is a function of loop gain and frequency offset from the discriminator center frequency.

Since spread spectrum systems (and especially direct sequence systems) require frequency tracking as close as possible, phase-lock detection often is more advantageous. Even so, FMFB demodulation is useful and can often be combined with phase-lock methods to the advantage of both.

PDM Demodulation

A demodulator for the SCPDM (suppressed clock pulse-duration modulation) technique described in Chapter 4 is shown in Figure 5.40. We note here that in spread spectrum systems in which PDM modulation is employed the PDM demodulator commonly follows another demodulator such as a Costas loop because the information is not in PDM form as transmitted and must be converted to it.

The SCPDM demodulator shown[241] has exhibited superior threshold characteristics to FM, and curves demonstrating this performance are given in Figure 5.41.

Frequency Hopping Demodulation

The demodulators discussed so far have been unsuited to frequency hopping systems. Few (if any) frequency-hopping links are able to

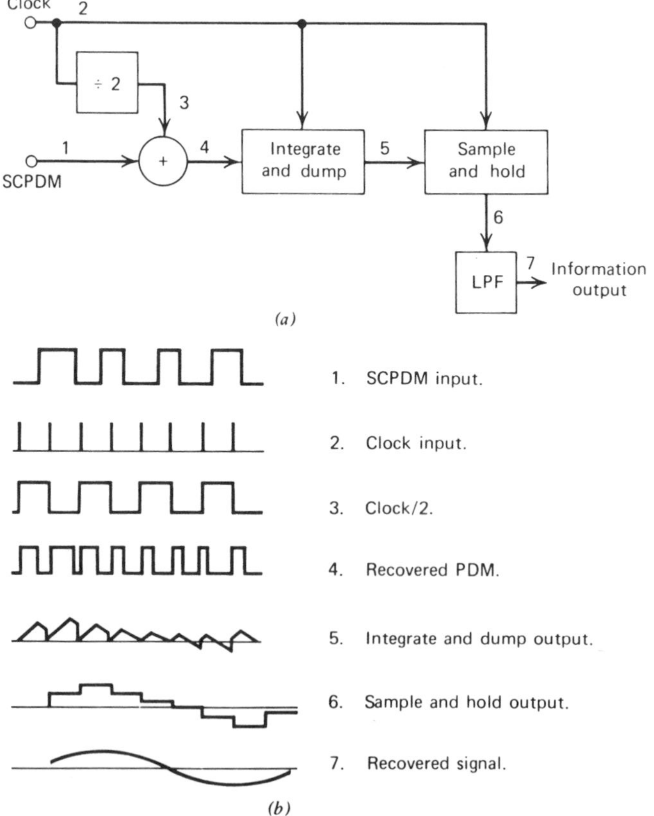

Figure 5.40 SCPDM demodulation process: (*a*) suppressed carrier PDM demodulator; (*b*) waveforms in SCPDM form.

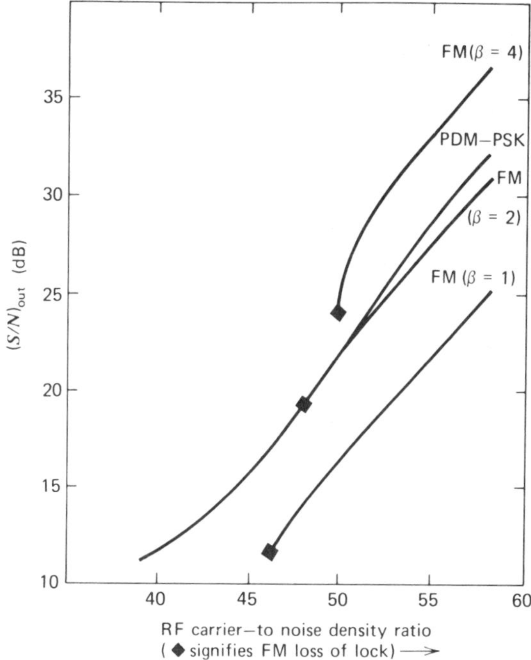

Figure 5.41 Performance of PDM demodulator compared with FM.

support coherence* so that the signal into the demodulator changes phase each time the system hops to a new frequency. In addition, the signal is likely to be pulsed due to sampling mark and space frequencies. Therefore, phase-lock and FMFB loops are not usually employed in frequency hopping demodulation. Instead, simple envelope detection, which does not care about input phase shifts and can respond rapidly to pulsed signals, is most often employed.

The most common method of transmitting information in a frequency hopping system is to send a frequency f_1 if the datum is a one and a frequency f_2 if the datum is a zero. Usually some transmission redundancy is employed so that one code-determined set of frequencies is used to convey a one and another set, to convey a zero. If more than one frequency is employed for redundancy, the transmitter has the choice of

*Here, and throughout this book the term "coherent" means carrier coherence when referring to frequency hopping systems. Frequency hoppers, like their direct sequence counterparts, must maintain code and code clock coherence (synchronization) in order to communicate. Carrier coherence is seldom found in frequency hopping systems because of the difficulty (note that we did not say impossibility) of maintaining it in an RF link. Direct-sequence systems, on the other hand, are almost always code, code clock, and carrier coherent.

transmitting all of them at once or one at a time. This choice has a significant impact on the receiver and demodulator.

Before proceeding, let us digress for a moment to list the demodulators most suitable for spread spectrum reception, together with the modulation format for which they are usually employed.

One or more of the demodulators we have discussed in this chapter are suitable for demodulating the forms of modulation typically used in spread spectrum systems. Matching modulation types with demodulators, we have the following:

Modulation Type	Detector
Carrier FM/FSK	Phase-lock detector
Clock FM/FSK	Costas or squaring-loop detector
PDM/PSK	Costas or squaring plus PDM demodulator
Frequency hopping (coherent)	Phase-lock detector
Frequency hopping (noncoherent)	Envelope and integrate and dump detectors

Of course, we can demodulate clock or carrier FM signals with a discriminator or double-sideband suppressed carrier by doubling and multiplying the result with the input signal, but for spread spectrum use these noncoherent methods offer little because of their threshold characteristics. Only the noncoherent frequency hopping systems can benefit from the use of demodulation of other than the phase-lock variety.

Figure 5.42 shows a typical frequency hopping demodulator in a noncoherent receiver. The block diagram is for a multiple-chip-per-bit receiver, in which chip decision is made on the basis of each sequential set of chips. This demodulator is designed for compatibility with sequential sampling of one and zero frequency channels; that is, the frequency synthesizer inserts into the receiver a frequency corresponding to a one being transmitted and follows it with the frequency corresponding to a zero transmission. Each of these frequency alternates is sampled for one-half the chip period. (we pay for the reduction in receiver complexity by doubling the effective bandwidth of the postcorrelation signal.)

In the demodulator in Figure 5.42 an input signal, a series of pulses, is passed through an envelope detector to an integrate-and-dump filter, which is dumped at the chip rate. Alternate chips are sampled and compared and the largest of each pair is assumed to be intended signal. Therefore, if a one is intended, then one-half a chip (say the first) would contain a pulse of RF, the second half would not. Comparison of the signal level in the two half chips would then allow the one/zero decision to be made for that chip.

We might wonder why this seemingly overcomplex chip decision process should be used when a level detector such as a Schmitt trigger

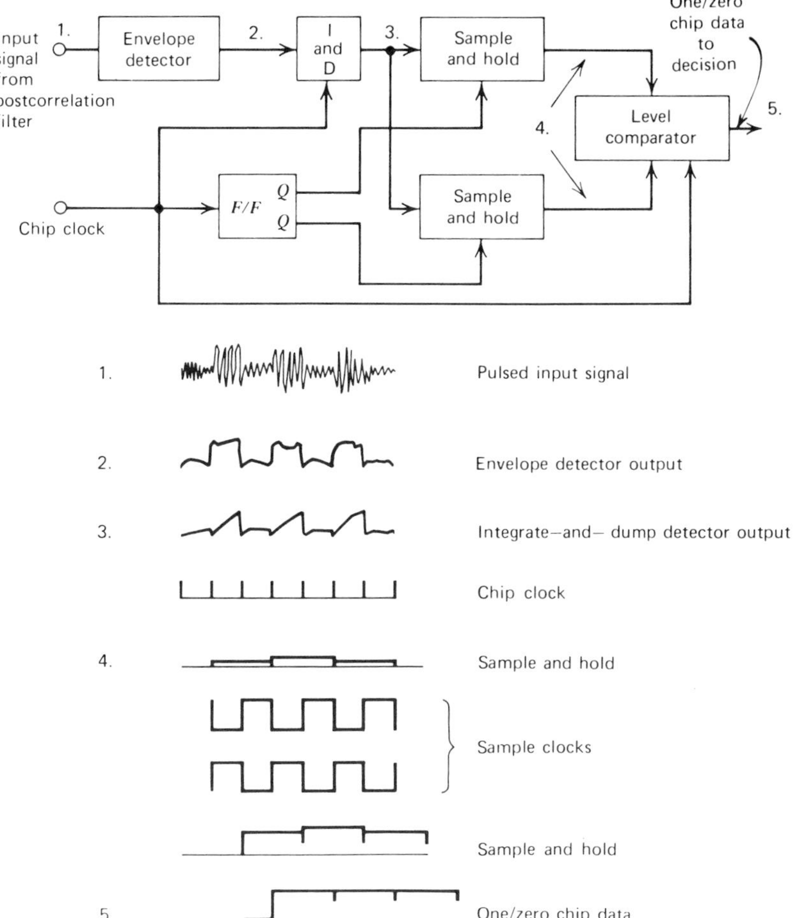

Figure 5.42 Frequency hopping demodulator (noncoherent).

placed at the envelope detector output could give the same information. The difference is that, when interference is added to the input signal, random hits by the jammer in complementary channels cause both half chips to contain an RF burst (again on a random basis). Integrate-and-dump operation, with subsequent comparison of signal level in the two half chips, allows us to decide which channel contains the largest signal rather than basing the decision on whether a signal exceeding threshold is there (which the simple level detector would be forced to do).

The output signal to the data decision circuit consists of n-tuples of chip data in which the number of chip-related bits is the same as the number of chips sent for each bit. For a two of three majority decision system each n-tuple consists of three chip-related bits. The decision circuit then determines on the basis of these bits whether a one or a zero was intended by the transmitter.

When an interfering signal is received with the desired signal, occasional hits in the unintended channel will occur (i.e., the interference will appear in the one channel when a zero is being sent and vice versa). When this occurs (and the probability is precisely the same as the ratio of the number of interfering carriers to the number of frequency-hopping channels available), the chip comparator must decide whether the real signal or the interferer is that sent by the transmitter. This decision is based on a simple comparison of the power in the two alternate channels sampled by the receiver; that is, if the interfering signal has more power in its channel than the desired signal has in the intended channel, the wrong decision will be made.

The threshold of the chip decision is just the decision threshold ϵ of the level comparator (Figure 5.42). Whenever either signal power or interfering power exceeds the other by ϵ, the chip decision will be made in favor of that channel. *For this reason a jammer does not increase his or her effectiveness by increasing his or her power at any one frequency once it exceeds the signal by* ϵ.

If an interfering signal does cause a chip error, the bit decision input gets as a signal a single zero in place of a one, or a single one in place of a zero. When the bit decision is based on two of three, as in our Figure 5.42, this single chip error will not cause a bit error. If two or more chip errors occur in the same set of chips, however, a bit error will follow. The chip clock/6 input to the two of three decision circuit in Figure 5.43 is used for output data because the actual data rate in a frequency hopping system,

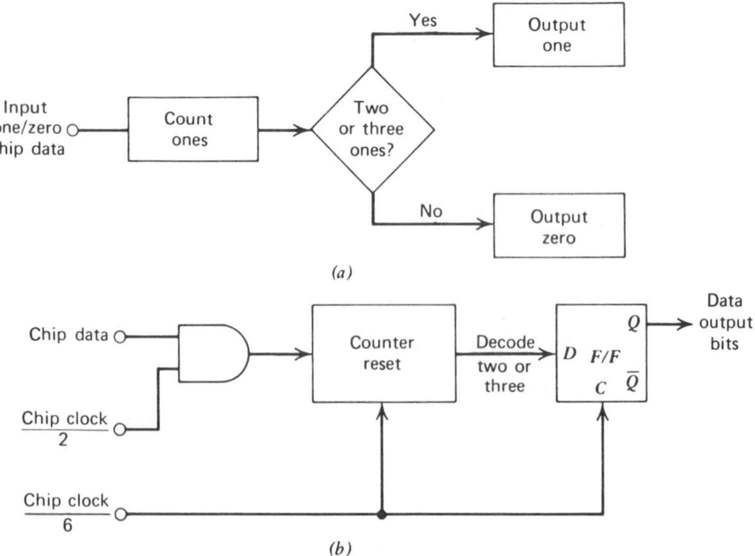

Figure 5.43 One/zero chip-to-data decision for two of three decision; (*a*) flow diagram for a two of three majority decision based on three chips per bit; (*b*) block diagram for a two of three majority decision.

with respect to the receiver, is slower than the receiver's chip rate (which is twice the transmitter's chip rate because it samples both one and zero channels) by a factor of two for chip sampling and a factor of three because there are three chips per bit.

Neither the three-chip-per-bit format nor majority decision as advanced here is the optimum. They are given only as representative of typical techniques.

Integrate-and-Dump Filters

An important type of filter, called "integrate and dump" (I and D), is shown in Figure 5.42 as part of the frequency hopping demodulator. It is matched to the period of a bit at the chip rate. Figure 5.44 shows two implementations of this filter—one for operation at RF or IF and the other for baseband (as in our frequency hopping illustration). The two are equivalent from the standpoint of their function, but here we emphasize the operation of the baseband filter.

Two conditions must be satisfied for any filter to be "matched":

1. Its impulse response must be a time-reversed replica of the signal to which it is matched.
2. Any particular unit of information must be independent of that preceding and succeeding it.

That integrate-and-dump filters meet these conditions will be evident as our description progresses.

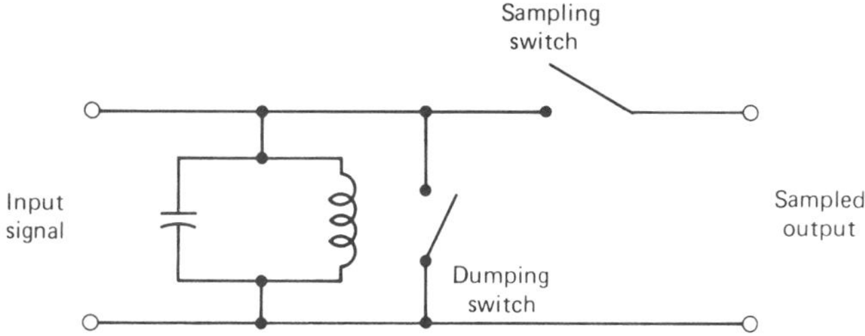

RF/IF I and D filter using high Q parallel–tuned circuit for integration.

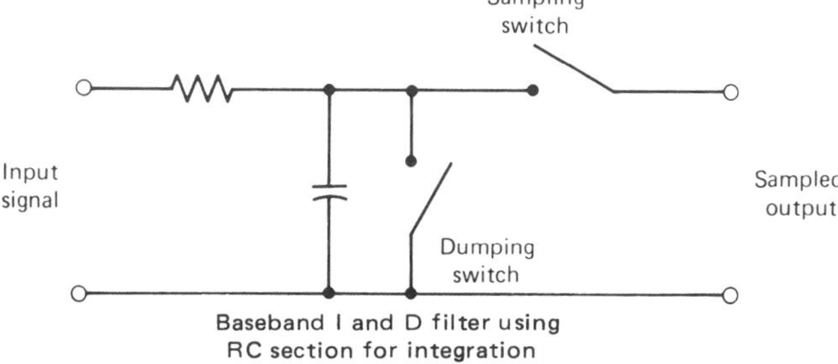

Baseband I and D filter using RC section for integration

Figure 5.44 I & D filters.

The signal input to the baseband integrate-and-dump filter, which would be most likely to be used in a frequency-hopping system, is just the envelope-detector output. Thus the input signal is a series of what is intended to be rectangular pulses but in the real world is distorted by noise and only approach rectangularity. At any rate, the series of pulses is applied to an *RC* integrator whose period is long compared with the pulse period ($T_{RC} = R \times C >> T_{\text{signal}}$); that is, the integration performed is restricted to the more linear part of the integrator's charge characteristic, where V_c closely approaches $V_t/RC = I_{pk}t/C$.

The integrator is allowed to charge for the period of an input pulse and is then sequentially sampled and dumped (discharged). If the signal existing at the input is a square wave, the integrator during Δt will charge to $V\Delta t/RC$. For any other signal shape, however (assuming that the peak voltage is the same), the charge at the end of Δt is lower and a decision threshold can be set on the basis of a desired signal being present in a given signal space. For the integrate-and-dump detector the process of discharging the signal stored is a matter of shorting across the integrator, preparatory to the start of accumulating energy during the following chip period. Figure 5.44 shows shorting and sampling switches connected to the integrator. (Of course, in practice, electronic switches are used.) To accomplish the purpose of signal detection and switches shown are closed (1) to sample the integrated signal and (2) to dump the integrator's charge.

Independence of signals is ensured by the dumping process, which isolates the energy in one chip from that in the others; it accomplishes one of the requirements of a matched filter. The second requirement—that its impulse response be a time-reversed replica of the desired signal—is also met.

Consider the *RC* integrator, with both its switches open. An impulse at its input charges the capacitor to a level $(I_{pk}/dt)/C$, which remains until dumped. This is exactly the same as the desired flat-topped square pulse. For the integrator that uses a high-Q tuned circuit the input impulse causes the circuit to ring at the tuned frequency—again forming the duplicate of a desired input signal. In both types of integrator, time inversion has no meaning, for square wave and sine wave pulses are the same when their leading and trailing edges are reversed.

Both types of integrate-and-dump detector, then, are matched filters, and the advantage of a matched filter is demonstrated for signal detection. When the baseband integrator charges to its highest level for a desired input and the lower levels for any other signal, the detection process is optimized.

The Dehopped Signal

A frequency hopping receiver must perform the same operations as its companion transmitter, except that it does them in reverse order. First, the spread spectrum modulation is removed by a chip-by-chip remapping of each received frequency into a narrowband filter and demodulator. The "dehopped" signal must then be processed to demodulate the information sent.

Frequency hopping receiver performance depends on how well the bandpass filter following the dehopping multiplier is able to select the desired signal from among other signals (and noise). In many frequency hopping receivers frequency shifts caused by instability, Doppler shift, or even nonoptimum signaling may force the filter bandwidth to be wider than the minimum signaling bandwidth, which in turn results in a loss in performance.

In a binary* frequency hopping transmitter data transmission (in this section only digital information transmission is considered) is performed by sending one frequency to signify a space or an alternate frequency for a mark. Whether only one or multiple chips are sent for each bit of information, one of two frequencies is sent for each chip. The receiver must then be able to observe both frequencies and decide which was intended. (It is often jammer strategy to try to send a signal in the complementary channel so that the receiver will make an incorrect decision.)

The receiver must be doubled (essentially two receivers) to observe both alternate channels at once.

Ignoring for the moment that if two receivers were used for data reception the system complexity and cost would be increased, let us look at the received signals. Figure 5.45 shows the output signals from a pair of mark/space data receivers. We see that the input to the mark or space band-pass filters is a series of pulses at the data rate (if multiple chips are sent per bit, the pulses are at the chip rate). These RF pulses are envelope detected and used to make a bit decision.

Observing that the mark/space pulsed-RF streams are complements immediately suggests that an alternate receiver without this redundancy might be possible. If we consider, for instance, that the absence of an RF pulse is a space, the mark channel receiver above could be used and only half the receiver or Figure 5.46 would be required.

Alternatively, if the synthesizer in the receiver were made to hop twice as fast as the transmit synthesizer—first to the mark frequency and then to the space frequency—it could sample both. Its block diagram and the waveforms generated are shown in Figure 5.47. Note that even though the output data are the same as in the preceding two cases, the signal input to the band-pass filter in Figure 5.47's mark/space sampling receiver

*There are also "m-ary" frequency hopping systems (of which one is TATS).[111] These transmitters choose from a set of frequencies corresponding to data inputs; for example, three binary bits could be sent for each 8-ary signal chip, with each of the eight levels corresponding to one of eight possible binary 3-tuples 000, 001, 010, 011, · · ·.

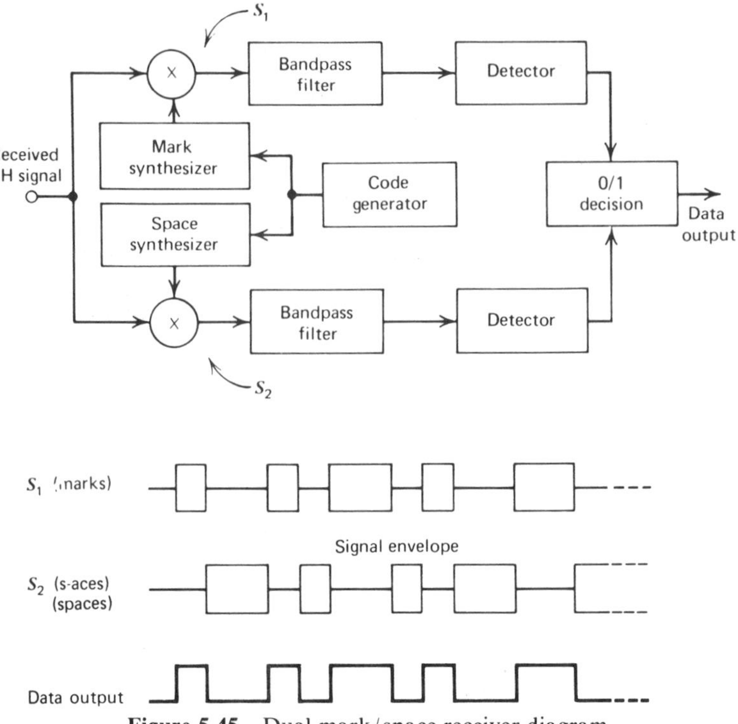

Figure 5.45 Dual mark/space receiver diagram.

consists of many pulses of half the duration of the transmitted frequency-hopping signal. Because the pulsed-RF input to the postcorrelation bandpass filter is of different duration, depending on the specific receiver implementation, we see that the filter must have a different bandwidth for proper reception. We note, however, that the data rate is the same for all three and therefore there must be a difference in receiver performance because of the different filter bandwidths.

Before considering this difference and the effect of it, we must point out one more factor—the coherence* or noncoherence of the frequency

*Coherence in a frequency hopping synthesizer requires that, when jumping from one frequency to another, the synthesizer output signal phase be well defined; for instance, one frequency burst might end at a zero crossing and the next start at a zero crossing. In a frequency hopping system coherence implies that *both* the transmit and receive frequency synthesizers maintain constant phase relationships so that the remapped single frequency correlator output is phase continuous. Such coherence, by the way, is no mean undertaking, for it requires not only a precise transmitter but uniform link (propagation) characteristics and identical receivers. There are those who speculate that overall system coherence in an actual operating over-the-air wideband frequency hopping is impossible to achieve.

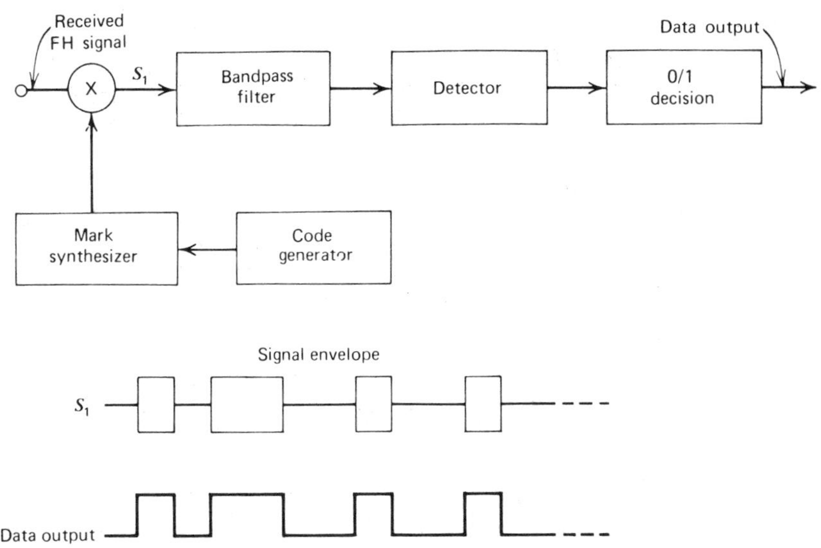

Figure 5.46 Alternate nonredundant receiver.

hopping synthesizers used. If the entire transmit and receive system is phase coherent, the postcorrelation bandpass filter output signal is phase continuous; that is, the pulsed-RF signal appears to be simply a pulse-modulated RF carrier. On the other hand, if the entire system is not coherent, the pulses are a series of RF bursts, each at a different phase, and the signal appears to be both amplitude and phase modulated.

What difference does this make? A significant difference. The bandwidth of the coherent signal is less than that of the random-phased noncoherent signal; thus the postcorrelation filter bandwidth may be decreased for a net gain in threshold and/or processing gain. When receiver sampling and multiple chips are used, the difference becomes even more pronounced. Figure 5.48 when the various chip envelopes are compared, shows that a three-bit-per-chip sampled signal, for instance, is very nearly a square wave. If that square wave RF envelope were considered to be 100% AM modulated by a square wave, it would readily be seen that the spectrum was that of an AM transmitter with modulation sidebands distributed according to the distribution of harmonic energy in the square wave. This distribution is

$$(\omega_m) = \frac{4}{\pi}\left(\cos \omega_m t - \frac{1}{3}\cos 3\omega_m t + \frac{1}{5}\cos 5\omega_m t + \cdots\right)$$

according to the Fourier expansion. The most significant power falls within the first and third harmonics; therefore, a filter that passed only

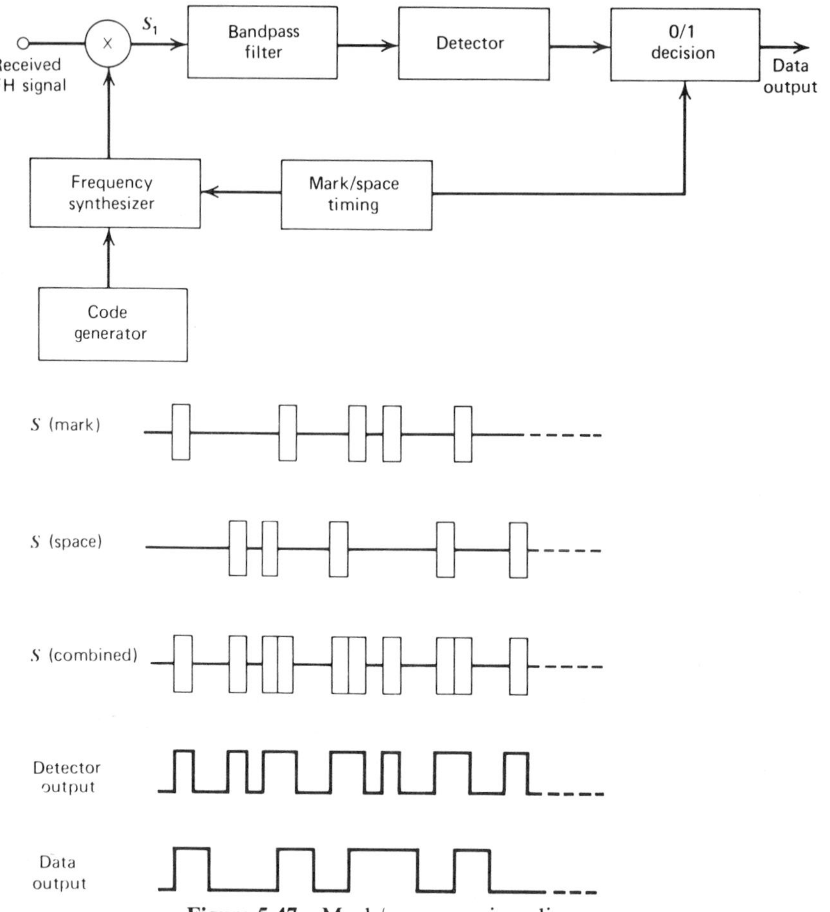

Figure 5.47 Mark/space receiver diagram.

these harmonics would output a reasonable RF pulse. If only the first pair of sidebands were passed by the filter, the output would be near-sinusoidal.

For noncoherent reception the phase of each RF pulse is different and the effective modulation rate is twice that of a coherent pulsed signal. This leads to a signal spectrum at least twice as wide as that of the coherent signal thus, requiring a wider filter and causing a loss in receiver performance. What is the expected frequency distribution for a randomly phase-shift-keyed signal? What else but $[(\sin x)/x]^2$? Figure 5.49 compares the spectra of coherent and noncoherent frequency hopping signals after dehopping. Note that the noncoherent signal main lobe occupies twice as much bandwidth as the coherent signal and its first sidebands.

Figure 5.50 is a photograph of an actual noncoherent dehopped signal's spectrum, which confirms our expected result.

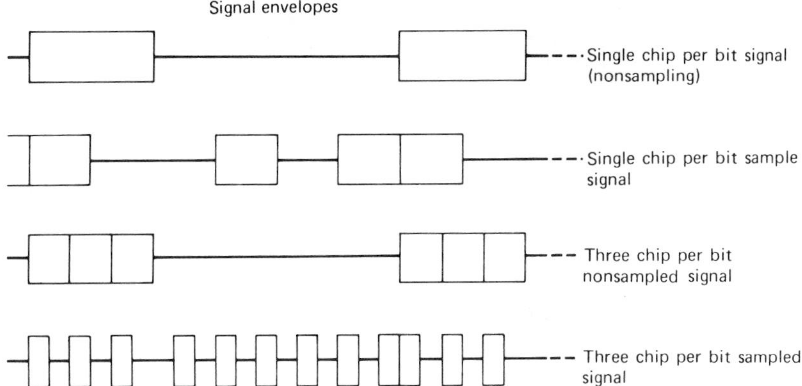

Figure 5.48 Comparison of RF envelopes of postcorrelation signals in a frequency hopping receiver.

Of course, there is nothing special about the $[(\sin x)/x]^2$ noncoherent signal that would prevent filtering it to the same $2\omega_m$ bandwidth as the coherent signal. This would result in a similar RF envelope from the postcorrelation bandpass filter but would give a decrease in signal power because of the different distribution of power. To be more exact, filtering to the ω_m bandwidth would result in a loss of 25% of the power in a $[(\sin x)/x]^2$ spectrum but only 10% of the coherent signal's power. This small difference (about 1 dB overall) could be signficiant in determining system operation or failure.

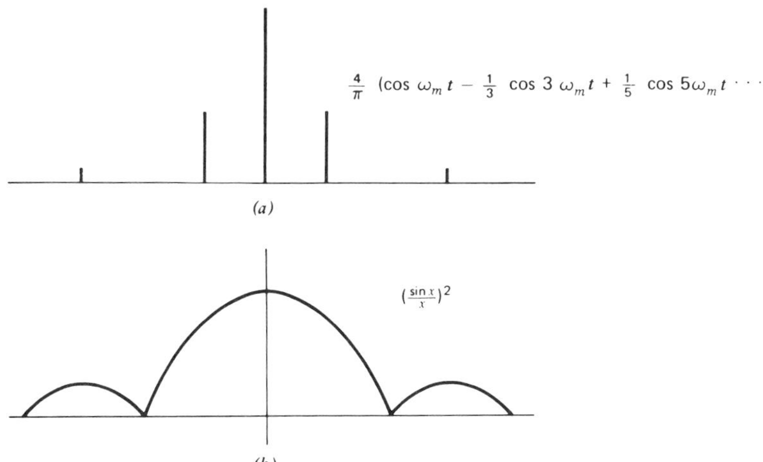

$\frac{4}{\pi}(\cos \omega_m t - \frac{1}{3} \cos 3\omega_m t + \frac{1}{5} \cos 5\omega_m t \cdots)$
(a)
$(\frac{\sin x}{x})^2$
(b)

Figure 5.49 Comparison of coherent and noncoherent frequency hopping signal spectra after dehopping; (*a*) coherent square-wave-pulse AM spectrum; (*b*) noncoherent square-wave-pulse AM spectrum.

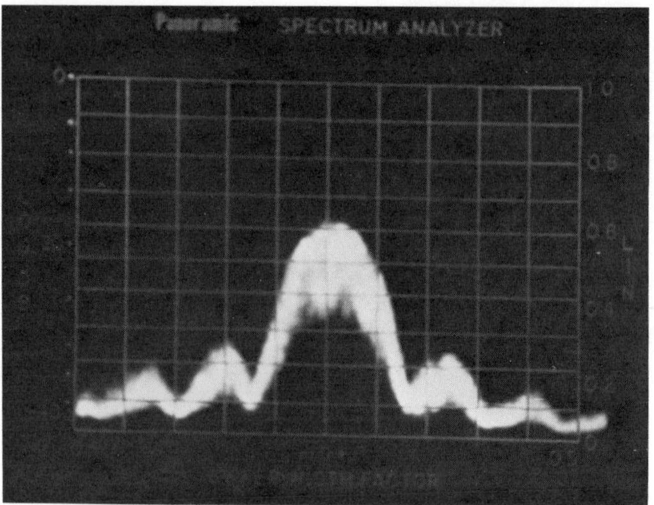

Figure 5.50 Photo showing actual noncoherent dehopped frequency hopping signal.

m-ary Detection

Detectors for *m*-ary signals are similar in principle to the alternate-chip-comparing detector already described, except that they simultaneously compare all possible signal channels and make a decision based on the largest signal detected. An *m*-ary frequency hopping signal is one in which the transmitter has a choice of *m* frequencies in each chip time rather than two as in our previous discussion. The transmission can then represent more than one bit of information; that is, an 8-ary transmission with one of eight frequency choices being transmitted at any given time would signify three binary bits by each chip. A 16-ary transmission would convey four binary bits per chip, and so on.

The *m*-ary transmitter can transmit on any of the frequencies made available by its synthesizer, of course, but allowing any set of *m* frequencies from a large set can pose a difficult demodulation problem for the receiver. To be exact, if any *m* of the synthesizer frequencies are used, the receiver must monitor all *m*, which forces the receiver to employ as many as *m* synthesizers or even an array of *m* receivers. The approach taken in TATS (tactical transmission system)[111,124] is to use a frequency frame like that shown in Figure 5.51 in which each possible frequency has a definite relation to $m - 1$ others.

In Figure 5.51 we see a set of frequencies shown for three different chip times, t_1 through t_3. Each set of frequencies consists of eight possible transmission frequencies clustered around a center frequency. The center frequency f_c changes at each chip time, then each *m*-ary set is hopped around and becomes a frequency-hopping transmission in itself. In addition,

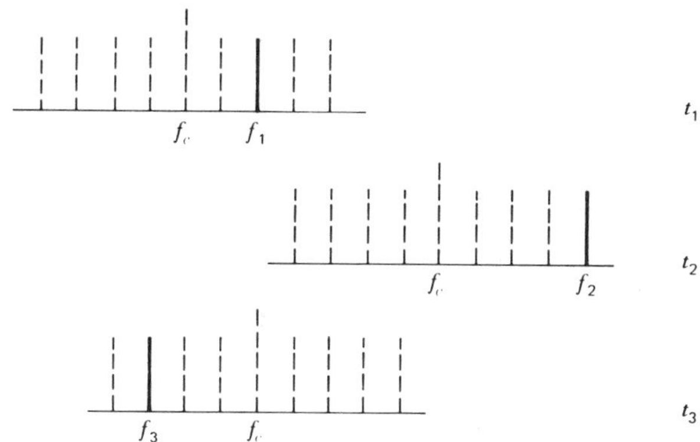

Figure 5.51 Typical transmission format for m-ary frequency hopping.

however, the information to be transmitted offsets the center frequency to $f_1, f_2, f_3, \cdots$, depending on the data being sent. Therefore, except for synchronization, the center frequency is never actually transmitted. Only the frequencies offset from the center by the input data requirements are transmitted.

In addition to the more obvious advantage that is afforded by m-ary transmissions over binary, that is, more data per chip, a simple receiver format is permitted. Only one receiver with separate detectors operating at baseband, rather than a separate receiver for each possible frequency, is satisfactory. The demodulator in Figure 5.52 is satisfactory for receiving such transmissions. This demodulator is a simple extension of those shown for frequency hopping systems except that all possible frequency inputs are sampled at once and the channel with the largest signal in a given chip interval is chosen as the correct one. If, for instance, the channel judged to have the largest signal corresponds to a 110 input, then 110 would be the demodulator output. Similarly, any of the other seven

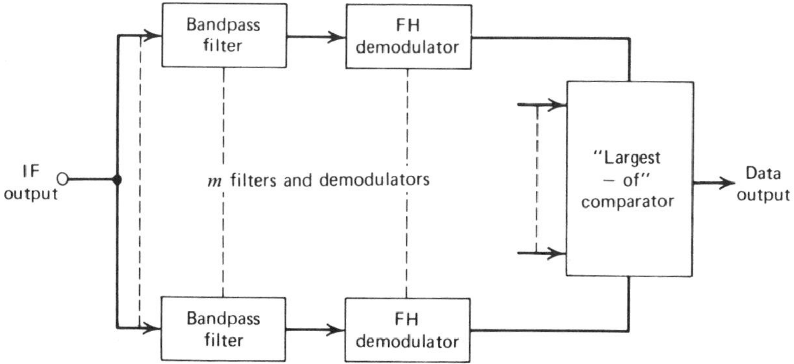

Figure 5.52 Block diagram of m-ary demodulator.

channels (referring to Figure 5.52) containing the largest signal would cause an output 000 through 111, depending on the 8-ary to binary correspondence.

One problem with the *m*-ary transmission format does exist; that is, a system's jamming margin may be reduced below the level that would be expected for a given number of available hopping channels. This is due to the extra vulnerability of *m*-ary transmissions to interference. If any one of the *m* channels other than that intended is hit by an interfering signal, an error is possible. In an 8-ary receiver, for instance, an interfering signal in any one of seven unused channels could cause an error and that error would represent three bits. To overcome this problem error correction coding can be used. The TATS system employs a Reed–Soloman encoding technique for this purpose. A determined jammer, however, could note frequency, retransmit that frequency offset by the channel spacing, and cause an error almost every time. Only the hopping rate of the desired signal and the geometry of the particular situation preventing the jamming signal from reaching the friendly receiver during the correct chip could avert this type of vulnerability.

When employing the *m*-ary format, it is common to employ one of two techniques to avoid the offset jamming signal vulnerability mentioned above. The first of these is to choose the *m* frequencies used from the complete set available, rather than restricting them to a contiguous set. The second is to increase the number of frequencies available by a factor of seven, which reduces the probability of a bit by the jammer.

PROBLEMS

1. Correlation filter output signal-to-noise for a 20-dB correlator interference-to-signal input and 30-dB process gain would be how much (neglect implementation loss)?
2. Draw a block diagram implementation of a correlator designed to represent the function

$$\int f_{(t)}\, g(t - \tau)\, dt$$

3. What is the overall probability of an error in a frequency hop chip if 30% of the available frequency channels contain an effective interfering signal?
4. Draw impulse response waveforms for the integrate-and-dump detectors in Figure 5.44.
5. Compare the binary and *m*-ary frequency hopping approaches.

6. What, in a direct sequence system, is the equivalent of m-ary information transmission?
7. What bandwidth would be expected of a common channel spread spectrum signal (i.e., one similar to the desired but employing a different code sequence) at the output of a correlating multiplier?
8. Given 30-dB process gain, what would be the output $(S + I)/I$ ratio in a correlator when input interference is 10 dB above the desired signal?
9. What repetition rate, or line spacing, would we expect to see at the output of a correlator's multiplier for a 4095-chip code at a 1 MB rate when multiplied with a CW carrier.
10. What is the phase angle between the input signal and reference in a phase-lock loop when the input is offset by 10 kHz from the loop internal oscillator and loop gain is 10^5?
11. Compare phase lock and Costas receivers for subcarrier FM or FSK demodulation.
12. What would be the effect of employing simultaneous carrier FSK and subcarrier FSK in a direct sequence system?

6

SYNCHRONIZATION

Now we must talk about the hardest part. Throughout this book we have assumed good synchronization (on the part of the code) between transmitters and receivers. By "good" synchronization we mean that the coded signal arriving at a receiver is accurately timed in both its code pattern position and its rate of chip generation (with respect to the receiver's reference code). So far we have assumed that the codes in the system we have talked about were already synchronized and would remain so.

What an assumption! More time, effort, and money has been spent developing and improving synchronizing techniques than in any other area of spread spectrum systems. There is no reason to suspect that this will not continue to be true in the future.

Code synchronization is necessary in all spread spectrum systems because the code is the key to despreading the desired information and to spreading any undesired signals. The overall requirements of frequency hopping and direct sequence systems with regard to synchronization are similar, but somewhat more latitude is possible in the frequency hoppers. Here we discuss the synch requirements of both types of system, describe ways of attaining synchronization from a cold start, and discuss ways to maintain synch once it has been acquired. Chirp systems are not considered because they do not require synchronization in the same sense as other types of spread spectrum systems. To be sure, chirp systems do synchronize, but synch is on a pulse-by-pulse basis because the most important property of a chirp matched filter is rapid total-pulse correlation.

Sources of Synchronization Uncertainty

In spread spectrum systems two general regions of uncertainty exist, with respect to synchronization. These are code-phase and carrier-frequency uncertainties and both require resolution before a spread spectrum receiver can operate. Code phase must be resolved to better than one chip, and center frequency, as seen at the receiver, must be resolved to the degree that the despread signal is within the aperture of the postcorrelation filter. Furthermore, the carrier frequency is often constrained to be accurate enough to work well with a demodulator.

Code clock rate must also be resolved, but because clock rate can be related to carrier frequency, if a system is so designed, it is not necessarily a separate source of synchronization uncertainty.

Most of this code-rate, phase, and carrier-frequency uncertainty may be eliminated simply by providing accurate frequency sources in both transmitter and receiver. Not all, however, can be overcome in this way. Doppler-related frequency errors often cannot be predicted and will affect both carrier frequency and code rate. In addition, mobile transmitters and/or receivers see a relative code-phase change with every change in their relative positions. Even fixed-position transmitting and receiving stations may see changes in code phase, center frequency, and/or code rate due to propagation-path-length changes. Such effects are especially pronounced in tropospheric scatter communications but may also be seen in all but the most stable direct line of sight links.

The amount of Doppler-frequency uncertainty applied to a received signal is a function of the relative velocity (range rate) of the receiver and transmitter and of the frequency transmitted. Consider a sinusoidal traveling electromagnetic wave. If that wave is observed passing a given point, it will repeat at a rate $f = c/\lambda$, where f is the frequency of repetition, λ is the wavelength, and c is the speed of light in the propagation medium* being considered. Thus, if neither the transmitter that sends the wave nor the receiver that observes it is moving, the frequency of the received wave is the same as that sent by the transmitter. Let the point of observation move with respect to the transmitting point, however, and the number of repetitions in a given period is changed; for instance, if the receiving point moves toward the transmitter, the rate of repetition of the electromagnetic wave at the receiver increases because it reaches the receiver at a rate higher by an amount exactly proportional to the receiver's approach velocity.

The "Doppler frequency shift" is seen to be equal to the number of wavelengths at the frequency of interest that the receiver covers in a 1 sec period. The amount of Doppler frequency shift then is just $D = V/\lambda = fV/C$ Hz, where V is the relative velocity of transmitting and receiving terminals and f is the frequency transmitted. When a given frequency f_t is sent, the received frequency is just

$$f_R = f_t \pm \frac{f_t V}{C}$$

*2.997925 × 10[8] m/sec in vacuum, per NBS Applied Mathematics Series 55, Handbook of Mathematical Functions, February 1971.

the plus sign for increased frequency is applicable to terminals approaching one another and the minus sign for decreased frequency is valid with departing terminals.

A rule of thumb for determining Doppler shift when frequency and velocity are known is that

$$\begin{aligned} D &= 3.336 \text{ Hz/MHz} \cdot \text{km} \cdot \text{sec} \\ &= 0.000926 \text{ Hz/MHz} \cdot \text{km} \cdot \text{hr} \\ &= 0.00148 \text{ Hz/MHz} \cdot \text{mi} \cdot \text{hr} \end{aligned}$$

Thus, Doppler effect must be considered in optimizing a communication system, for it is often the major contributor to frequency uncertainty when stable transmitting and receiving frequency elements are employed.

Often the frequency sources used in a communication system are not so

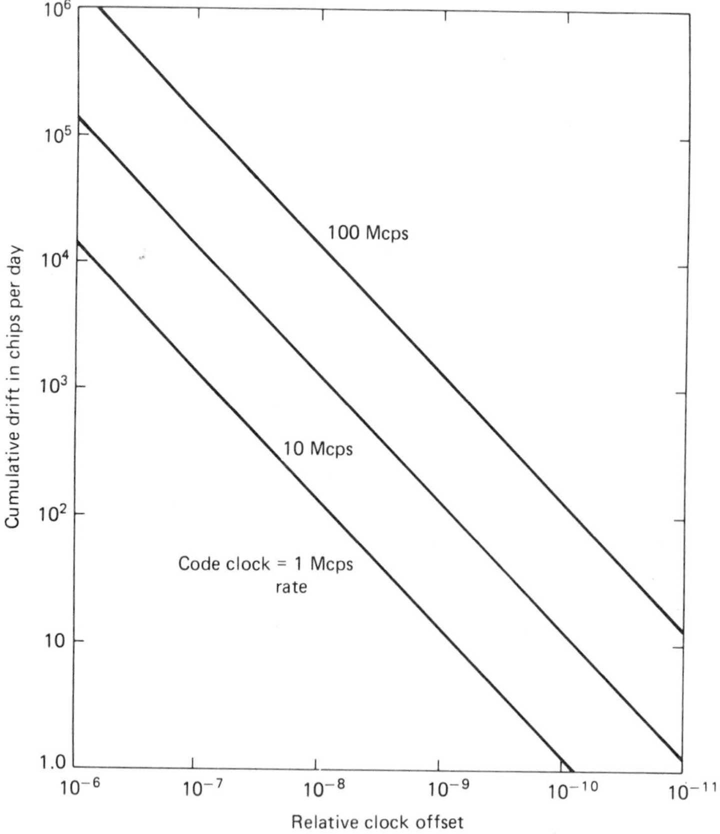

Figure 6.1 Curves showing cumulative daily drift for various code rates and relative code-rate errors.

stable as we might wish and their contribution to frequency uncertainty cannot be ignored. In spread spectrum systems another consequence of frequency uncertainty is also apparent—any clock rate offset is cumulative in code-phase offset; that is, a 10 Hz offset in clock rate for a code generator would translate to a 10 chip per second cumulative code error, which in an hour would cause a code-phase drift of 36,000 chips. Factors that contribute to oscillator frequency instability are temperature, attitude of the crystal (i.e., vertical, horizonal, or otherwise), amplitude stability, noise contributions from the amplifier stage, and others. Oscillator instability has been characterized in detail by Lindsey and Lewis.[1729]

The prime sources of uncertainty for synchronization, then, are those that are time* or frequency dependent. If the receiver could know range and transmitter time accurately enough, there would be no synchronization problem, or if both transmitter and receiver possessed oscillators accurate enough, they could derive the timing needed. Furthermore, that same accurate oscillator could act as a source of precise frequencies to remove all frequency uncertainty except for that produced by the Doppler effect. A highly stable frequency source is to be desired for any spread spectrum system. Figure 6.1 illustrates the amount of code drift that occurs between a pair of systems with various relative code rate differences.

6.1 INITIAL SYNCHRONIZATION

The initial synchronization problem is the most difficult of all. When synchronization has already occurred, it is often possible to base a subsequent synchronization on the knowledge of timing gained. A cold start without prior knowledge of timing, or at best very minimal knowledge, is the usual rule rather than knowing the proper time frame for achieving synchronization with a desired transmitter or receiver, however. Many techniques for achieving synchronization have evolved, some with simple requirements, others with complex implementation implications. Several are described in this section. Which is best depends on the application, the amount of time allowed, and the amount of uncertainty involved.

Certainly, if the time of day (and in some cases the date) is known to an accuracy of sufficient degree by both transmitting and receiving stations, the code generators used to determine the order of system functions could be set to coincide and the only uncertainty would be that caused by unknown range; for example, if two identical code sequences were started at precisely the same time and the clocks used to generate their chip rates

*Time uncertainty here includes any propagation time uncertainty due to unknown range.

continued from that time to give identical code rates, then at any point in time after the original start the two sequences would be at the same state. A transmitter and receiver employing this pair of sequences could be synchronized in real time and would remain so forever if no outside influence disturbed them. They still could not talk together, however, unless the distance between them were less than $1/Rc$ (R = code bit rate, c = speed of light), the distance corresponding to one code chip in range. Otherwise, the propagation time delay for a signal modulated with the transmitter's code sequence to reach the receiver would be such that the receiver's sequence could have reached a new state (and so would the transmitter), whereas the signal sent would be in transit and the coded signal as received would not be in synch with the receiver's code.

The obvious cure for this small difficulty is to move the transmitter's sequence ahead in time by an amount equal to the signal propagation delay, and therein lies a new problem; that is, the distance from transmitter to receiver may not be known well enough to time the code sequences satisfactorily. Thus we see that even perfect knowledge of time and perfect timekeeping does not completely solve the synchronization problem. This is especially true when one or both of the terminals communicating are mobile, for a poorly defined distance uncertainty is the result.

Even for fixed terminals with known range and precisely known time it is possible for some phenomenon such as multipath to require a special synchronization procedure. Bearing in mind, therefore, that even the stablest of code sequence generators does not completely overcome either initial synchronization or tracking problems, we now launch into the subject of synchronization methods.

The "Sliding" Correlator

The simplest of all correlation techniques uses a "sliding" correlator, so called because the receiving system, in searching for synchronization, operates its code sequence generator at a rate different from the transmitter's code generator. The effect is that the two-code sequences slip in phase with respect to each other, and if viewed simultaneously (say with an oscilloscope), as in Figure 6.2, they seem to slide past each other, stopping only when the point of coincidence is reached. The flow diagram of Figure 6.3 illustrates this process.

Figure 6.4 shows spectra typical of those seen at a direct sequence demodulator's input. Figure 6.4*a* shows the noise spectrum, band limited by the predemodulator IF, generated by the cross-correlation of the receiver's internal code generator with any unsynchronized input signal. This is the typical spectrum seen any time the receiver is not synchronized.

At the instant that the receiver's code sequence comes into synch with an incoming signal, a coherent output like that in Figure 6.4*b* appears.

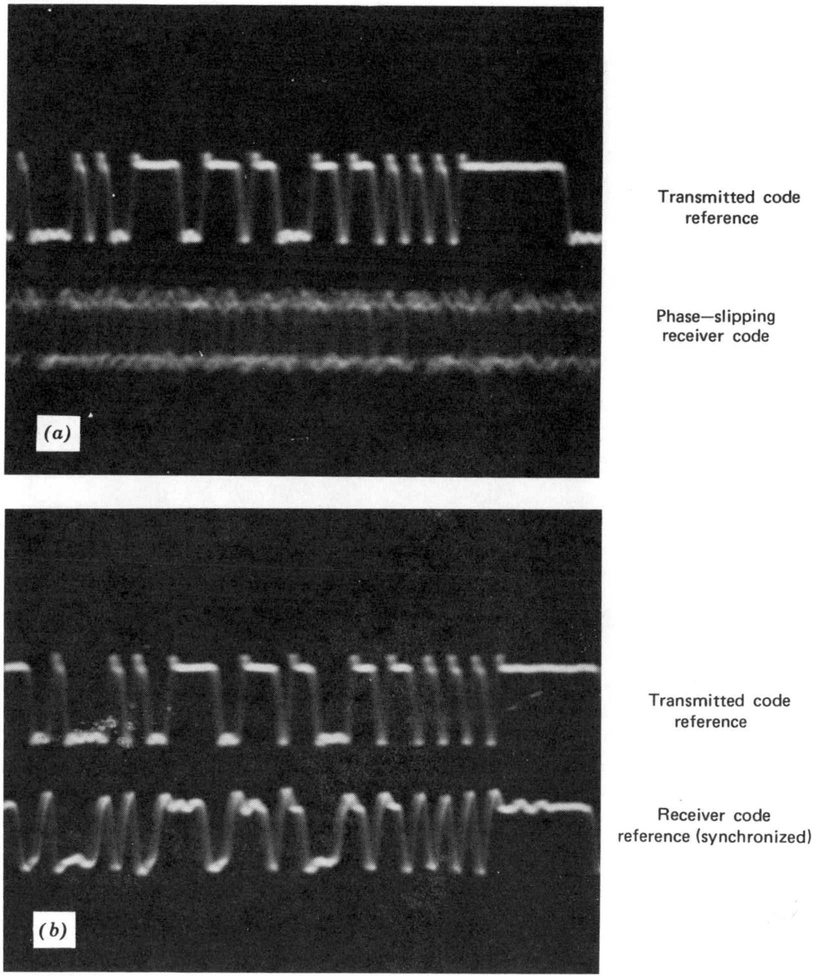

Transmitted code
reference
Phase—slipping
receiver code
(a)
Transmitted code
reference
Receiver code
reference (synchronized)
(b)

Figure 6.2 Sliding correlator action: (*a*) sliding process in operation; (*b*) synchronized codes after acquisition.

Loss
of
synch
Retard
clock
Is
receiver
locked?
Yes
Stop
search
Enter
track
mode
Continue
search
Track

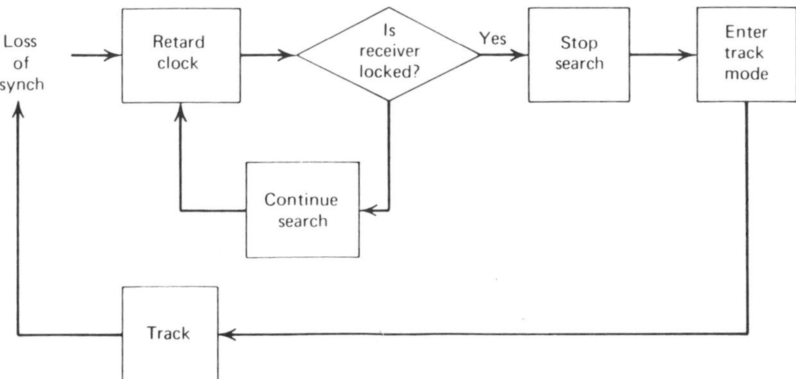

Figure 6.3 Flow diagram for sliding correlator synchronizer.

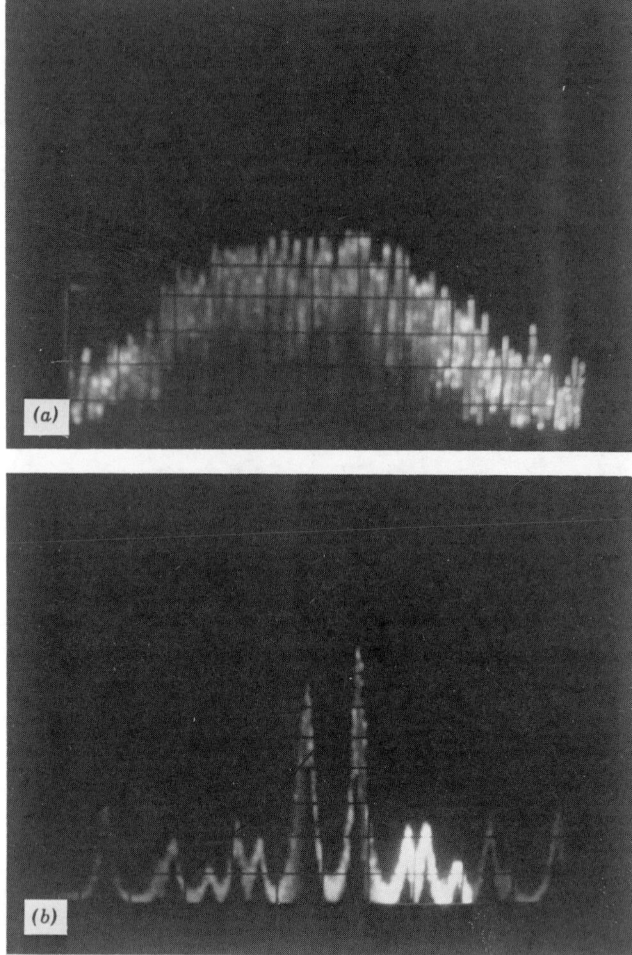

Figure 6.4 Signals seen at input to direct-sequence demodulator: (*a*) noise input to demodulator, unsynchronized condition; (*b*) DSB, 2500-bps signal after synchronization, 25-dB J/S ratio, CW jamming.

The demodulator then has the task of acquiring carrier lock on the signal. The signal shown in Figure 6.4*b* is double-sideband suppressed-carrier modulated by a 2500-bps square wave. Also, the input jamming to signal ratio for the receiver in Figure 6.4*b* is 25 dB.

The advantage of the sliding correlator is its simplicity in that nothing more is required than some way, on command, of shifting the code clock of the receiver to a different rate. The difficulty of using a simple sliding correlator for synchronization, however, is that, when a large degree of uncertainty is encountered, examination of all possible code-phase positions is impractical because of the time involved. Recognition of synchronization, which must occur to stop the sliding or "search" process at or near the point of synchronization, is limited in response time by the bandwidth of the system's postcorrelation receiver. Thus the rate of searching, or sliding through all of the possible code-phase offset positions, is also limited by that bandwidth. Therefore, to give an example, a system with a postcorrelation (baseband) receiver bandwidth of say 1 kHz could recognize synchronization in approximately 0.35 msec (applying the general rise time-to-bandwidth relation $T_r = 0.35/\mathrm{BW}$), and the time used to slide through the point of correlation (remember that the correlation function is two chips in width) should be a minimum of this amount. The maximum search rate then would be approximately $2/T_r$ or 5.7 kbps. Because it is possible under some circumstances for a receiving code sequence generator to be offset by trillions of bits, we can easily see that hours, days, or even years might be consumed in the initial synch process if the sliding correlator alone were used. As a rule of thumb, one can expect to search at a rate equal to the data rate for which the receiver has been designed.

In practice, sliding correlation is almost always employed, though the technique is often augmented by other methods for restricting the area of search. Most of the remainder of this chapter (at least that part having to do with synchronizing) discusses methods of initial synchronization that, in general, employ techniques to supplement sliding correlation; that is, these other methods bring the two code streams of interest into a range in which sliding correlation can reasonably be used.

Synchronization Preambles

One of the most effective techniques for making use of a sliding correlator employs special code sequences, short enough to allow a search through all possible code positions in some reasonable time but limited in how short they can be by correlation requirements. A well-chosen code sequence (called a preamble when used for synchronization) is a good solution to almost all synchronization problems. Range estimates are unnecessary when a preamble is used, and acquisition time is set by the length of the preamble. Consider an avionics system, for example, in which no knowledge of direction, speed, or distance can be derived before receipt of a signal. A synchronization preamble sent at the beginning of each transmission would allow normal push-to-talk system operation without having prior knowledge of relative positions. Of course, any time required for the synch preamble acquisition process would be added at the beginning of each transmission. A reasonable alternate might be to transmit a preamble code for acquisition only at the beginning of a series of transmissions, which would synchronize all link users. Thereafter synchronization would be maintained by the system clocks, broken only by transmission of an "out" signal.

Typical synchronization preambles range in length from several hundred chips to several thousand, depending on the specific system's requirements. When the information being sent lies within a specific band, it is advisable to select the preamble length such that its repetition rate does not fall within that band. Spectral components from code repetition × interference occur at intervals of R_c/L Hz, where R_c is the code chip rate and L is the code length in bits. If the preamble repetition rate is within the information bandwidth of the system, two or more of these spectral components will lie in the receiver's IF bandwidth and could appear to the demodulator as a pair of modulation sidebands, thereby interfering effectively with the desired signal. Though the preamble code repetition rate can act as an interference-enhancing element, it cannot be chosen to produce less than one spectral component in the information band. No matter how far apart the spectral components may be, a variation in interference frequency can cause one component to fall into the signal band.

At the opposite extreme we may choose the preamble length and code rate such that the repetition rate causes multiple spectral components to lie in the information band (this implies longer codes and/or lower code rates). If the code-produced frequency lies below the information band, the interference produced can be acceptable.

Figure 3.39 shows the line spectrum of an 8191-chip code when cross-correlated with a CW jamming signal.

Preamble synchronization methods have one significant weakness that comes about as a result of the very code property that makes them work well; that is, the relatively short sequence length which allows rapid synch acquisition tends to be more vulnerable to false correlations and to possible reproduction by a would-be interferer. With the exception of the possible vulnerability problem, however, preamble synchronization is by far the least critical, easiest to implement, least complex, and best for all around use.

It is possible to employ short synch preambles if they are hidden by transmission well below the level of a masking signal. For example, suppose that a long code and a short code modulate a pair of orthogonal carriers, but the short-code-modulated signal is attenuated by at least 10 dB. Then, the two signals are summed together. The preamble-modulated signal is then 10 dB or more below the longer code level, and cannot be demodulated.

Bounds on the preamble's length are generally set by four criteria: minimum preamble code sequence length is bounded by cross-correlation and interference rejection requirements, whereas the maximal code sequence length is set by maximum available acquisition time. The fourth criterion, that of code repetition rate, may tend to bound preamble code length in either direction.

Frequency Hop Synchronization

The comparatively slow code rates employed in frequency hopping systems give rise to greater ease of synchronization than those of direct sequence systems, because at the slow code rates used a given clock error accumulates much more slowly; that is, given a clock error of one part in 10^6, a 50-Mcps direct-sequence generator would accumulate a one-chip offset in 50 msec, whereas a typical frequency hopping code generator at 5 kcps would drift only one chip in every 500 sec. Thus the area of code-phase uncertainty is significantly different for frequency hopping systems than for direct sequence systems. For this reason direct sequence systems may employ a frequency hopping preamble to facilitate more rapid synchronization.

Frequency hop synchronization is implemented by allowing for two code sequences which are generated at different rates, one for the direct sequence mode and one for frequency hopping during synch acquisition. The codes must be related, in that frequency hop code synch allows for decreasing the direct sequence code phase uncertainty; for example, assume that the direct sequence code to be synchronized is 10^6 chips long* and its chip rate is 10 Mcps. We then can take a slower 1000 chip code sequence at say 10 kcps, synchronize the two codes so that their starting points coincide, and arrive at a pair of code sequences for which each chip in the slower code corresponds to 1000 chips in the higher-speed code.

If the slow code is then used for synch acquisition, the synch uncertainty which would exist could be no worse than 1000 chips instead of the 10^6 chip uncertainty possible with the higher-rate code. Therefore, an initial synchronization operation can be carried out in one-thousandth the time.

Initial acquisition of the slow code, however, does not complete the acquisition process. Because synchronization to a slow code chip resolves the high-speed code only within 1000 chips, serial code search must be carried out to synchronize the high-speed code within 1 chip. So we see that by searching over a 1000 chip uncertainty twice we can resolve a 10^6 chip ambiguity.

A slow code, unlike a high-speed code, is not adapted to direct sequence transmission. A 10 cps code could hardly be expected to give a useful amount of spread spectrum processing gain in a direct sequence modulator/demodulator. The solution, therefore, is to use the slow code to drive a frequency hopping synthesizer during an initial synchronization period, with a switch to the high-speed code and direct sequence operation following frequency hopping synch acquisition.

*A number chosen here for convenience; a more realistic number would be $2^{20} - 1$, or about 1.05 million.

Frequency hop synch provides for rapid synch acquisition but does not allow for simplification when direct sequence modulation has been chosen to avoid a frequency hopping synthesizer requirement. When a flexible frequency synthesis technique already exists, frequency hopping synch methods are a good alternative to direct sequence preamble synch methods.

Transmitted-Reference Methods

When a receiving system must be the simplest possible, a transmitted reference[1502,1506] may be used for initial synch acquisition, tracking, or both. Transmitted reference receivers employ no code sequence or other local reference generator. Instead the coded reference is generated at the transmitter and transmitted at the same time as the intended information-bearing signal. Both frequency hopping and direct sequence systems are amenable to transmitted reference methods. Figure 6.5 shows a transmitted reference system.

The operation of a transmitted reference receiver is precisely the same as that of any other receiver using an offset local reference signal. The difference is that the local reference is generated in the transmitter and sent to the receiver along with the signal to be demodulated. Their carrier frequencies are offset by an amount equal to the first IF in the receiver, and when mixed, a correlated IF signal is produced.

Transmitting the reference is obviously advantageous when the receiver must be limited in cost, weight, or size because there is no need for a code sequence generator, search or tracking circuits, or any of the code-related mechanisms. Of course, if the reference is transmitted only to achieve rapid synch acquisition with a subsequent switch to normal direct sequence or frequency hopping demodulation, we must also still include these subsystems.

If the transmitted reference technique is to be employed when interference or active jamming is a threat, some other synch provision may be necessary. Otherwise any two identical signals offset by the amount of the IF would produce false synchronization in the post-correlation circuits. One method of protecting the IF in this situation would be to frequency hop one of the signals transmitted, which would require the receiver IF to hop with the same frequency pattern, or a frequency hopping local oscillator would be needed to act as a local reference for the frequency hopping subsignal. In either case the synchronization time would be increased to account for frequency hop synch acquisition.

One other disadvantage exists, and this is that noise introduced in the reference due to being transmitted through the link (rather than locally generated) degrades the receiving system. When interference appears at both desired and reference-signal frequencies, correlator output products

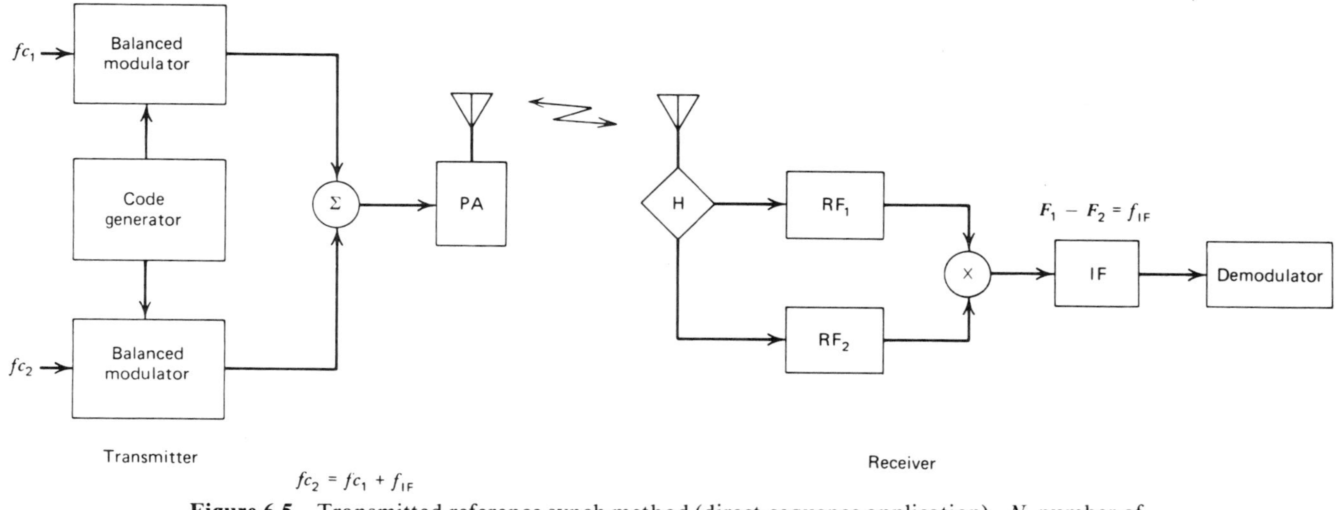

Figure 6.5 Transmitted reference synch method (direct-sequence application). N, number of stages in shift register; T_c, code clock period; L, code period. Loading time = NT_c; no noise acquisition time = NT_c, Integrate for LT_c and make decision.

are produced which can fall into the IF band, thus also degrading signal demodulation.

Universal Timing

When accurate, dependable, universally known time is available, it is possible to devise a system that never loses synchronization. (This does not necessarily mean, however, that a search procedure is not needed for synch acquisition. Even if, in real time, transmitting-system and receiving-system code sequence generators are in perfect alignment, a signal sent from one to the other would not be synchronous when it reached the receiving system because of the time required for signal propagation. For this reason, and others, some search-and-track capacity is always required, even when code chip rates are precise and distances are well known.)

Consider that every communication terminal knows the time of day and the date within 1 μsec. Then at prearranged times, say at each hour on the hour, all code generators could be set to a new state derived from that time and date information. Resetting to a new state could, of course, be done more often when, with minimum waiting, entry to the network by new users is desired. If the code sequences in question have rates of 1 Mcps or less, universal 1 μsec timing ensures real-time synch to within 1 chip time. In any case, however, some search-and-track capability would still be required.

Universally timed spread spectrum systems have been used principally in satellite communications in which ephemeris data provide accurate distance information and ambiguity can be reduced to the point at which search processes may be minimized. Future spread spectrum systems for all uses will probably employ universal timing concepts for synchronization and tracking, as frequency sources for mobile use improve and as computerized position estimation methods provide more accurate information for range compensation.

Burst Synchronization

The idea behind burst synchronization is to transmit a short, high-speed message that provides the spread spectrum receiver with enough information to gain rapid access to a subsequent direct sequence or frequency hopping signal. It is, of course, hoped that any would-be interferer is to be caught napping by the sudden appearance and disappearance of the synchronizing signal. A determined interferer might leave his or her interfering signal on continuously, however, which would negate any element of surprise. On the other hand, the high peak power available to a low-duty-cycle burst transmitter could cause our would-be interferer to require an inordinately large amount of continuous RF power, thus providing interference protection.

No information, other than code and / or carrier synch, is normally sent during a synch burst. Information to be transmitted then follows a switch to direct sequence or frequency hopping.

Sequential Estimation

Much akin to burst synchronization methods are those of sequential estimation, in which the receiver to be synchronized simply demodulates the incoming coded signal, inserts the data (code) as a vector input to its own sequence generator, and makes a trial check for synchronization. The code being received may or may not be easily acquirable. The received signal is demodulated in exactly the same way we would demodulate any PSK data stream (this technique is not readily applicable in frequency hopping). Once the local code generator is loaded with a code input taken from the received signal, it must race ahead (search forward) by a number of bits $n + T_p/f_k$, where n is the code generating register length in bits, T_p is the receiver's processing delay time, and f_k is code clock rate.

Sequential estimation (see Figure 6.6) has a real advantage as a rapid synchronization method. Ward's results[65] show a time improvement factor of 23 for replacing a sliding correlator with a sequential estimator. Sequential estimation, however, does have the drawback that it is more vulnerable to interfering signals than some other techniques. This is because the demodulator demodulates the code received without benefit of processing gain.

Average acquisition time recorded by Ward shows 90 msec at −6 dB, 300 msec at −10 dB, and 1 sec at −12.5 dB signal-to-noise ratios. (A

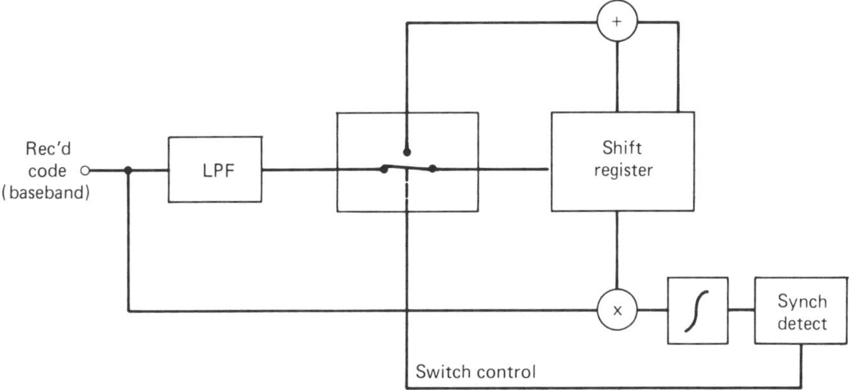

Figure 6.6 Illustration of sequential estimation synchronizer.

white-noise source was used for these measurements.) Although this performance against noise may not appear to be impressive when deliberate interference is expected, those spread spectrum applications in which the system contends only with natural disturbances—and there are many of them—could benefit by synch acquisition through sequential estimation techniques.

Special Coding for Synch Acquisition

Special codes, specifically designed for synch acquisition, have been employed to advantage in the quest for more-rapid spread spectrum system lockup. Short preamble codes might be considered special codes, but in general the only difference between them and the code sequences used after acquisition is length. Here we are more interested in codes that have special properties to make them more easily acquirable than the maximal sequences or to give some implementation advantage that can be translated to a simpler or more-rapid synch procedure. Emphasis in this section is placed on use of these codes rather than on the codes themselves. Those who are interested in code properties and code generation in more detail are referred to Chapter 3 and the references listed in Appendix 2.

The best-known easily acquirable code type is the JPL component code,[219,330] made up of shorter codes. These shorter codes, each of different length are combined to result in a code with a length $(2^m - 1)(2^n - 1) \cdots (2^r - 1)$, where $m \neq n \neq r$, and with a special correlation property, which is different from the maximal linear sequences in that the m-sequences have only one point of autocorrelation. The component codes, on the other hand, have one more correlation possibility than there are components in the code (i.e., if a code is made up of three component subcodes, there are four autocorrelation points associated with the overall code). Moreover, all but one of these autocorrelation levels are associated only (and separately) with the individual codes making up the composite. The highest correlation level corresponds to total composite code synchronization.

Synchronization, by using the JPL component codes, is accomplished by first cross-correlating one of the component codes with the composite code. Once this component code reaches the point of synchronization with its mate, which is embedded in the composite code, a partial correlation occurs. The partial correlation is then the signal for the second component code cross-correlation to be initiated, which causes the partial correlation level to be increased. This process continues until all the component codes making up the overall composite code are individually synchronized with their counterparts in the received signal. When all are individually synchronized, the correlation is the same as if the process had simply synchronized the composite code. The advantage of this technique is that it provides for rapid synch acquisition without the use of a preamble or anything other than the composite code itself. When the component codes are, say, 200, 500, and 1000 chips in length, separate search processes over these individual lengths (a total of 1700-chips search) can be accomplished much more rapidly than a search of the composite 10^8 chips.

This advantage to lock-in time is paid for by a decrease in the signal-to-noise ratio in the correlator output when all code components are not locked. Therefore, when interference is a problem, the partial correlation levels combined with code × interference noise could reduce the correlated signal-to-noise ratio below a usable level. Also, interference could cause any correct decision regarding correlation level to be difficult to make.

The Gold[312,313] codes previously discussed are also specially formulated for synch acquisition, though they are not normally expected to reduce synch time. It may be, however, that use of a Gold code makes it possible to reduce the overall code length, which in turn reduces search time. Gold codes are actually sets of nonmaximal linear codes whose correlation properties are uniform and well defined over the entire set.

Anderson[31] has also defined a set of codes with well-defined correlation properties that may be employed for synchronization. Although they are not really special codes, short sectors of longer codes or even very short codes (say 63 chips) are sometimes sent as synch preambles to be detected by a delay-line matched filter. This technique is used, as all the other special encoding techniques usually are, to reduce synch acquisition time. The delay-line matched-filter signal detector is discussed in a subsequent paragraph.

Is a special synchronization code generally useful? No. But can it be used for a specific purpose? Definitely. Special codes are useful for specific purposes. Most are intended to reduce acquisition time but may pay for this by providing reduced interference rejection, poor cochannel-user rejection, reduced ranging capability, or some other performance loss. As in many other areas, treatment of a special synch code is a matter of matching it to a specific bounded use.

Matched-Filter Synchronizers

The matched-filter synchronizers referred to here are not the same sort of matched filters already discussed—the integrate and dump detector—although both are generically matched filters (Figure 6.7). A matched filter is any device that in effect generates a time-reversed replica of its desired input signal when its input is an impulse. As Turin[1711] states, "The transfer function of a matched filter is the complex conjugate of the signal to which it is matched."

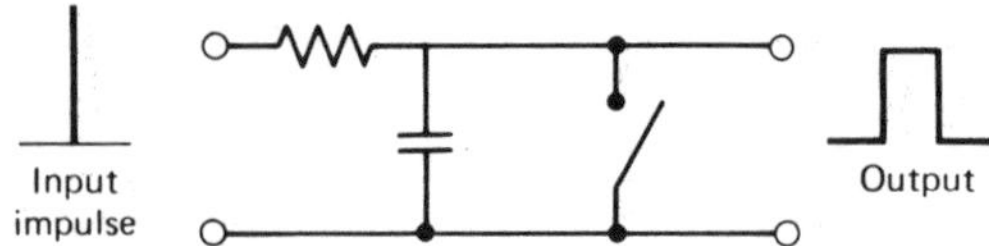

Figure 6.7 Integrate-and-dump filter showing impulse response.

Both IF and baseband matched filters have been implemented and for comparable filter lengths give similar performance. IF matched filters tend to be the surface acoustic wave type, with lithium niobate, lithium tantalate, or quartz as base materials, depending on the length and temperature range required. Baseband matched filters have been implemented as digital integrated circuits by using normal integrated-circuit techniques. The length of acoustic matched filters is limited by the material and by the acceptable loss in the devices themselves; the longest acoustic matched filter presently known is made up of a series of short lines to reach 4000 total chips. Digital baseband matched filters (see Figure 6.8*a*) are limited only by the ability to sum the correlated signal properly and by the power dissipation of the circuits employed. (Digital baseband matched filters employ active circuits, as opposed to the surface wave filters, which are strictly passive.)

Matched-filter synchronizers, generally made up of delay elements, are quite different from integrate-and-dump detectors and chirp filters. They do, however, act as conjugate signal generators to the proper signals, as do truly matched filters. Figure 6.8*b* illustrates the typical delay-line matched filter that one would expect to find in a matched-filter synchronizer, although in a usable filter there would undoubtedly be more delay elements.

The delay-line matched filter is intended to recognize a code sequence and, in a specific configuration, recognize a particular sequence and that sequence only. Each delay element has a delay equal to the period of the expected code clock so that each element contains energy corresponding to only one code chip at any one time. Still considering Figure 6.8, let us assume an incoming signal, biphase-modulated by a seven-chip code sequence, 1110010. This signal, with its phase reversals occurring at each one/zero transition in the code, travels down the delay line until chip one is in T_7, two is in T_6, and so on. When all delay elements are filled and the signal-modulating code corresponds to the filter delay elements, the signal phase in T_2 is the same as that in T_5, T_6, and T_7. Also, delay elements T_1, T_3, and T_4 all contain the same signal phase.

Summing the sets $\{T_2, T_5, T_6, T_7\}$ and $\{T_1, T_3, T_4\}$ separately, inverting the summed $\{T_1, T_3, T_4\}$ output, and summing both together, we find that the signal-energy contained in all seven delay elements adds in phase and the total output level is seven times greater than the unprocessed signal level. This phase addition property is identical to that of the chirp

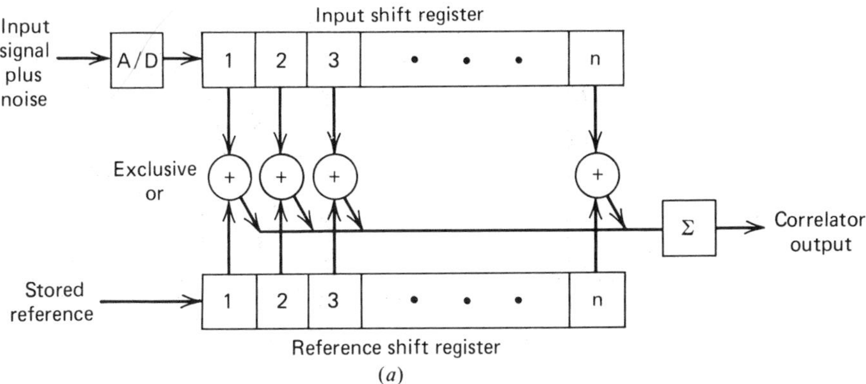

(*a*)

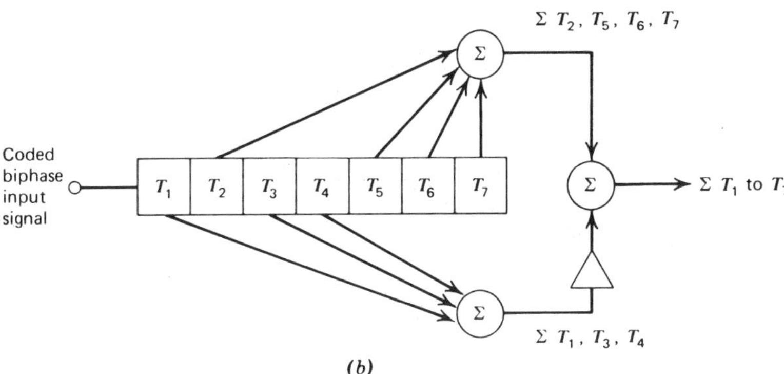

(*b*)

Figure 6.8 (*a*) Baseband, digital matched filter. (*b*) Delay-line matched filter.

filters in Chapter 2, though here the signal is a code-determined discrete phase-shift-keyed signal, whereas a chirp signal exhibits continuous phase change.

The signal enhancement afforded by this technique may be increased by adding delay stages, and process gain available in this fashion is simply

$$G_p \text{ (delay line matched filter)} = 10 \log n$$

where n = number of delay elements summed.

Practical limits to the maximum length of a delay-line matched filter exist, although recent strides have been made toward avoiding size limitations. Acoustic surface wave delay lines appear to be excellent from the size standpoint, since they have been constructed with as many as 200 delay elements in less than 1 in.[3] The prime practical limit, however, appears to be the accuracy with which the delay elements can be constructed, especially in the smaller types of lines. A second limitation that exists is that of length in time. For production units the length appears to be in the range of 20–40 msec, which means that the process gain is limited to approximately $40 \times 10^{-6}/(\text{chip rate})$ or 36 dB for 100-Mcps codes. Also, the necessity for matching to a particular code length means that the tap structure must be changeable if more than one code is to be accommodated. This has not been accomplished in phase-coded matched filters at lengths greater than 128 chips.

A significant point in the use of delay-line matched-filter synchronizers is that they must accurately represent the clock period of the code sequence to be detected. When the delay-line elements exactly match an input signal, the signal summation is perfect. If, however, the code chip rate does not match the delay line, only partial correlation between delay line outputs occurs. Doppler shift, clock frequency drift, or any other cause for clock offset can cause the delay-element periods in the filter to be mismatched with the incoming signal. For this reason it may be necessary to employ an array of delay-line matched filters with graduated delay periods whenever a clock-rate uncertainty exists.

How does a mismatched input signal affect a matched filter of this type? A code input in which the code is correct but the code period is incorrect has only partial signal phase addition because the signal phase shifts do not occur at the proper time. This means that signals overlap into adjacent delay periods and thus do not add properly. If the mismatched input is a CW carrier, it is summed in phase for both parts of the delay line (Figure 6.6); that is, $\{T_2, T_5, T_6, T_7\}$ adds and so does $\{T_1, T_3, T_4\}$. Inversion of $\{T_1, T_3, T_4\}$ before summing with $\{T_2, T_5, T_6, T_7\}$, however, causes the overall summation of $\{T_1, T_2, \cdots, T_7\}$ to have a net output equal to the input. The delay-line matched filter, therefore, has an advantage over a CW interfering signal of approximately N, where N is the number of delay elements. As another example of the utility of linear maximal sequences, we should note that the balance between the two sets of signals summed in our example is due only to the one/zero balance in the sequence used.

An equivalent but more flexible approach than the phase-coded matched filter is the surface acoustic wave "convolver." In such a convolver a wave is inserted into each end of a slab of surface wave material. The two waves then propagate toward one another just as the waves in a pond travel. If the two waves match, then the surface displacement is maximum, and if the energy can be integrated over the surface, a pair of matching waves generates peak energy.

In a surface wave convolver, the two waves entered from the opposite

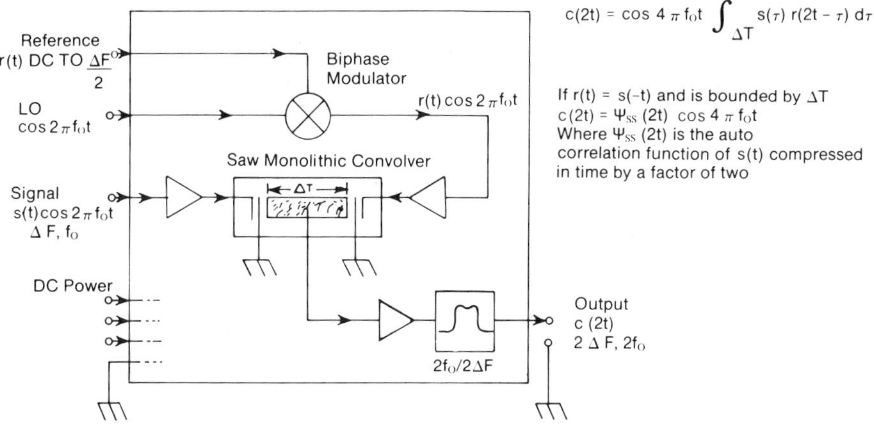

Figure 6.9 Convolver (courtesy Anderson Labs).

ends of the surface wave material (see Figure 6.9) are typically modulated by a code and the mirror image in time of that code. The mirror-image code is generated as the local reference in the receiver. (With linear maximal codes as a base, this is easily accomplished.) Thus the two coded signals generate a maximum correlation when they are propagating in opposite directions in the surface wave device. All that remains is to integrate the energy over the length of the wave interaction. The best approach to the energy integration task is to employ a configuration called an "elastic convolver" in which a piezoelectric material (usually zinc oxide) is deposited on the surface of the surface wave device, with electrodes attached. Then, as the surface wave material vibrates, a signal is generated that is a function of the material motion. Thus, as two matching phase-coded signals travel past each other, they generate a correlation peak that is identical in shape to the autocorrelation of the code but with half the period. That is, if a code at p chips per second is being convolved with its time reversed replica, the correlation function generated is $1/p$ sec wide (at the base). The code's autocorrelation is $2/p$ wide, however. A typical convolver output is shown in Figure 6.10.

Convolvers have been constructed with process gain as high as 40 dB. They are limited in length and dynamic range, just as the phase-coded matched filters are, however. This means they must be employed in receivers having either automatic gain control or some alternate means of restricting their input signal range.

Acoustooptical correlators called Bragg cells also have promise. These devices employ a surface wave device combined with a coherent light source (a laser). The laser is aimed at the surface of the surface wave device at the "Bragg angle." If the laser beam is modulated by an input signal and the surface wave device is modulated by an identical reference signal, then the reflected wave from the surface wave device is

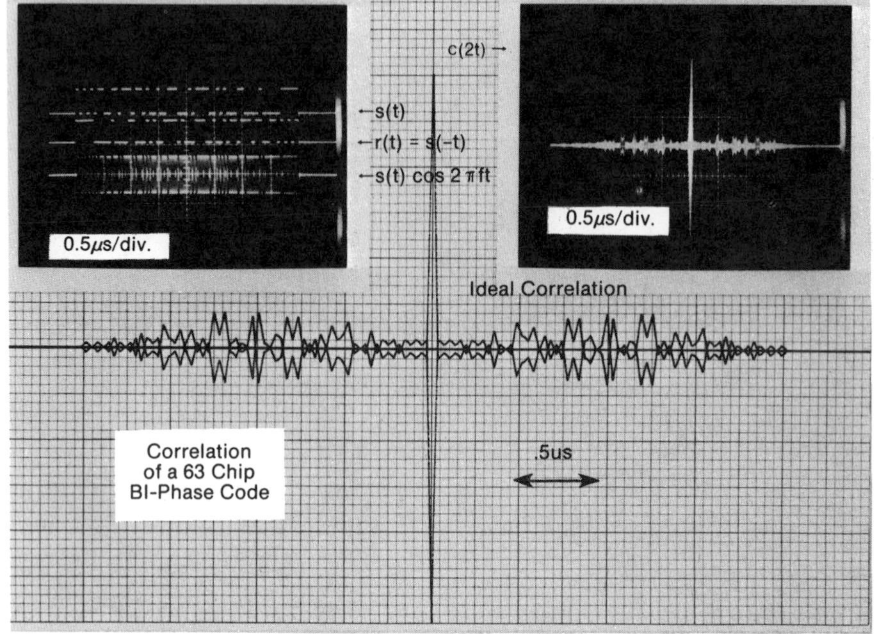

Figure 6.10 Typical convolver output.

concentrated and can be detected by a sensor. If the input and reference are not identical, however, the laser beam is scattered. Thus it is possible to recognize the difference between a particular modulating code pattern and any other. Bragg cell detectors are still limited by the same thing that limits phase-coded matched filters and convolvers; that is, their internal surface wave device limits the period of the correlation practical at this time.

Synch Recognition

Recognizing synchronization is an inherent part of the synch acquisition problem in a spread spectrum system because the receiver must recognize that it is synchronized, stop the searching process, and start its local code to tracking before the received and local codes drift apart. To complicate the process all this must be done with a minimum of misses or false starts, even in the face of deliberate interfering signals.

The search process in most spread spectrum receivers is implemented by causing the receiver's code to slip in phase with respect to the received code. This phase-slipping process may be accomplished by simply offsetting the receiver's code clock rate by a given amount and holding that offset until the proper point of synchronization is reached. This, however, is seldom the case. Instead the method of search normally used phase shifts the receiver's clock periodically by some small increment so that the two codes in question "slide" past each other in jumps. Figure 6.11 illustrates this effect on the correlation function. Instead of having a smooth triangular shape, the correlation function becomes a double staircase in which the number of steps in the stepped-search correlation function is equal to $2s - 1$ (s is defined as the number of phase increments used to shift one clock period). The exact shape of the staircase function varies, depending on the basic clock phase relationship. Figure 6.12 shows the staircase correlation functions generated by phase stepping one sequence by $\frac{1}{10}$, $\frac{1}{8}$, $\frac{1}{5}$, $\frac{1}{3}$, $\frac{1}{2}$, and 1-chip increments with respect to a second sequence. In the first column the clocks generating the two codes are synchronous, only the codes are offset. In the second column the code clocks are offset by a constant equal to $s/2$, or one-half the phase increment taken in the search process (i.e., if the search process causes one code to move $\frac{1}{5}$ chip per shift, with respect to the other, the two code clocks are offset by $\frac{1}{10}$ chip). This is the worst case from the standpoint of generating a maximum correlation value, although the peak value generated does dwell for a longer period of time. The question is, "Is it better to employ a large number of phase increments (small steps) or is it better to reduce the number of steps while increasing the time spent at peak correlation?"

The answer to this question depends on the type of signal detector used to demodulate the input signal fed to the synch recognition circuits. Where the demodulator is a phase-lock type, threshold is minimized but lock-in time is a factor. If, on the other hand, the demodulator is an envelope detector, there is minimum lock-in time but the threshold is higher.

Let us examine the searching process to see if there is an advantage to be gained in manipulating the manner of search. Figure 6.11 shows a

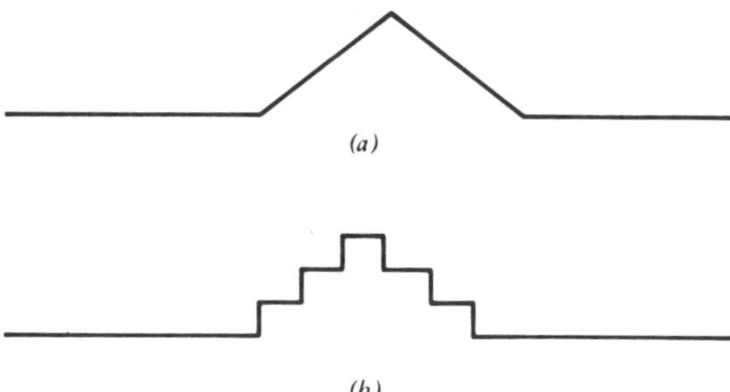

Figure 6.11 Comparative correlation functions for continuous and stepped search processes: (*a*) continuous correlation function for a frequency-offset clock; (*b*) incremental correlation function for a phase-shifted clock.

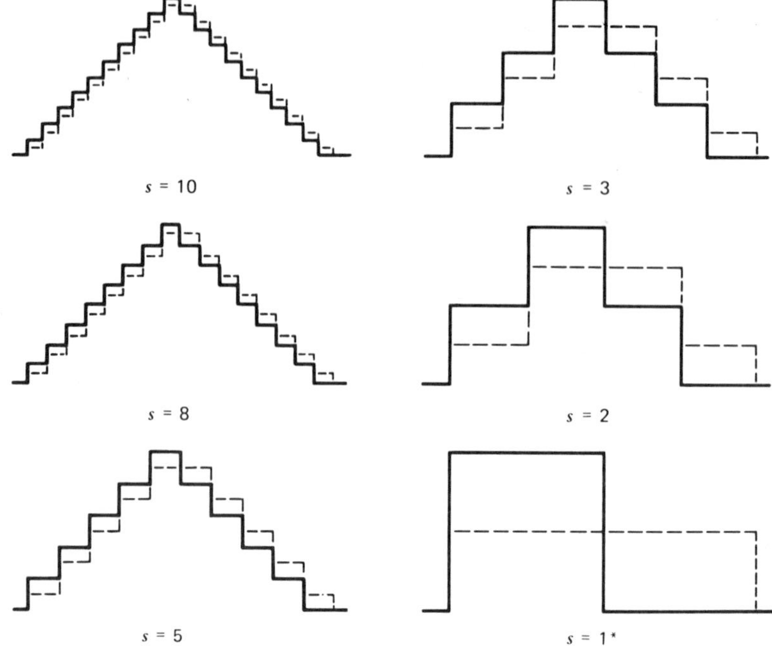

Figure 6.12 Staircase correlation functions generated by incrementally stepping code phase in the code search process.

continuously searched code correlation function with height

$$H = \int_0^T f(t)\,f(t-T)\,dt \quad \text{and} \quad \text{basewidth} = \frac{2}{Rs}$$

where R_s is the search rate for synchronization. The threshold point for synchronization is taken to be the 50% point on the curve. This is the point where the signal-to-noise ratio is high enough to allow the carrier-tracking loop to lock, which in turn triggers the synch-recognition detectors.

For the correlation curve of Figure 6.13, the total area under the curve is simply

$$\frac{1}{2} \times \frac{R}{Rs} \times H = \frac{H}{Rs}$$

The area above the threshold point is

$$\frac{1}{2} \times \frac{1}{Rs} \times \frac{H}{2} = \frac{H}{4Rs}$$

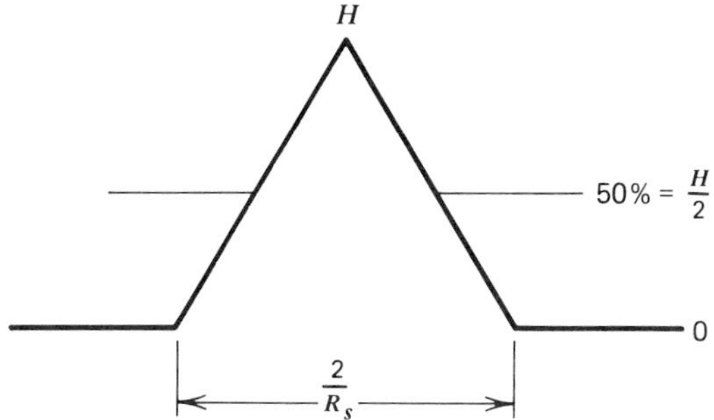

Figure 6.13 Continuous-search correlation function due to constant offset Rs in receiver code clock.

and the time at or above the 50% point is

$$\frac{1}{Rs}$$

For stepped search, as in Figure 6.12, the correlation functions are somewhat different, having a shape developed by the number of steps taken and the step rate. Observance of Figure 6.12 will show that the highest correlation is reached when the initial offset of the codes is an exact multiple of the chip period. The longest dwell time at the high point occurs, however, when the initial offset is exactly half the step size, plus an integral number of chip periods.

The time offset, in other words, in chip periods is

$$\frac{U_t}{T} = \text{number of chip periods } p + r$$

where r = remainder,
U_t = total time uncertainty,
T = (code chip rate)$^{-1}$.

The correlation function has a shape that is determined by

$$\frac{U_t}{T} = p + r$$

and if $r = 0$, then the highest correlation occurs. If r = (step size)/2, then the longest correlation at a high point occurs. Continuing, the number of steps in the correlation function with $r >$ (step size)/2 is $2s - 1$, but for $r <$ (step size)/2 $\neq 0$, the number of steps is $2s + 1$.

The worst case maximum height for the correlation function occurs when $r = 0$, and is

$$H\left(1 - \frac{1}{2s}\right) = \text{worst case height}$$

but the width at the worst case height is always 2/(step rate), where the step rate is just the search rate times the number of steps per chip (s).

The smooth and stepped correlation functions can be compared with respect to their areas. A summary of this comparison follows:

	Smooth Curve	Stepped Curve
Total area:	$\frac{H}{Rs}$	$\frac{H}{Rs}$
Area at or above 50% point:	$\frac{H}{4Rs}$	$\begin{cases} \frac{H}{4Rs}, & s \neq 1 \\ \frac{H}{4Rs}, & s = 1, 0 < r < \frac{\text{step size}}{2} \\ 0, & s = 1, r = \frac{\text{step size}}{2} \end{cases}$
Time at or above 50% point:	$\frac{1}{Rs}$	$\begin{cases} \frac{1}{Rs}, & s \neq 1 \\ \frac{1}{Rs}, & s = 1, 0 < r < \frac{\text{step size}}{2} \\ \frac{2}{\text{step rate}}, & s = 1, r = \frac{\text{step size}}{2} \end{cases}$

Thus we see that the stepped-search approach is equivalent to the smooth-rate search approach if the area of the correlation function is taken as a criterion.

A typical synch-recognition lineup is shown in Figure 6.14. (That shown is for a direct sequence system. Frequency hop synch recognition arrangements are similar, however.) A received signal is multiplied in the correlator with a local reference whose code is slipping in phase with respect to the incoming signal. During this time the signal output for the correlator is the input signal convolved with the nonsynchronous local reference which produces pseudonoise whether the input is desired or undesired. (The IF and demodulator cannot discern the difference before the occurrence of code synchronization.)

When the point of code synchronization is reached, the correlator output is transformed into a coherent narrowband signal which is passed by the IF and demodulated, and a signal-presence indication (carrier level, etc.) is fed to the synch decision block. A positive synch decision then causes the search process to stop and the clock tracking to start. All operations must be carried out, preferably while the receiver's code sequence is at or near the point of maximum correlation. For this reason

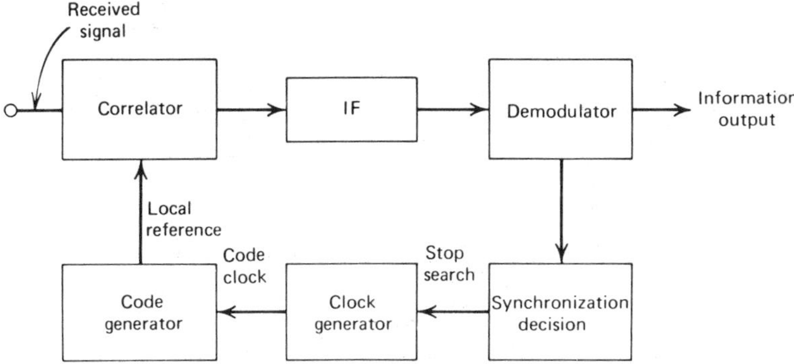

Figure 6.14 Typical direct-sequence synchronization-recognition lineup.

careful attention must be paid to the relation between the signal detector used and the time allowed for recognizing synchronization—especially when the signal detector is the phase-lock type. Laboratory observations have shown that the lock in time for a phase-lock loop increases by as much as 8–10 times over its high signal-to-noise lock-in time when its input signal is at threshold. This increase must be taken into account in the design of a synch recognition subsystem.

One other consideration that bears heavily on synch decision circuitry is that the noise output from a spread spectrum system's correlator is often not random, but pseudorandom. This difference, though not apparently significant, forces design provisions that must compensate for this lack of randomness. No spread spectrum synch recognition system can perform at its best without taking the true correlator output noise statistics into account. This especially true when relatively short-code sequence lengths are used, because the shorter the code, the less it resembles real noise.

Figure 6.15*a* shows the lack of randomness in a short pseudorandom sequence by cross-correlating an 8191-chip sequence with a CW carrier. The short-term correlation peaks in this signal tend to cause subsequent synch recognition circuits to register false correlations. Worse yet, each time the CW carrier or the code is changed, the pattern of short-term partial correlation changes and, therefore, it is difficult to design a correlation detector to be optimum for all the possibilities. For the purpose of comparison Figure 6.15*b* also shows the same cross-correlation between the code and a CW signal except that the code employed in Figure 6.15*b* is many millions of chips long and its cross-correlation with a CW signal appears to be random. We see from this comparison, then, that code length can play an important part in the design and performance capability of a spread spectrum system, especially when the system must be designed to reject large cross-correlating signals.

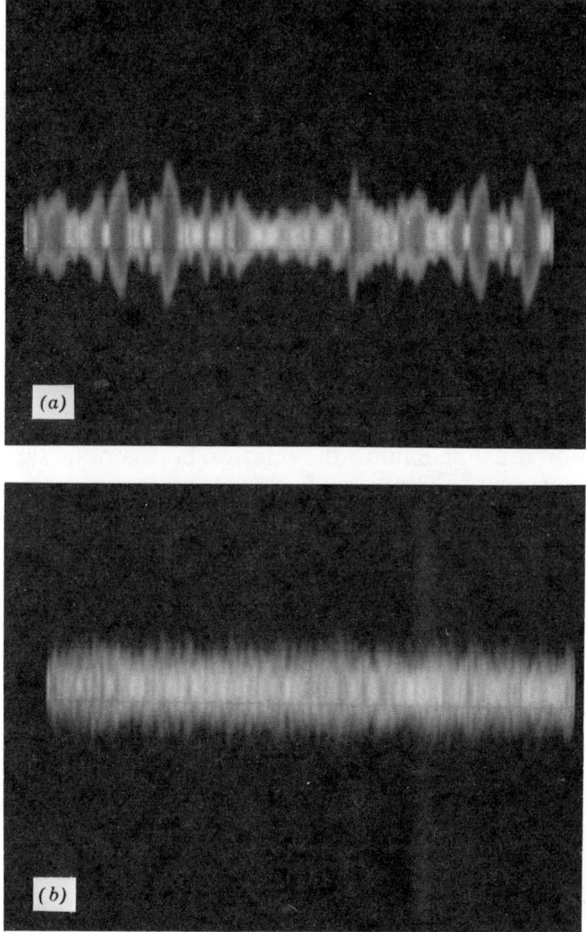

Figure 6.15 Comparison of cross-correlation of short- and long-code sequences with a CW signal: (*a*) typical cross-correlation of a short synchronizing code with a CW signal (note pattern repetition); (*b*) typical cross-correlation of a much longer code under conditions identical to Figure 6.7*a*.

The synch-recognition function is performed by some device capable of detecting the presence of a coherent signal. The desired signal must then exceed the synch-recognition-circuit's threshold sensitivity, whether the synch recognizer is preceded by a phase-lock loop, envelope detector, or whatever. Typical synch-recognition circuits are shown in Figure 6.16. It is easy to perceive that the only physical difference in the two synch-detector configurations in Figure 6.16 is that one employs a phase-lock loop and the other an envelope detector. Both the phase-lock loop and the envelope detector perform the same function insofar as the threshold detector is concerned; that is, they output a signal that is a function of the input signal level. The integrator or low-pass filter that follows, in

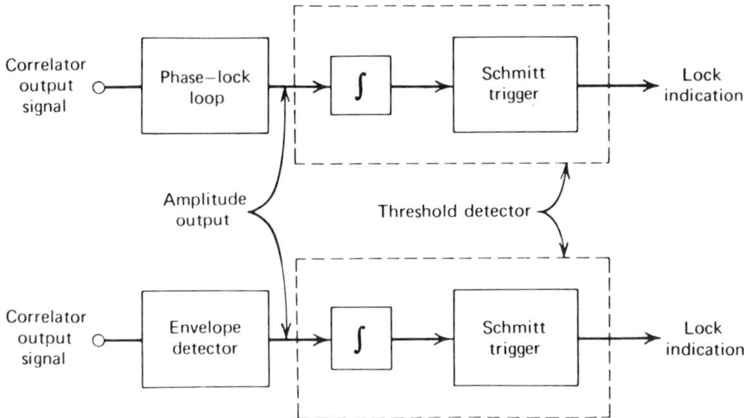

Figure 6.16 Typical synch-detector configurations.

combination with a Schmitt trigger as a level sensor, then smooths the envelope-generated signal and, if the smoothed signal exceeds the Schmitt's triggering level, a signal is recognized. This, in turn, stops the search operation.

Rise time of the synch-detection circuit is set by the integrator, except for those cases in which a phase-lock-loop's lock-in time is actually the determiner of recognition time. When phase-lock time is not a limiting factor, the integrator portion of the threshold detector determines lock-in time. In fact, this detector is most often not one but a bank of detectors, each having different integration times. The purpose of such a bank of threshold detectors, combined to make up a single synch detector, is to optimize the probability of detecting synchronization if it occurs while holding the occurrence of false alarms to a level as low as possible. Figure 6.17 is a block diagram that shows a typical multiple-integration-time synch detector. In the synch detector of Figure 6.17 a DC level corresponding to an input signal's envelope would be applied to all three integrators whose function is to determine the firing rate of the Schmitt triggers shown as threshold-detection devices. The problem is to design the filters (integrators) and threshold detectors so that the probability of detecting the desired signal, when it appears, is maximized, while minimizing the probability of recognizing the wrong signal.

Error rate, and thereby the probability of detecting a desired signal, is set by the first detector, which has the fastest rise time (or widest prefilter bandwidth). The third detector sets the synch-false-alarm rate. It is the final decision maker which starts the tracking process, allows data to be demodulated or marks a range measurement. (See the flowchart and timing diagrams in Figure 6.18.) The second detector's prime purpose is to bridge the gap between first and third detectors; that is, because the third detector is delayed for some time before its decision is made, the first detector can be extinguished by noise before the third can fire. This would

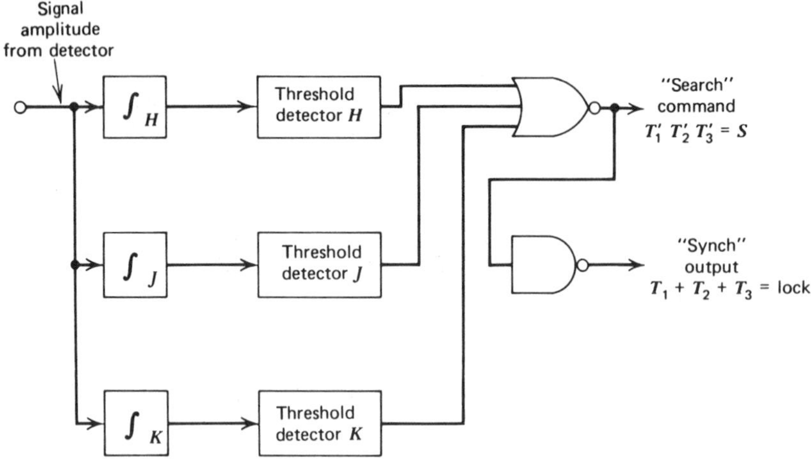

Figure 6.17 Multiple-integration synch detector.

permit the search process to be restarted even after synchronization has occurred, thus aborting an otherwise successful acquisition. The second detector is then designed to have a probability of signal detection lower than the first detector but with an attendant lower error rate so that its synch decision can be made and held until the third decision occurs.

Figure 6.19*a* illustrates the relationship between detector-threshold setting, noise level, and signal-plus-noise level. This relationship holds true when a signal level is to be detected by a threshold-sensitive device that might be used with the signal of Figure 6.19*b*. When the signal-and-noise density functions and the detector threshold are known, it is possible to determine probability of signal detection and false alarm for a given signal-to-noise ratio. Referring again to Figure 6.16*b*, the probability of signal detection, $P(d)$, where a detection threshold setting prevails, we can find the probability of detection by integration

$$P(d) = \frac{\int_{\gamma}^{\infty} f(s+n)\, dV}{\int_{0}^{\infty} f(s+n)\, dV}$$

and the probability of false alarm

$$P_{(fa)} = \frac{\int_{0}^{\infty} g(\text{noise})\, dV - \int_{0}^{\gamma} g(\text{noise})\, dV}{\int_{0}^{\infty} g(\text{noise})\, dV}$$

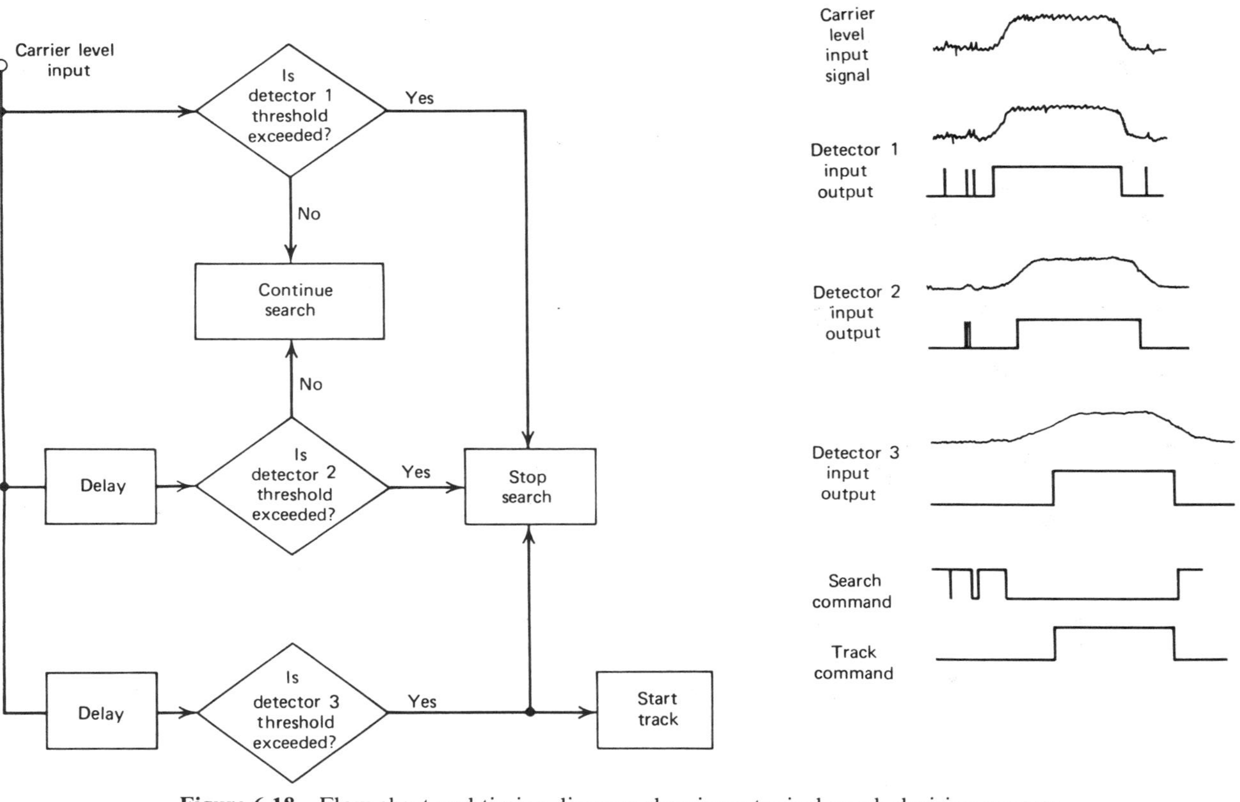

Figure 6.18 Flow chart and timing diagram showing a typical synch-decision process.

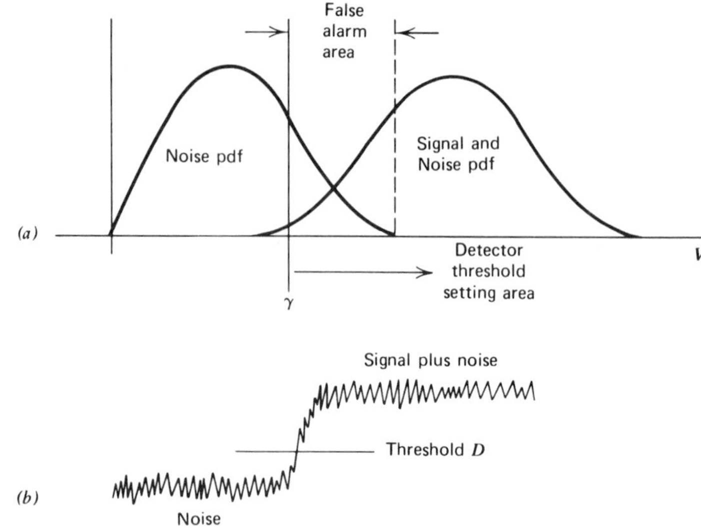

Figure 6.19 Relation of signal and noise with regard to detector-threshold setting.

These derivations are obvious because any signal greater than γ (lying to the right of the threshold setting γ) will trigger the threshold detector, whereas any noise exceeding γ gives a false alarm. Determining the areas in the signal and noise density functions with relation to γ then gives a measure of false alarm rate and probability of detection. Note that the lower the setting of γ, the more of the signal density function will lie above the threshold, thereby giving higher assurance that a signal will be recognized. Note also, however, that lowering γ increases the noise lying above threshold, which increases the false alarm rate.

Figure 6.20 plots the probability of false alarm versus signal-to-noise ratio for various detection probabilities. This curve shows that for a 0.95 probability of detection and 10-dB signal plus noise-to-noise ratio, the probability of false alarm is 2×10^{-2}. Thus it is seen that we can determine the probability of signal detection and false-alarm rate for the synch-detection subsystem by setting synch-detector threshold(s) to the proper level.

Referring once more to the three parallel synchronization detectors shown in Figure 6.17, we find that this configuration is used to maximize the probability of recognizing a correct desired signal when it occurs. An alternate configuration in which the three detectors are placed in series (i.e., Schmitt trigger H must fire before J can fire and J must fire before K can fire) tends to force a more positive identification of the desired signal while minimizing any tendency to false alarms in the final code track-start decision process. It is interesting to note that this configuration does not differ significantly in the rate of search stoppages because they are

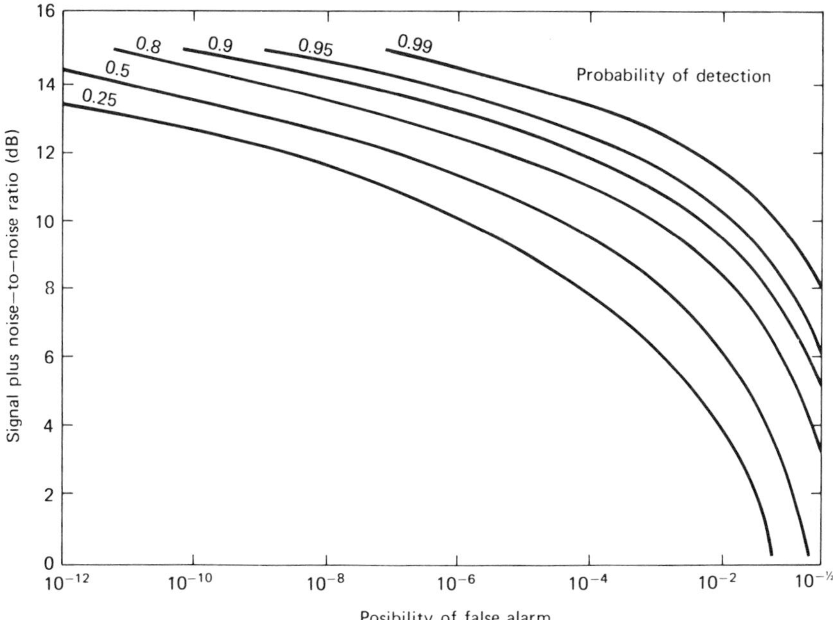

Figure 6.20 Plot of false-alarm probability and detection probability versus signal plus noise-to-noise ratio when signal and $S + N$ pdf's are Gaussian.

determined primarily by the first decision level (H) and there is little or no difference in this decision circuit, whether the series or parallel configuration is employed.

There is, of course, nothing sacred about three synchronization detectors. Any number can be used, depending on the requirements for a particular system with respect to recognizing synchronization. The whole idea is to make sure that the search process stops when a possible desired signal is present, that the receiver looks at the signal for a long enough period to allow its demodulator lock to the desired signal if it is truly present, and that the search process goes on about its business with minimum delay when there is no desired signal present. In the proper setting one detector would work well. In others, we might need many more than three.

One other latitude of control over synch detection is in the area of the filters preceding the threshold detectors. These filters delay the decision as to whether the signal desired is present and also as to its loss. Such delayed decisions can be advantageous in preventing recognition of a false signal as well as in holding synch when a desired, but marginal, signal is present. If, for instance, a synch decision is delayed for a period longer than the period of the code being used, it is certain that a partial correlation cannot trigger a false indication of synch. Therefore, at least one synch-detector circuit is almost always set to require code synch for a time much longer than a synch code period. This same circuit, with a long decision delay, is effective in preventing loss of synch when short signal losses or interference bursts occur whose duration is shorter than the detector delay time.

One way of providing an adaptive integration time based on signal likelihood is an approach called "sequential" searching (having nothing to do with "sequential detection"). Figure 6.21 illustrates the difference between serial search (the simplest form of sliding correlation) and sequential search, with a pair of flow diagrams. In serial search, the code is slipped in time by an increment (usually less than one chip interval). Then, the receiver integrates for a fixed interval, usually as long a time as possible, and samples the integrator's output. If the signal is deemed to be present (the integrated output is above a set threshold), then "synch" is declared and the receiver starts to track the incoming signal.

Sequential search is different is that a second threshold level is set, lower than the "synch" threshold, which is considered to be what we will call a "maybe" threshold. If at the sample time the signal is below the maybe threshold, the search processor takes another step. If it is above the maybe threshold, however, the same code position is held and the receiver integrates for another period. This technique allows the receiver to increase selectively its integration time when the likelihood of a synchronized signal being present is high, and to go on to more fruitful

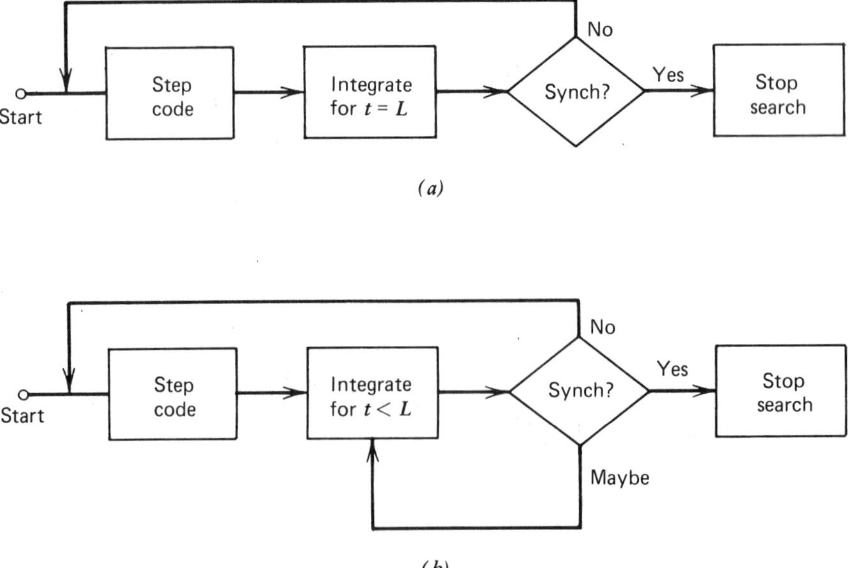

Figure 6.21 (*a*) Serial and (*b*) sequential search flow diagrams.

territory when it is low. Comparing the time required for serial and sequential search:

Serial search,

$$\text{synch time} = \frac{U_t}{T\text{ chip}} \times \text{steps per chip} \times \text{integration time}$$

where U_t = time uncertainty

$$T\text{ chip} = \frac{1}{\text{code rate}}$$

Sequential search,

$$\text{synch time} = 1 + p(f.a.) \times \frac{U_t}{T\text{ chip}} \times \text{steps per chip} \times \text{integration time}$$

where $p(f.a.)$ is usually 0.1–0.2 integration time (sequential search) $\leqslant$ [integration time (serial search)]/10. Comparing the two, then,

$$\frac{\text{serial search time}}{\text{sequential search time}} = \frac{1.2}{10} = 0.12$$

if integration time is $\frac{1}{10}$ for sequential search and the false-alarm rate is 20%. Even if the false alarm rate is 50% and the integration time is one-half that of serial search, the synch time would be 0.75 for sequential search. In typical sequential search implementations, the rate of improvement in search rate is a factor of 3–5 over serial searching.

Now let us discuss a typical synchronization detector design, assuming that we have a typical sliding correlator synchronizer consisting of a signal detector, a code-rate offset device, a coded signal multiplier (correlator), and a synchronization detector. In the first detector in the triad we choose to have a 0.95 probability of detecting the desired signal, which gives a $10^{-1/2} = 0.314$ probability of false alarm at a 6-dB $(S + N)/N$ ratio. In our second detector we allow a detection probability of 0.8, for a 0.09 probability of false alarm, and the third detector we set for a 10^{-3} false-alarm rate and detection probability of 0.25. The overall probability of signal detection, because any one of the parallel detectors can register the signal's presence, is 0.995.*

*The probability of occurrence of one or more events from a set of independent events is ΣP_n minus the sum of all the possible combined probabilities; that is, $P_{(\text{overall})} = P_{(a)} + P_{(\text{b})} - P_{(\text{a})} P_{(\text{b})} = P_{(\text{a and/or } b)}$, $P(\text{a}) + P(b) + P(c) - [P(a)\,P(b) + \text{P}(a)\,P(c) + P(b)\,P(c) + P(a)(b)(c)] = P(a \text{ and/or } b \text{ and/or } c)$.

Considering the time elements now, the first detector would have no filtering other than that provided by the demodulator preceding it. The second detector's filter would delay its decision for a period such that the dropout probability of the first detector does not prevent the third decision from occurring before synch is lost, and the third detector delays until enough time has passed to ensure that a large portion of the receive code has been synchronized with the incoming signal before making its decision.

The multiple parallel decision network approach is useful in many spread spectrum systems, providing for synch recognition and rejection of spurious cross-correlated signals under all conditions.

6.2 TRACKING

The second part of the synchronization problem follows initial acquisition. Once a spread spectrum receiver has synchronized with a received signal it must continue to operate in such a way that it remains locked, with its code reference exactly tracking the coded incoming signal; that is, once the code sequences are matched in time, the receiver must cause its own code chip rate to match the incoming code chip rate as precisely as possible (usually so that the codes remain matched within a small fraction of a clock period).

Without some active method to force the receiver's code clock rate to track an incoming signal's code rate, the two will not maintain synchronization. The two most common techniques for doing this, actually very similar, are called "tau-dither" (or "tau-jitter") and "delay-lock." A third technique, is called "carrier lock." For frequency hopping, a method called split-bit tracking is also feasible.

Tau-Dither Tracking

Tau-dither code tracking is an error-signal-generating method similar to those used in some servo systems for position sensing and correction. Those familiar with servo systems will recognize the word "dither" as describing a small amplitude, usually high-frequency, vibrating signal, applied to such things as hydraulic actuators to reduce friction and provide positioning information.

Tau-dither clock tracking in spread spectrum systems makes use of the triangular code correlation function possessed by binary codes. The idea is to cause the code phase in the receiver to remain as close as possible to a received code phase so that the timing of the two codes will remain essentially at the peak of correlation. The way this is done is to degrade the correlation deliberately by a known amount, observe the effect, and employ the information gained to improve the degree of correlation.

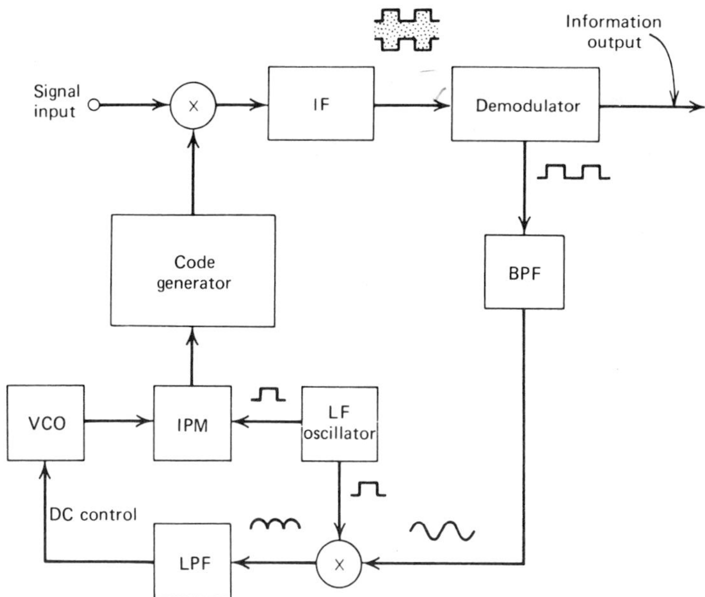

Figure 6.22 Tau-dither clock-tracking loop.

A typical tau-dither clock-tracking loop is illustrated in Figure 6.22. In a tau-dither loop a phase modulator capable of shifting the phase of an input signal by some increment—say one-tenth clock period—is used alternately to set the clock that drives the receiver's code reference back and forth between shifted and unshifted positions. When this occurs, the degree of correlation between the received and local codes also changes, which causes a shift in the amplitude of the signal input to the demodulator. (The information demodulator, usually for PSK or some other form of angle modulation, is not affected by a small amount of amplitude change, because of the orthogonality of angle-modulated and amplitude-modulated signals.) As the clock phase is shifted back and forth, the signal is amplitude modulated (see Figure 6.23) at the phase shifting rate.

By observing Figure 6.24 it is easy to envision the tau-dither operation. Consider that we have a pair of identical code sequences, which we can vary (with respect to their phase relationship) in any way we wish. Now, as we approach the point of perfect synchronization and pass it, we generate the familiar triangular function. Let us, however, set the code phase relationship so that we are at point 1*a* on the curve. This means that our two codes are shifted approximately one-half chip apart in a direction we call negative. The signal output from a correlator in which these two codes are inputs would be about half its maximum value.

Now assume that the code phase is moved (closer to maximum correlation) to a more positive point 1*b*. The correlation between our two

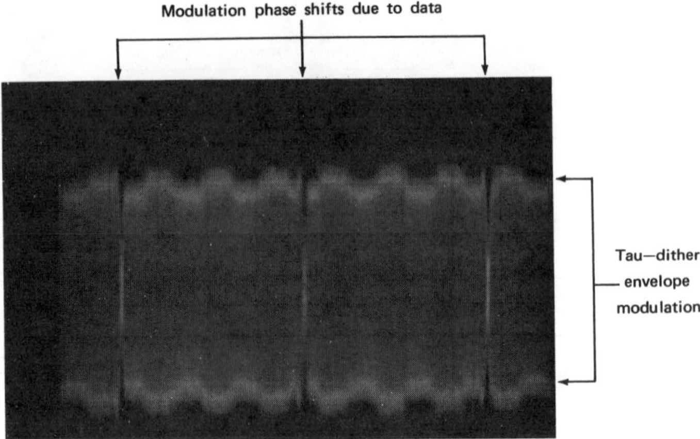

Figure 6.23 Correlator output signal showing envelope modulation due to code phase shifts.

codes is now greater. Similarly, for point 2*a* on the positive side of the correlation function the correlation between the two codes is greater than at point 2*b*.

The correlator output signal is also greater when code phase is at points 1*b* and 2*a* than at 1*a* or 2*b*. We have, then, the basis of a coherent error generation and detection loop.

Let us postulate a square wave, which, when true, causes a code sequence to be advanced by a fraction of a bit, and when false, causes it to go back to its unadvanced state. Assume that the relative code states at the beginning of our operation are such that we are at point 1*a* and that the advance in phase moves us to point 1*b*. Our square wave then causes us to go back and forth between 1*a* and 1*b*, thereby amplitude modulating the correlator output at our square wave rate. Once this amplitude-modulated signal is passed through a detector, filtered, and fed back, the amplitude modulation gives an indication of the direction in which the code pair must be moved to reach perfect synchronization. (In the region

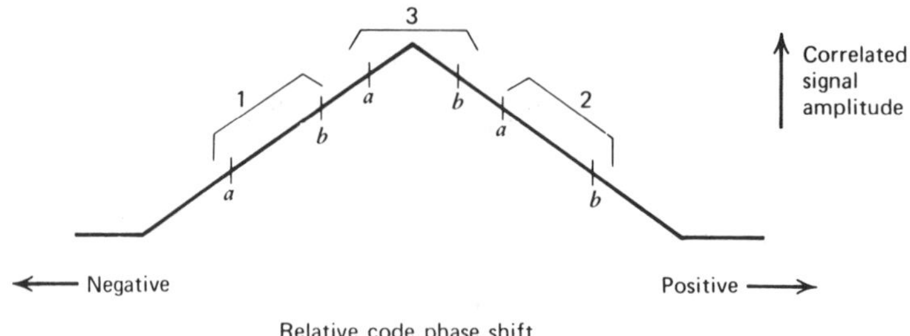

Figure 6.24 Correlation function showing phase-shift increments.

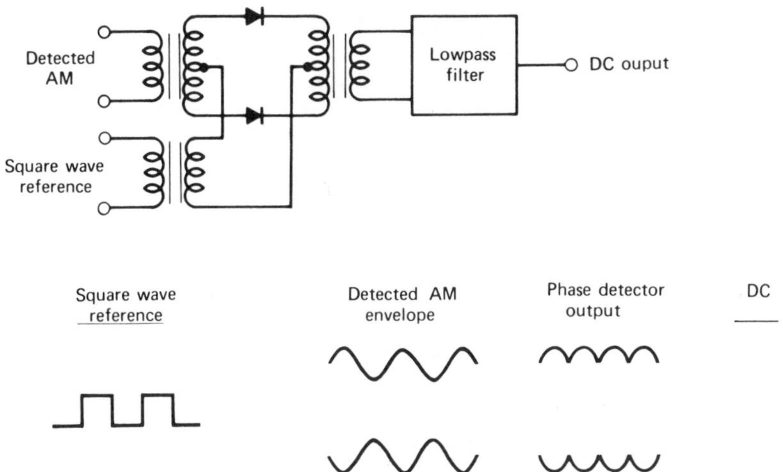

Figure 6.25 Phase detector operation.

of the correlation peak the amount of amplitude modulation is also an indication of how far the two codes are offset. This effect is discussed further in a later paragraph.) It should be obvious that the sense of the amplitude modulation (i.e., upward or downward modulation) produced by phase shifting between points 1*a* and 1*b* is opposite to that produced by going from point 2*a* to 2*b*. Detection of this amplitude modulation in a phase detector then allows us to generate a DC signal which has the proper polarity (and amplitude) to set the VCO of Figure 6.22 to run faster or slower, so that the code sequences in which we are interested speed up or slow down, as required for correct synchronization. Phase detector operation is illustrated in Figure 6.25, and Figure 6.26 shows signals typical of an actual operating tau-dither tracking loop.

Referring again to Figure 6.24, let us examine area 3 in which the code pair being examined has relative phase shifts ahead of, and behind, perfect synchronization; that is, if the overall incremental phase shift over which the codes are shifted is $\frac{1}{10}$ chip, the shift in area 3 of Figure 6.24 is $\frac{1}{20}$ chip negative and $\frac{1}{20}$ chip positive, and we see that the tau-dither operation centers on the peak of correlation so that it is never more than $\frac{1}{20}$ chip from perfect synchronization.

When the code loop is tracking so that the phase shift does straddle the correlation peak, as in area 3, the phase jumps produce no modulation* of the correlated signal. As the codes drift apart, the modulation increases from zero at perfect phase synch to a maximum that is a function of the total phase shift. If the total incremental phase shift used is $\frac{1}{10}$ chip, the maximum modulation will be limited to that produced by a $\frac{1}{10}$-chip phase

*In practical systems some residual modulation remains.

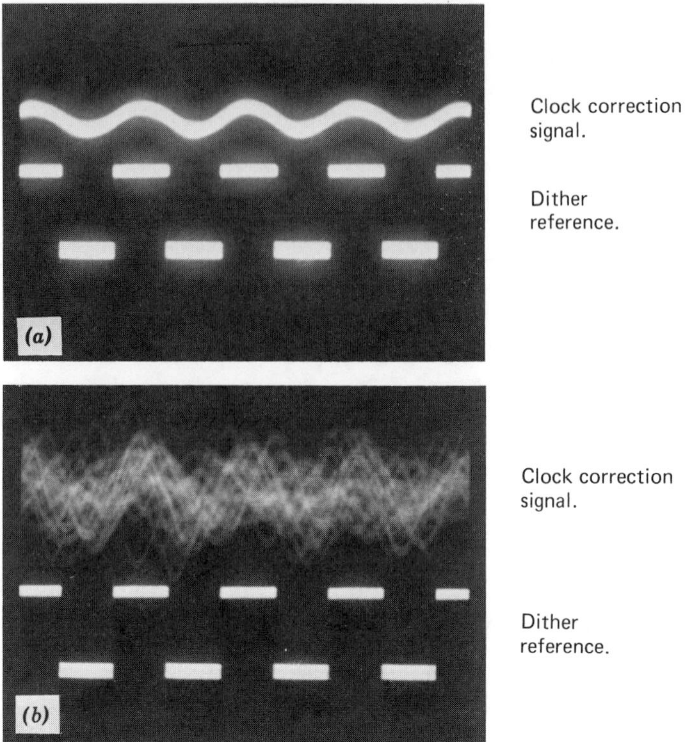

Figure 6.26 Dithered clock loop waveforms: (*a*) no interference; (*b*) 15-dB interference.

shift. Figure 6.27 plots the level of modulation amplitude as a function of code phase offset over the region of synchronization, where the phase increment used for tau-dither error generation is $1/P$ chip.

Considering that our two codes are synchronized and that the tau-dither phase shift is centered on the peak of correlation, what is the effect of the tau-dither operation? Remember that the two-position ahead-and-behind phase shift operation always holds the codes at a position slightly offset from the optimum, so the correlated signal is never at its best. The amount of correlation loss, however, is only $1/2P$th of the maximum, so that when the tau-dither phase jump is $\frac{1}{5}$ chip, the maximum correlation loss is $\frac{1}{10}$ or 10% (about 1 dB). Thus we see that a receiving system is degraded by no more than $100/2P\%$ by a tau-dither phase shift of $1/P$ chips.

Delay-Lock Tracking

The second common type of tracking loop is called a "delay-lock" loop, because two separate correlators are used which are driven by two local

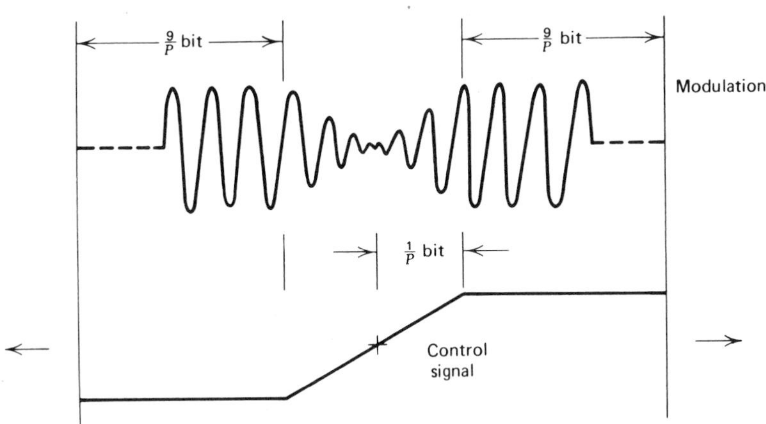

Figure 6.27 Modulation and control signal amplitude in the region of ±1-chip synchronization for tau-dither loop.

code reference signals—identical except that the one is delayed in time from the other. The amount of delay between the local codes is usually one or two bits, either of which is readily available, because adjacent shift register stages (if properly chosen) have outputs delayed by one chip. Figure 6.28 is a block diagram of a delay-lock loop. Except for the details of error generation, a delay-lock loop is identical to a tau-dither loop. (Here we will not conjecture which type preceded the other.)

In a delay-lock loop two local reference signals are generated. Identical except for the time delay mentioned in the preceding paragraph, these references are used for comparison with a single incoming signal in two separate correlators. Just as before, the correlation between an encoded

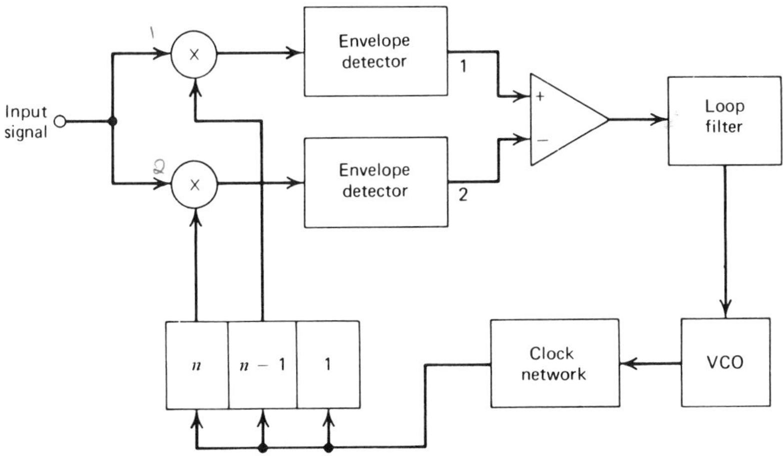

Figure 6.28 Delay-lock tracking loop.

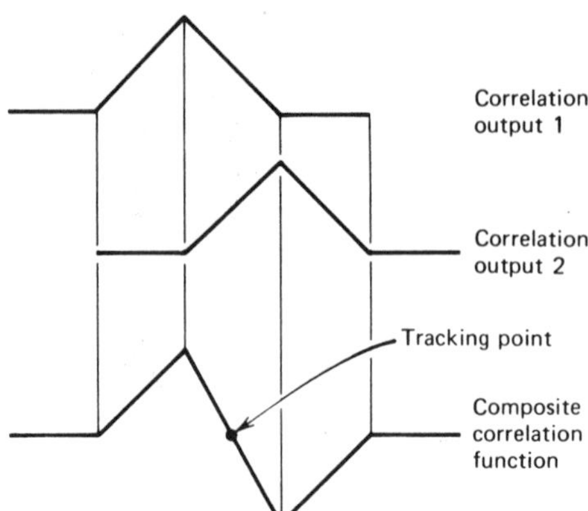

Figure 6.29 Correlation waveforms in delay-lock loop.

incoming signal and the local reference is a triangular function 2 chips wide, but now there are two correlated signals available (one at the output of each correlator), each with a triangular correlation function but with their peaks of correlation offset by the amount of delay between local reference signals. Thus the composite correlation function for a delay-lock loop has a double-peaked triangular shape (see Figure 6.29) after summing, in which one-half of the double triangle is inverted, so that the composite correlation has a linear region centered around the point halfway between the two correlation maxima.

Now, if the summed correlator outputs are filtered and used to control the receiver's clock source (the VCO shown), the receiver's code will track the incoming code at a point halfway between the maximum and minimum of the composite correlator output. (If the composite waveform is viewed as a discriminator response, the delay-lock loop is quite similar to a common AFC loop.) Because the receiver's two code references are forced to track the incoming code with an offset equal to one-half the delay between them, it may be necessary to decrease the delay to some value much smaller than 1 chip. Otherwise the code correlation would never be better than half its peak value. An alternative to decreasing reference code delay would be to take a separate $\frac{1}{2}$-chip-delayed code to a third correlator. This would provide an optimally correlated signal output for information demodulation. Figure 6.30 illustrates this concept in block form.

Coherent Carrier Tracking

Both RF carrier and modulating signals making up any communications signal are acted on in exactly the same way by the path from transmitter to receiver; that is, if either the transmitter or receiver is in motion, the Doppler effect will be equivalent (though of different magnitude) on modulating signals to whatever it may be on their carriers. Because

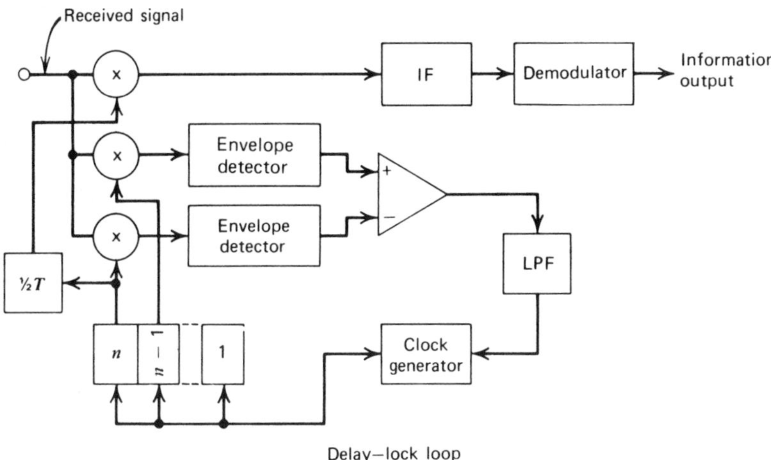

Figure 6.30 Delay-lock loop with half-chip-delayed correlator for signal demodulation channel.

modulating signals are, in general, at low frequencies, the Doppler shift seen is usually insignificant. With spread spectrum systems, however, and especially direct-sequence systems, the high modulation rates are prone to significant Doppler frequency shifts. Because the carrier and modulation signals suffer from identical shift mechanisms, it is possible to correct for both carrier and modulation (code) uncertainties with a single tracking loop.

In carrier-coherent systems the mechanism used is just what the name implies. In other words, the code sequence clock rate is designed to be a submultiple of the carrier frequency or both carrier and clock are derived from the same source. Common sources must be used in the transmitter and receiver (i.e., each must have common code rate and carrier sources—not implying that both transmitter and receiver have the same source).

When the modulation rate and carrier frequencies are coherently related, tracking one will also provide for the other; for instance, consider a transmitter in which a 50 Mcps code sequence direct sequence modulates a 500 MHz RF carrier. The carrier is derived as a 10-times multiple of the code clock. At the receiver the local oscillators are also derived from the code clock, and the postcorrelation signal is translated down to a convenient demodulation frequency by mixing with these code-rate-related oscillators. Thus the transmitter's 50 Mcps code clock rate is precisely related to the carrier signal seen at the receiver's demodulator, and a single phase-lock demodulator can be used to track both code and clock and carrier-frequency shifts. It may be necessary to provide a method for resolving the phase ambiguity which can result from locking to a frequency related to code rate but different from it. A simple ±1-chip correlation search is sufficient to satisfy the difficulty and once the proper phase is chosen only frequency need be tracked.

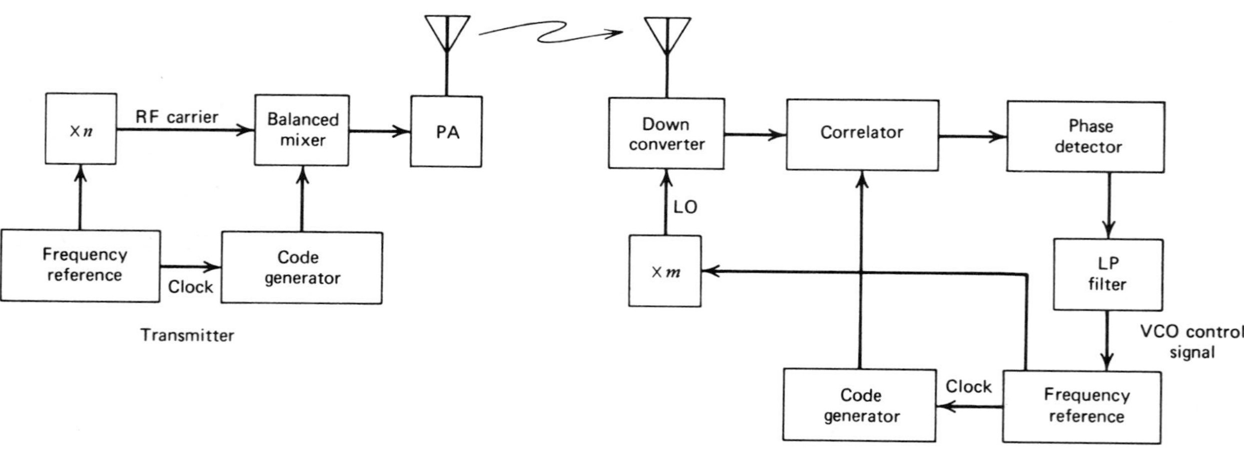

Figure 6.31 Carrier-lock tracking system.

A carrier-coherent tracking system is illustrated in Figure 6.31. It is interesting to note that whenever the frequency reference in a carrier-lock receiver is adjusted to compensate for an input signal frequency change the frequency of all local oscillators also changes. This fact can be put to use by designing the receiver so that local oscillator changes tend to cancel input signal frequency changes. The overall effect therefore is that of combining a frequency feedback loop with a phase-lock loop. In effect, the demodulator's tracking range is reduced by the amount of frequency feedback cancellation, which often permits less complex demodulator design.

When it can be employed, the carrier-lock tracking method is quite useful. It is limited, however, in that the transmitter from which it receives signals must have code clock rate and carrier phase coherence for the receiver to operate. Otherwise unrelated frequency drifts between clock rate and RF carrier would cause the receiver to lose lock.

Split-Bit Tracking

Split-bit tracking takes advantage of the envelope slope of the signal at the output of a frequency-hopping-receiver's correlator. This envelope is constant in amplitude, though not in phase (phase shifts occur at the hopping rate) when the receiver is synchronized perfectly. If, however, the receiver is ahead in time, then the correlator output is high during the first part of the hop period but low during the second part (see Figure 6.32).

When the receiver is behind the transmitter in time, the correlator output is low during the first part of the hop period and high during the

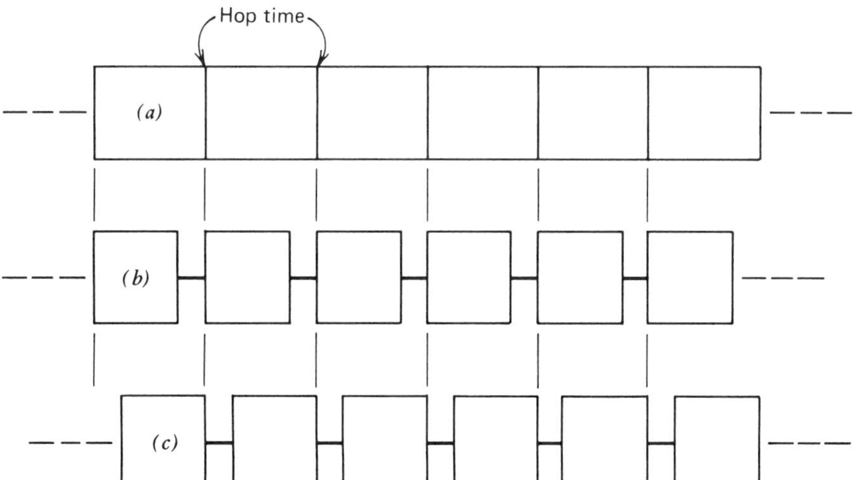

Figure 6.32 Illustration of timing for split-bit tracking loop: (*a*) perfectly synched signal envelope; (*b*) receiver ahead in time; (*c*) receiver behind in time.

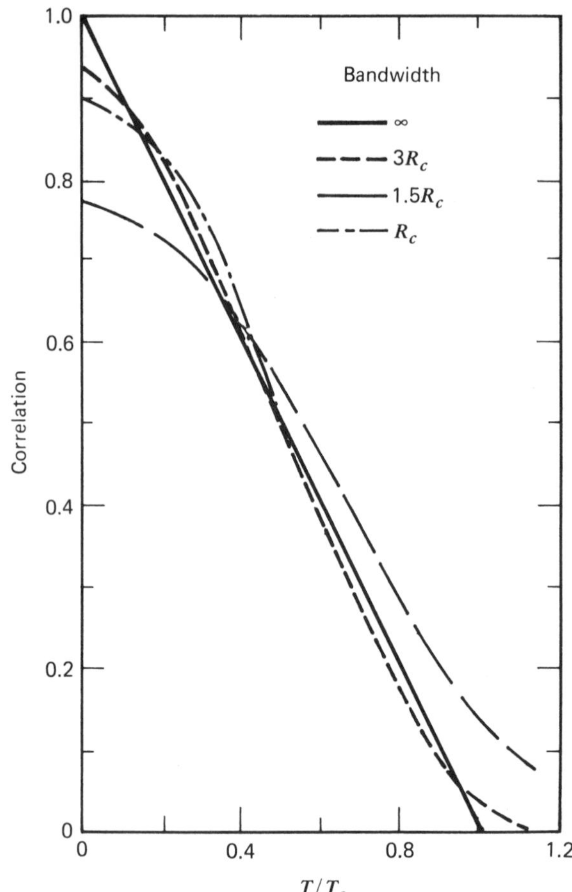

Figure 6.33 Correlation function loss as a function of bandwidth.

second part (which is just the opposite of the case when the receiver is ahead of the transmitter). Therefore, by integrating over the two halves of the hop time and comparing the energy present, a loop can be constructed that causes the receiver to adjust its code timing so that the energy is equal in both halves of the chip time. Thus the frequency-hopping receiver can be caused to track the incoming signal timing.

Correlation Loss Due to Bandpass Filtering

The correlation function generated by the delay lock and other tracking loops can be degraded by bandpass filtering of the transmitted signal. This is illustrated in Figure 6.33 (from C. R. Cahn). Figure 6.33 shows the changes in the shape of the correlation function as filter bandwidth is varied from the chip rate of the code to 3 × chip rate, and then the filter is removed. Note that only with no filtering is the function sharply peaked.

Loss expected is as follows:

Filter Bandwidth	Loss (dB)
2 R_c	0.45
1.6 R_c	0.5
1.2 R_c	0.75
R_c	1.1
0.8 R_c	1.6

Tracking error loss as a function of phase jitter caused by jamming is:

Error	Loss (dB)
$0.1T$	0.72
$0.2T$	1.51
$0.3T$	2.37
$0.4T$	3.34
$0.5T$	4.42

where

$$\text{loss} = 10 \log 1 - \sqrt{\frac{2}{\pi}} \frac{\sigma}{T}$$

and σ = phase jitter,
T = clock period.

PROBLEMS

1. What synchronization uncertainty might one expect from a 10 Mcps code generator whose chip-rate accuracy averages 1×10^{-9} after $2\frac{1}{2}$ days?
2. What if the above system employed a 100 kcps code generator?
3. Given a system with 35 dB process gain, 20 MHz RF bandwidth, and a 4095 chip synchronization preamble, what minimum time must be allowed for synch?
4. For a probability of detection of 0.9, what probability of false alarm might be expected of a detector whose input signal plus noise-to-noise ratio is 10 dB?
5. What system loss might be expected from the use of a tau-dither loop for clock tracking, whose total time shift is $\frac{1}{2}$ chip?
6. A direct sequence system with a 10 Mcps code clock rate is traveling at a relative velocity of 980 mph away from us. What is the Doppler shift on his code clock as received by us?

7. The system turns and approaches us at 580 mph. What is the Doppler shift on the clock now?
8. The carrier frequency of the transmissions in question is 375 MHz. What is the carrier frequency that would occur due to Doppler?
9. A frequency hopping system that is 112 miles away transmits to us. Each of the terminals is synchronized in real time to within 10 nsec. Hop rate is 100 khps. How far must our receiver search to gain initial synch?
10. We reply without resetting our code timing. How far must the other terminal search to synchronize with our signal? What if we reset to our original timing before transmitting?
11. A delay-line matched filter is to be constructed to match a 1023-chip code sequence. Code rate is 5 Mcps. What are the major delay characteristics required?
12. What signal-to-noise ratio would be required to give a probability of detection of 0.99 for a probability of false alarm of less than 10^{-2} (for a single detector)?
13. Compare chirp signals, from the standpoint of synchronization, to direct sequence and frequency hop signals.

7

THE RF LINK

Spread spectrum modulators and demodulators are dependent, in general, on transmitting in a frequency band capable of supporting a wideband signal, in which enough bandwidth can be allocated. In the usual case modulation and demodulation are performed at an intermediate frequency and translated up or down, to and from, the transmission frequency. Some systems have performed the code modulation and demodulation functions at the transmission frequency. Implementation problems, however, with regard to the required signal multipliers and filters have almost invariably forced these operations to be carried out at lower, fixed intermediate frequencies.

Careful spread spectrum system design also extends to the propagation medium. Even the atmosphere can limit the band of usable frequencies. When a broad band of frequencies is to be used, both phase shift and amplitude distortion can be introduced on the signal just by passage through a supposedly benign atmosphere. This chapter outlines some of the special problems that must be addressed in designing spread spectrum receivers and transmitters—actual bandwidth requirements, effect of poor receiver noise figure, AGC, and dynamic range, and other considerations peculiar to spread spectrum systems. Its most important purpose is to show that even though spread spectrum processing is, in general, transparent the task of actually constructing a workable system involves more than adding a code generator and modulator to an existing conventional system. To be sure, such appliques are possible and even practical, but their performance is always limited by the system to which they are grafted.

7.1 NOISE FIGURE AND COCHANNEL USERS

Noise and undesired cochannel signals (also noise) are just more interference, or jamming, to a spread spectrum receiver. They are added to any other undesired incoming signal and processed in the same way.

Our first inclination, then, is to expect a spread spectrum receiver to have greater sensitivity than a conventional receiver. Not so.

Except for narrowly defined and even (some might say) picayune gain made possible by a combination of bandwidth manipulation applicable only in direct sequence systems and noise × code-reference spreading, little is gained. Because the effective gain can be up to about 2 dB, however, there are situations in which this gain could be useful.

Any receiver's sensitivity is limited by the amount of noise it must take in along with its desired signal, including its own internally generated noise. These sources of noise are additive, as is any other interfering signal. Noise power into the receiver is kTB watts, where k = Boltzman's constant = 1.38×10^{-23} J/K; T = temperature of the receiver input termination in degrees kelvin; B = noise bandwidth of the receiver.

In a spread spectrum receiver the receiver bandwidth must be commensurate with the desired signal bandwidth. Even though a given baseband signal is spread over an RF band many times its own bandwidth, the receiver must necessarily accept more noise in exactly the same ratio as the increase in bandwidth; that is, if a 10-kHz (two-sided bandwidth) signal is spread over a bandwidth of 10 MHz (yielding a processing gain of 30 dB), the increase in noise intake would also be 30 dB.

For a given information bandwidth, the sensitivity of a narrowband radio is defined as that point where:

$$\text{input } \frac{S}{kTB_{\text{IF}}\,NF} \geqslant \frac{S'}{kTB_{\text{IF}}\,NF} \text{ (minimum)} = \frac{S'}{B_{\text{info}}\,NF} \text{ (minimum)}$$

where $S' = CS$,

C = conversion factor depending on modulation type and index,

NF = noise factor.

With a spread spectrum radio, sensitivity is still defined at a given output signal-to-noise ratio, but the receiver's processing gain enters the picture:

$$\text{input } \frac{S}{kTB_{\text{RF}}\,NF} \geqslant \text{output } \frac{S'G_p}{kTB_{\text{RF}}\,NF} \text{ (minimum)}$$

but

$$G_p = \frac{kTB_{\text{RF}}}{B_{\text{info}}}$$

therefore,

$$\text{output}\ \frac{S'kTB_{\text{RF}}}{kTB_{\text{RF}}\ NF\ B_{\text{info}}} = \frac{S'}{B_{\text{info}}\ NF}\ \text{(minimum)}$$

which is identical to the narrowband result. Thus we see that a spread spectrum receiver should have similar sensitivity to a narrowband receiver, for the same information bandwidth.

By judicious choice of receiver bandwidth, however, we could reject more of the potential noise than signal and produce some gain. Figure 7.1 illustrates cumulative power distribution in a direct sequence signal's main lobe and compares it with the cumulative power distribution for flat noise such as atmospheric or receiver noise. For receiver bandwidth equal to main-lobe bandwidth the receiver would accept all the noise and the desired signal. Narrowing receiver bandwidth to 60% of the main-lobe bandwidth, on the other hand, would reduce the desired signal to 90% of its original power, a loss of 0.46 dB. The noise contribution, however, would be reduced to 60% of its original power, down 2.2 dB. This simple expedient (reducing receiver bandwidth) thus produces a signal-to-noise ratio gain of approximately 1.7 dB.

For frequency hopping receivers, bandwidth manipulation does not give an advantage because the frequency hopper's power distribution is flat, as is the noise distribution. To be sure, if the frequency hopping

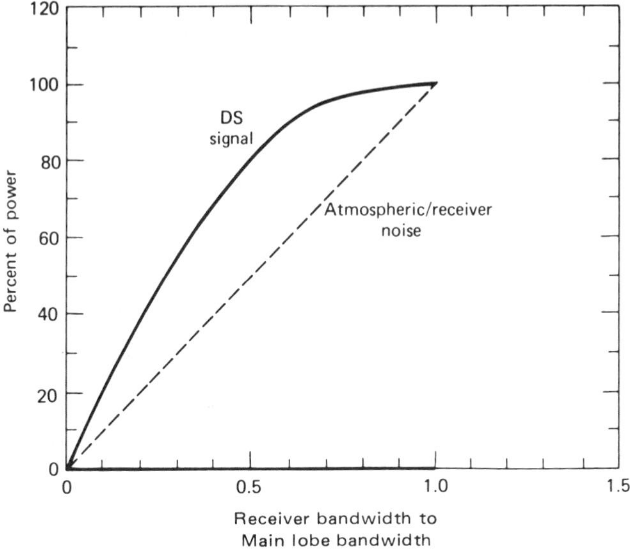

Figure 7.1 Comparison of power versus bandwidth for direct sequence and flat noise signals (normalized to direct sequence mainlobe bandwidth).

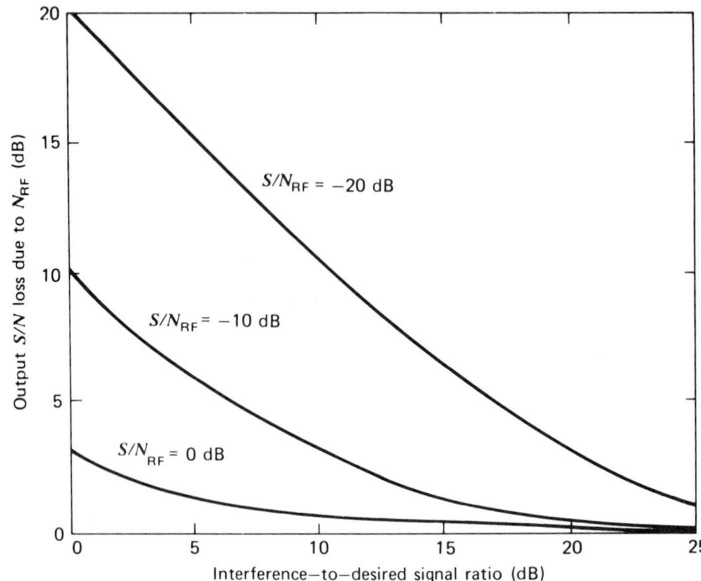

Figure 7.2 Loss in receiver output S/N ratio versus J/S for various RF input S/N ratios.

signal is truncated, as a filter might, the receiver's error would increase, because all the power in any chip transmission lying outside its filter passband would be completely lost.

Assuming that we have chosen a bandwidth for our receiver, what effect has our receiver's noise figure, or cochannel interference? These two noise sources have nearly the same effect on receiver performance. When noise figure or cochannel interference alone increases, then, if output quality is to be preserved, the desired signal must increase a like amount.

When jamming margin is the important factor, neither noise figure nor cochannel interference is really dominant, even though both contribute to system performance. Figure 7.2 shows the loss in output signal-to-noise ratio suffered by a spread spectrum receiver as input interference (jamming, etc.) is varied. Various signal-to-noise (exclusive of the jamming signal) ratios are shown as parameters. As an example, consider that N_{RF}(N_{RF} = cochannel interference + atmospheric + front-end noise) is 10 dB above the desired signal. Then, as the jamming signal is increased, the loss due to N_{RF} decreases to the point at which it is negligible. Thus we realize that even when cochannel interference and receiver noise power equal that of the maximum expected (design level) jamming, the degradation in jamming performance is no more than 3 dB. This is a simple but important result.

It should be pointed out that in most cases a spread spectrum receiver should not be expected to give maximum rejection for undesired signals and simultaneous operation on a minimum RF signal. Maximum

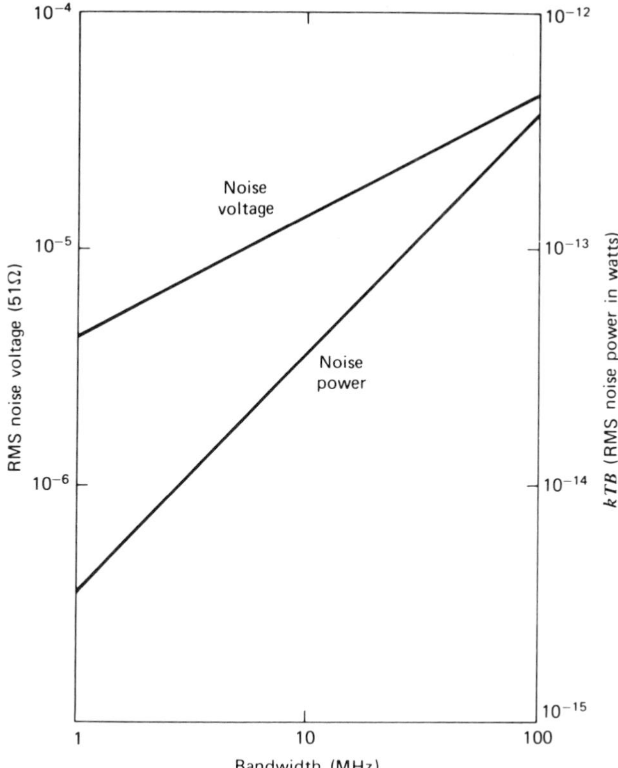

Figure 7.3 Noise input to receiver for common spread spectrum bandwidths.

sensitivity implies that the desired signal is small enough that combined atmospheric and front-end noise fall just short of jamming the receiver. Therefore, a spread spectrum receiver should be expected to give optimum interference rejection only when its input signal is at least 3 dB above the minimal level.

Figure 7.3 is a plot of kTB for bandwidths encountered in spread spectrum systems, to help the designer determine minimum signal operation. Receiver noise factor* directly multiplies this kTB.

7.2 DYNAMIC RANGE AND AGC

The dynamic signal range a particular receiver must handle depends largely on the type of environment in which the receiver is expected to

*Noise factor is the number of times a receiver front end multiplies the noise (kTB) at its input Its relation to noise figure is: noise figure = 10 log noise factor (dB).

operate; for example, a satellite-to-ground link is much more stable than an aircraft-to-aircraft or aircraft-to-ground link, for variations of geometry in the aircraft link are extreme and in general unpredictable, whereas satellite location is predictable at all times.

In addition to variation in relative transmitter–receiver location, any earthbound receiver must contend with variations in signal strength due to the propagation medium. Because the earth's atmosphere is far from homogeneous, it offers various discontinuities that cause diffraction, refraction, and occasional ducting, all of which can produce large variations in signal strength regardless of link geometry, even for fixed-terminal configurations. Even the angle of propagation through the atmosphere can make a significant difference in signal strength and quality; for example, signal strength commonly varies over a range of six or more orders of magnitude in an airborne UHF link. Other types of communication link may vary as much, but few have such rapid variations as the aircraft link. This means that any receiver whose demodulation system is sensitive to input level must, for optimum reception, incorporate some method of compensating for changes in RF signal level seen at its antenna.

The total signal range for a receiver can be estimated as

$$10 \log \Delta P - 10 \log \Delta M + 20 \log D_{\text{max}} + 10 \log L_{\text{max}} = \text{signal range}$$

where ΔP = signal maximum/signal minimum,
ΔM = antenna gain/null depth,
D_{max} = longest operating distance,
L_{max} = maximum atmospheric losses allowed under design conditions.

These can easily reach up to 100 dB or more, which means that AGC range must be similar. The front end and all stages up to the correlator must have at least 30 dB headroom above the desired signal level to remain linear and provide 30 dB process gain.

One method of holding signal level constant is to design the receiver system in such a way that clipping or limiting action takes place prior to the demodulator at all levels down to the design threshold. This method is obviously precluded, however, if any amplitude information is to be expected from the signal, for the limiting action reduces all incoming signals to a single value. The limiting approach is restricted to a few types of demodulation systems of the frequency- or phase-modulated variety.

All demodulators, even those designed to demodulate only frequency information, are sensitive to signal level to a certain extent, for example, the phase-lock FM demodulator. Because this device is essentially a servo loop that employs a phase detector as an error-sensing device, any change in the level of input signal varies its effective open-loop gain, thereby affecting the closed-loop bandwidth and damping factor. When the phase-lock loop is used as a threshold-extending demodulator, as is typically the case in spread spectrum communications equipment, these parameters must be held within narrow limits in order to achieve the optimum threshold and processing gain for the overall system.

Direct sequence receivers suffer losses up to 6 dB in jamming margin when operated with limiting front ends[13] and degrade even more when the jamming signal is coherent with the pseudonoise carrier. Indeed, any nonlinearity ahead of a direct sequence system's code-matching correlator can produce spurious interfering coorelator inputs and consequent self-jamming. Although limiting amplifiers have been employed in certain direct sequence communications systems, this has been done to avoid the added complexity of an AGC system and at the cost of less than optimum performance.

When signal limiting is not allowable, two general methods are available for controlling the transfer of signal from the antenna to the demodulator in modern communications systems. These are (a) controlled gain amplification and (b) controlled loss attenuation.

Gain control techniques rely on devices in which a DC circuit parameter may be varied to cause a change in the small-signal AC circuit parameters. As an example, a transistor with a nonlinear beta characteristic can be used in a circuit in which the DC operating point may be moved by changing the circuit's DC bias voltage.

This change in bias would cause a smaller change in output current at one bias point for a given change in input current than would result from the same amount of change in input current at another bias point. Various transistor types have beta curves that may provide an increase in gain as collector current is increased or decreased with current, depending on the particular device type.

Other variable-gain devices of note are tetrode transistors, both field effect and bipolar types, which have been used successfully in many communications systems. However, tetrode field-effect transistors have not yet been developed for the higher-frequency bands and bipolar tetrode transistors have poor noise figures. Both are being supplanted by other means. Attenuating control devices, for instance, are used as adjustable shunt impedances which cause a variable amount of loss in a voltage-dividing network. Some of the devices that may be used in this way are unijunction transistors and pin diodes, both of which vary inversely in resistance as their DC bias current is varied. Pin diodes have been employed as attenuators at frequencies up to and into the microwave range.

A gain control system, whether of the variable attenuation or variable gain type, is considered mandatory for use in most spread spectrum systems. The following block diagram (Figure 7.4) shows a typical receiver with AGC which controls the signal level into its correlator and thereby the demodulator input level.

In addition to the problems presented by their requirement for linear

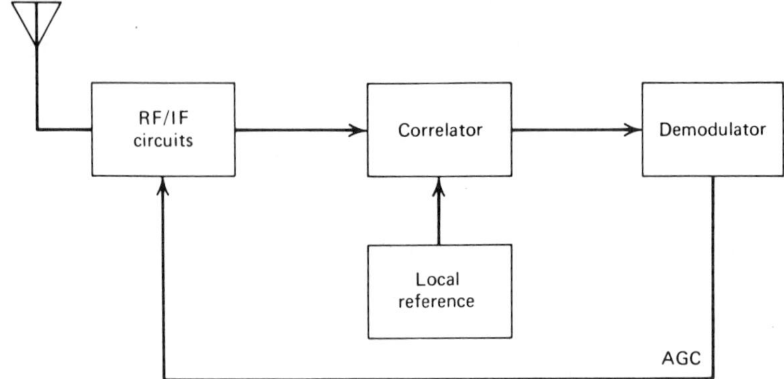

Figure 7.4 Correlation receiver with AGC loop.

transfer function over a large range of values of the desired signal, spread spectrum systems, and antijam systems in particular, are required to control their gain not only as a function of a desired signal but must also have control over the value of all signals that fall within their frequency aperture, including interference. This means that interference present in the absence of a desired signal is allowed to control the system gain. When a desired signal is present, however, interference must not be able to control system gain. If an interfering signal's effect in the presence of a desired signal is not minimized, the desired signal could be depressed simply by increasing the interference level, even within the jamming margin of the equipment. (Interference power may exceed signal power several hundred times within this margin.) This situation could force the desired signal down in level to the point of demodulator thresholding or cause amplitude modulation of the synchronized signal, which could, in turn, make the jammer even more effective.

The ideal gain control system for a spread spectrum system must not only control the signal level fed to the detector but must also operate under the following conditions:

1. It must control signal level to prevent false synch recognition when the desired signal is not present but a large amount of noise due to interference is being applied.
2. It must control levels only as a function of desired signal input when the desired signal is present, ignoring all other signals.
3. It must provide near-critical damping to prevent pulse-stretching effects in a pulsed interference environment.

Conditions 1 and 2 imply the use of two different AGC detectors, one before synchronization and a second after synchronization, because the bandwidths and gains under the two conditions are necessarily different.

Thus the problem of disengaging presync AGC and engaging the postsync AGC is presented. The changeover from unsynchronized to synchronized AGC control must be accomplished with minimum transient overshoot or undershoot, thus requiring that the AGC be properly damped under both noise and signal conditions. Proper damping is also required to prevent stretching of pulsed interference effects, for any overshoot or undershoot in the AGC loop could prevent synchronization over a larger percentage of time than necessary, giving the interference larger effective duty cycle; for example, assume a pulse large enough to preclude synchronization during its "on" time and a pulse width less than the fallout time of the sync recognition channel so that once synchronization has occurred the "on" pulse will not cause loss of tracking. Then synchronization can occur at any time when the pulsed signal is in its "off" state, and the probability of success for a single transmission will not be less than 1− interference duty cycle. The effects of interfering pulses may be stretched by improper AGC loop damping, however, and sync probability will be significantly lowered.

Reference to Figure 7.5 points out how a pulsed signal's effect may be stretched by nonoptimal AGC processing. Because the AGC waveform is ideally a reproduction of the envelope of the pulsed signal, any gain controlling signal larger than required suppresses a desired signal unnecessarily and degrades threshold, whereas control less than optimal permits overdriving of level-sensitive detection circuits and causes false indication of synchronization.

Processing of the presynchronization AGC envelope by periodic sampling could be used to detect the presence of pulse interference and cause the search process to be speeded up by an amount equal to the interfering signal duty cycle, which could result in a system that searched

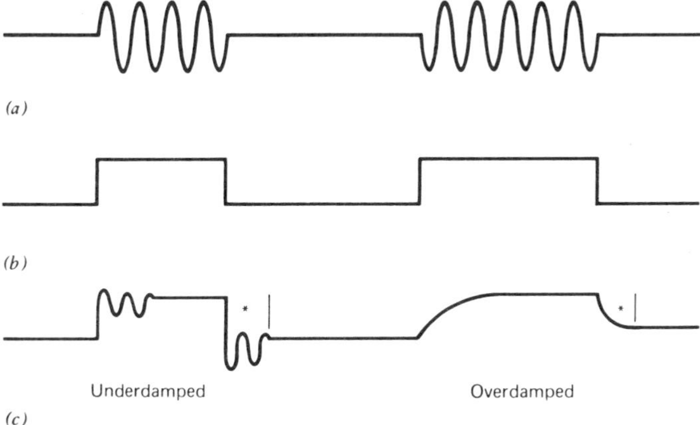

Figure 7.5 Pulse-jamming AGC waveforms: (*a*) pulse-jamming signal; (*b*) ideal AGC waveform; (*c*) poorly damped AGC waveform.

for synchronization only when the jamming signal was in the off state, thereby completely ignoring the interference. Again, however, near-critical AGC damping would be mandatory.

For a frequency hopping system special consideration must be given to the presynchronization AGC control system. Direct sequence systems permit AGC control by a detector operating within the bandwidth of the signal demodulator. Because all signals applied to the correlator are spread at least to the bandwidth of the local reference signal, a part of the noise power is forced to fall within the bandwidth of the demodulator where it may be processed to generate AGC voltage.

In a frequency hopping system little or no energy may fall within the band of the demodulator because the local oscillator will hop to one of many discrete frequencies, some of which will be far enough from the center frequency so that no energy will fall into the demodulator passband. Frequency hopping signals at the input to the correlator (see Figure 7.4) lie in one range of frequencies, whereas the local reference signals fall in another range. The difference frequency output then lies in a range centered at the difference frequency, but twice as wide as the local reference signal, for an unsynchronized frequency hopping input signal. When the input is continuous wave, however, the correlator output is a replica of the local frequency hopping reference translated down to the same band but centered around the difference frequency between the jammer and the local reference. Once synchronization occurs, the input and reference signals correlate and their constant frequency difference produces an output that is entirely within the detection bandwidth of the receiver.

One way to meet the requirements set forth for frequency hopping system automatic gain control, then, is to generate control voltage from a detector that immediately follows the correlator and samples its output over a bandwidth equal to twice the frequency hopping bandwidth. This may be termed a wideband AGC and requires fixed gain IF's following the correlator or a separate AGC loop for control of the stages following the correlator. It would be possible, however, to generate AGC from the frequency hopped energy which does fall in the demodulator information passband at the cost of greatly reduced AGC loop gain. Alternatively, if signal levels permit it, the wideband amplifiers preceding the correlating mixer can be left without AGC. In this case narrowband AGC alone, taken from a point following the correlator, is satisfactory. Narrowband AGC is generated as a function of the signal within the information bandwidth.

Figure 7.6 illustrates a two-loop gain control network for frequency hopping systems that could meet all the stated requirements. Control signal AGC_1 is generated as a function of the total signal power that falls anywhere within the desired signal band $f_F \pm f_{FH}$. This control signal would set the level of signals applied to the second IF amplifier. AGC_1

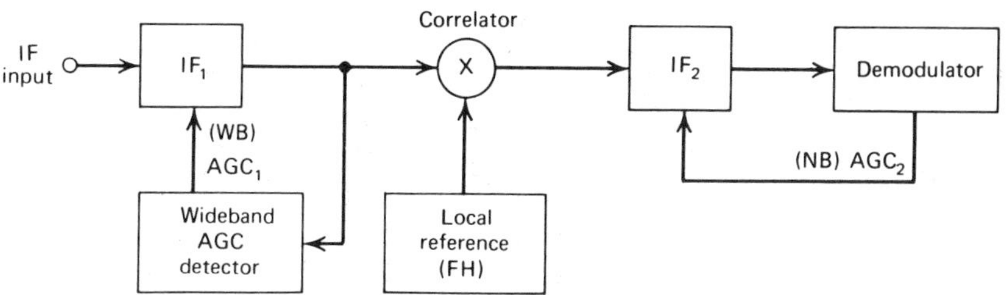

Figure 7.6 Wideband/narrowband gain control network for frequency hopping receiver.

loop gain would be set to allow maximum linear output signal level, and IF dynamic range must be such that a desired signal plus the maximum expected interfering signal may be handled simultaneously and linearly.

Control signal AGC_2 is generated as a function of the signal power falling within the desired information bandwidth.

Figure 7.7 shows a single-loop AGC network for frequency hopping. In this configuration no AGC is applied to the wideband receiver sections and an input signal-plus-noise level up to the maximum signal, plus the expected maximum interference, must be tolerated. This implies that IF_1 gain should be low because more than minimum gain would force the frequency hopping local reference to be supplied at a high level, which could be difficult to maintain and could also produce system self-interference.

In the case of a combined frequency hop and direct sequence system a combination of detection methods might be ideal in which a wideband detector would act to give precorrelator control with a narrowband AGC giving postcorrelator control. This would permit separation of the AGC function and good control of interaction, which is a must for a satisfactory spread spectrum AGC network.

In summary, automatic gain control for spread spectrum systems must provide the following

1. Linearity over the full range of desired and undesired signal inputs.
2. Effective noise AGC.

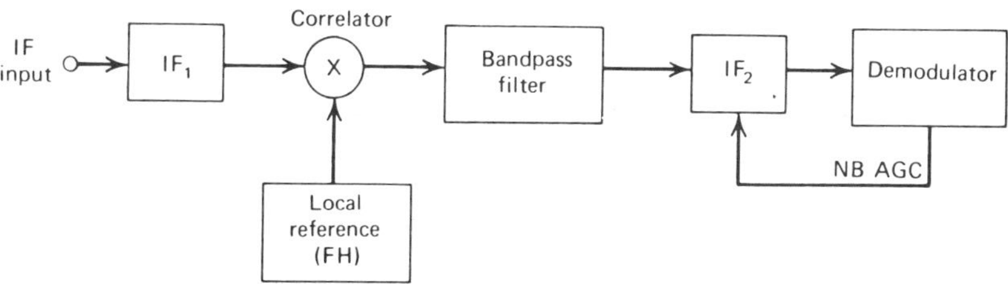

Figure 7.7 Alternative frequency hop gain control.

3. Close control of demodulator input signal level.
4. Proper signal versus noise AGC takeover characteristics.
5. Near-critical damping of AGC loop response.

Spread spectrum systems using these AGC configurations have been successful in operating in environments in which conventional modulation methods were not satisfactory.

7.3 THE PROPAGATION MEDIUM

Propagation for a spread spectrum signal is much the same as for any other signal, with the exception that the wide bandwidths employed sometimes restrict system use. The wideband modulation that can cause a system to be restricted in some areas can be a boon in others. This section lists some of the advantages and disadvantages of spread spectrum systems when faced with the common perils shared by all RF transmissions.

Line-Of-Sight Loss

Path loss for all kinds of spread spectrum signals is the same as for a narrowband signal operating at the same center frequency. The same physical laws apply. Therefore, if, for instance, an AM signal suffers a given amount of loss in traveling a certain distance, a direct sequence or frequency hopping signal would suffer that same amount of loss.

For convenience, a nomograph of path loss versus distance for frequencies between 100 MHz and 10 GHz is given in Figure 7.8. This nomograph is based on the well-known equation for path loss between isotropic antennas,

$$L = 37 + 20 \log \text{distance} + 20 \log \text{frequency}$$

and does not account for loss except that due to simple reduction in power density with distance.

We note here that for very wideband spread spectrum signals some compensation may be necessary to overcome differences in path loss across the band used. Such path loss differential would cause a linear decrease in signal strength at a 6 dB per octave rate, which could seriously distort the received signal spectrum and would especially hamper a frequency hopping system in which bit decisions are made on the basis of chip power comparisons. A simple solution is the insertion of predistortion in the transmitter or postdistortion in the receiver before the correlator. It is fortunate that the distortion is such that a simple 6 dB per octave high-pass section can compensate for this frequency-related loss.

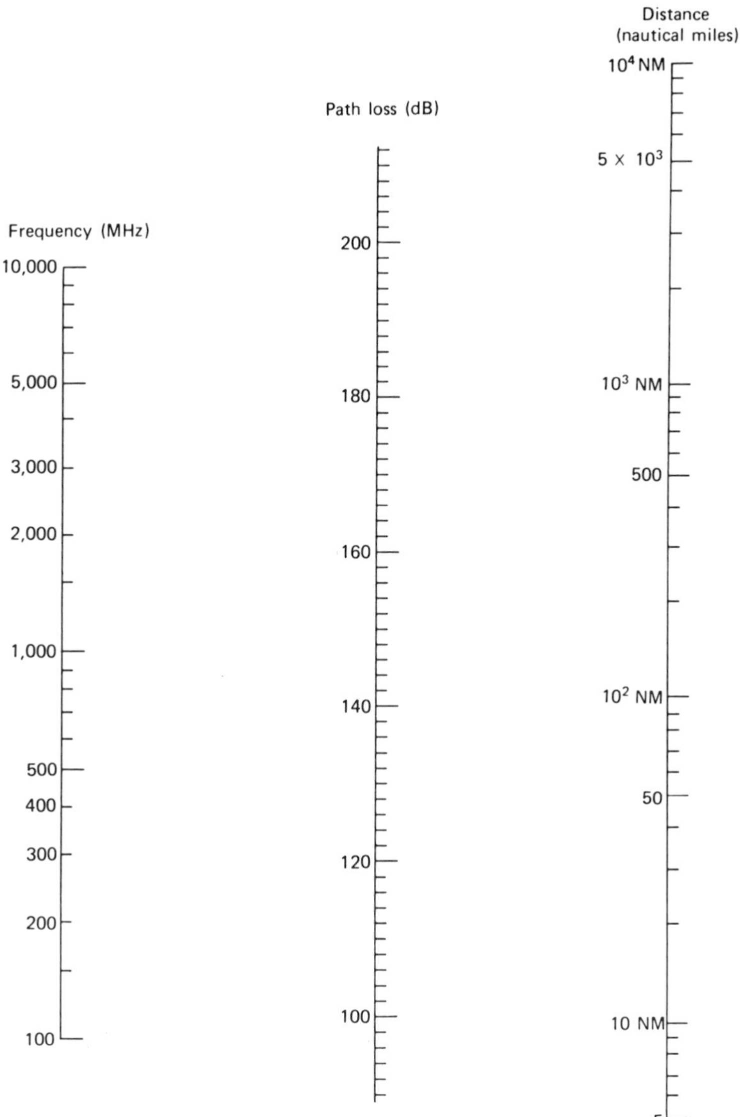

Figure 7.8 Path-loss nomograph.

Absorptive Losses

At certain frequencies signal propagation is sharply attenuated because of atmospheric constituents.[1302] These frequencies should be avoided unless it is required that for some reason the signal be limited to short range. The frequencies exhibiting highest attenuation are 23 GHz, where oxygen absorption is predominant, and 60 GHz, where water vapor absorption prevails.

Except for a bit of extra care to ensure that the broadband signals common for spread spectrum systems do not overlap one of the highly absorptive regions, there is no difference in spread spectrum and conventional designs as far as the absorptive frequencies are concerned. A broad band signal could, however, suffer differential amplitude loss in the vicinity of the oxygen and water vapor absorptive regions due to the slope of attenuation in the area.

Differential Phase Delay

Differences in the velocity of propagation with frequency can cause severe phase distortion, an effect that is pronounced in the ionosphere. Such phase distortion degrades digital signals so that their one–zero transitions are slowed (see Figure 7.9).

Eckstrom,[1408] Staras,[1417] and others have considered the transionospheric phase distortion effect and have shown that the RF bandwidth of a signal should be less than 5–10% of the carrier frequency to avoid serious differential phase distortion. Sunde[1418] has studied the problem of delay distortion in high-speed data transmission systems (A direct sequence transmitter is certainly a high-speed data transmission system, whether or not the high-speed code is recovered as data.) Sunde's results show that there is less than 2 dB transmission loss for biphase modulation up to delay differentials (from center frequency to first null) of 1.6 times the transmitted pulse period. This is true for both quadratic and linear distortion. Quadriphase modulation in linear distortion is somewhat worse; it exhibits 2 dB loss for a delay differential of $0.5T$ (where T is the period of a data bit).

In the frequency hopping world, differential delay prevents the transmission of a coherent frequency hopping carrier. The difference in delay seen at the various frequencies in a frequency hopped spectrum causes the dehopped carrier at the receiver to be randomly phase modulated at the hopping rate. This phase modulation in turn prevents the use of a phase-lock tracking loop in a frequency hopped system. It also causes the received signal bandwidth to be approximately 0.88 times the hopping rate, which can cause a loss in the receiver of up to 3 dB over a coherent signaling method.

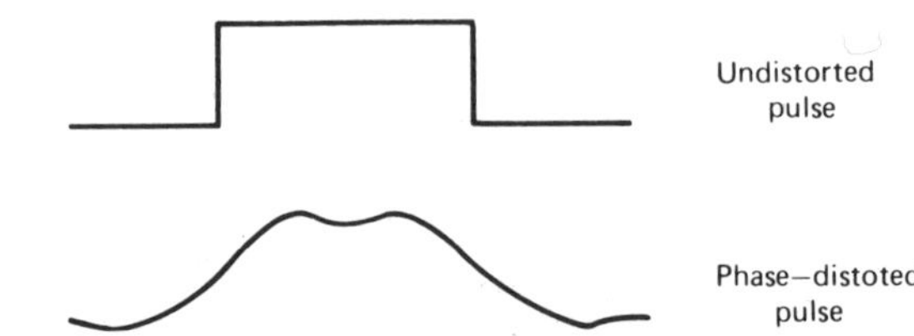

Figure 7.9 Distortion of square pulse due to differential phase delay.

This is not to say that coherent frequency hopping cannot be accomplished, but it requires careful design in at least three areas:

1. Precise phase coherence in the transmitter's frequency synthesizer, with no differential phase shift in the power amplifiers.
2. Precise phase coherence in the receiver's frequency synthesizer, with no differential phase shift in the receiver's RF and precorrelation IF stages.
3. Careful, accurate modeling of the link between transmitter and receiver, with precise delay compensation.

If these can be accomplished, then coherent frequency hopping can be a reality. Until all three can be achieved simultaneously, however, only noncoherent frequency hopping is practical for most applications.

Multipath

In addition to the direct signal path there often exists a reflected path for a signal to follow from a transmitter to a receiver. The receiver then sees at its input two signals—one delayed by the extra path length. These signals add according to their phase relationship to produce a signal that contains the modulation of both and has an amplitude determined by the phase relationship. In television receivers they cause "ghosts" or under extreme conditions complete loss of picture synch. It is possible for a receiver to receive reflected-path signals that are larger than the direct-path signal, especially in aircraft environments in which an antenna may be partly shaded by part of the aircraft structure.

The result of multipath reception is that a moving vehicle is subjected to rapid fading and peaking of its input signal as the direct and reflected signals cancel and reinforce one another. Stationary receivers are also subject to multipath but of a more stable variety, for the only variable is the reflecting surface. (Who has not seen the result of an aircraft flying between his television receiver and the transmitting station?)

Spread spectrum systems are less subject to multipath signal variations than conventional systems. In a direct sequence receiver, if the reflected signal is delayed (compared with the direct-path signal) by more than one code chip, the reflected signal is treated exactly the same as any other uncorrelated input signal. The higher the code chip rate for a particular system the smaller its multipath problem. With 5 Mcps codes, for instance, the reflected signal path must be 200 ft or less different from the direct path to have any effect, and as code rate is increased the path length for which the reflected signal interferes is further reduced.

Frequency hopping receivers avoid multipath losses in two ways: (a) by transmitting the same information at more than one frequency and (b) by

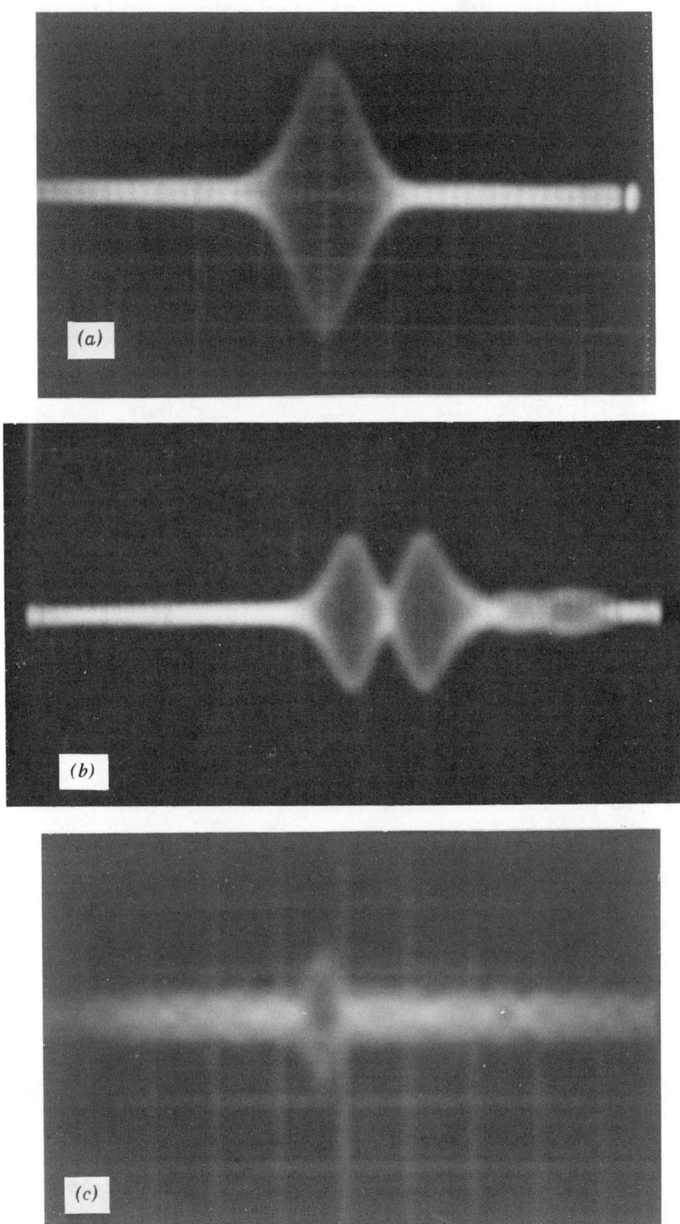

Figure 7.10 Matched filter signals. (*a*) Undistorted matched filter receiver output. (*b*) Receiver output with specular multipath. (*c*) Receiver output with diffraction due to hill in path.

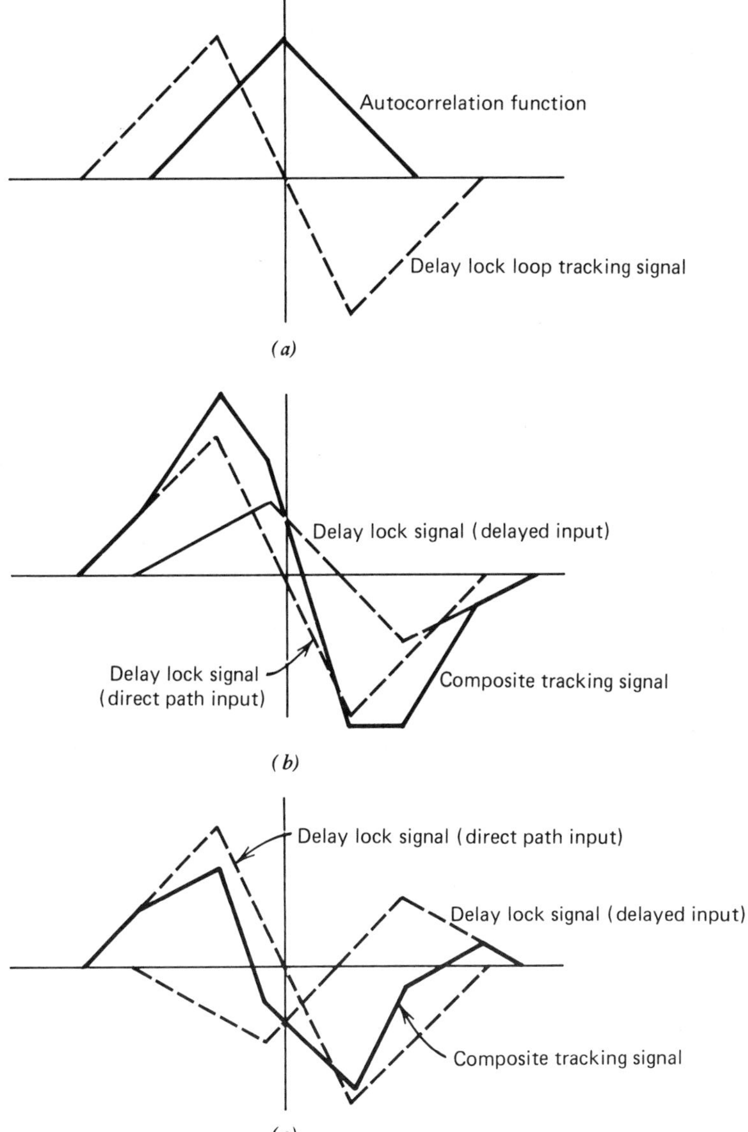

Figure 7.11 Effect of multipath: (*a*) no multipath; (*b*) in-phase multipath; (*c*) out-of-phase multipath.

comparing the power in alternate mark/space channels for each bit decision. (This considers that frequency hopping systems are used for data transmission. An analog-modulated, frequency hopping system would not have these advantages against multipath.)

When a delay-line matched filter or some similar device (chirp filters or digital matched filters, for instance) is used by a receiver, multipath can cause more than one filter output pulse; that is, one output signal is due to detecting the desired input and another is due to late-arriving input generated by multipath delay. This paired output from a matched filter under strong specular multipath conditions is shown in Figure 7.10*b*. Provision can be made in the matched-filter receiver to overcome any difficulty caused by double-pulse reception by designing the receiver to accept only the first-arriving pulse, which effectively nullifies the effect of late-arriving multipath signals.

Under conditions of diffuse multipath, a matched-filter detector's output is likely to exhibit noncoherent (in time) output such as that shown in Figure 7.10*c*. In this photograph, the signal was diffused by a hill in the signal path, between the transmitter and intended receiver.

Figure 7.10*a* shows a normal signal output from the same matched-filter detector as that used in 7.10*b* and 7.10*c*, where there is no significant multipath signal present. It should be evident that the receiver's jamming margin may be reduced under such conditions, as the matched-filter output level is reduced by division into one or more signals.

Multipath conditions can also drastically alter the shape of the discriminator function developed in a delay lock tracking loop. Figure 7.11, from C. R. Cahn, shows the effect of multipath on the discriminator function. Tau-dither loops are also affected by multiple signals offset in time. In either case, the tendency of the code clock tracking loop is to follow the energy centroid of the arriving signals. The shift in the center (zero crossing of the discriminator curve) is

$$\frac{2\Delta t}{1 + \dfrac{S_1}{S_2}} = \text{clock offset}$$

where Δt = time shfit between signals,
S_1, S_2 = signal amplitudes for signals S_1 and S_2.

The clock offset is forward or backward in time depending on whether the multipath signal is in phase or out of phase with the desired signal.

7.4 OVERALL TRANSMITTER AND RECEIVER DESIGN

Design of transmitters and receivers for use with spread spectrum signals is not basically different from design for any other system in the same frequency range. Bandwidths are obviously wider, however, and other considerations must be made because of the expectation that large signals will be encountered simultaneously with small ones.

Let us trace a signal through a typical system, observing the perils and degrading influences to which it is subjected on its way.

The Transmitter

We assume here that we have a modulated intermediate frequency and carry on from there. A popular intermediate frequency for communications modems (modulator/demodulators) is 70 MHz, which is acceptable for spread spectrum signals as long as their modulated bandwidths are not greater than about 20 MHz. Higher intermediate frequencies are required for the higher code chip rates and/or wideband frequency-hopping; 700 MHz IF's are common when wider bandwidth modulated signals are expected.

The transmitter must take a modulated fixed-frequency input signal and move it to the intended operational frequency, amplify it to the desired level, and output it to the antenna. The operation of moving the modulation from an intermediate to a transmitted frequency may be performed in two ways: translation and multiplication. Translation (or conversion) mixes the modulated signal with another signal set to the desired output frequency, plus or minus the intermediate frequency, and selecting (by filtering) the desired sum or difference frequency. The translation method can be used with direct sequence or frequency hopping modulated signals with equal results. Multiplication to the desired output frequency must be employed with a great deal more care because multiplication can change the basic modulation.

What happens to spread spectrum signals when they are multiplied? Frequency hopping modulation is increased in bandwidth and freuqency spacing by exactly the same ratio as the multiplication; that is, a 100-channel frequency hopper with 1 kHz channel spacing before multiplication by 3 would still have 100 channels but a 300-kHz bandwidth and 3 kHz channel spacing after multiplication. In many applications this multiplication of frequency spacing is quite useful because it allows a frequency hopping synthesizer to be constructed with narrow spacing and overall frequency range. That same synthesizer, multiplied, could supply almost any bandwidth or channel spacing. The same is not true of direct sequence-modulated signals. Multiplication of a direct sequence signal is practical only for certain specific multiplication factors. Others modify or completely remove the direct sequence modulation.

Some care must be exercised when multiplying any angle-modulated signal,* and this care also applies to phase-shift-keyed direct sequence signals. The action of a multiplier is such that not only is the center frequency multiplied but so is the rate of any modulation. Thus a biphase-modulated signal $\cos a \pm 90°$ when doubled becomes $\cos 2a \pm 180°$ (for the term of interest), which is a twice-frequency carrier without modulation because $+180°$ is identical to $-180°$. Multiplication by 5, on the other hand, produces $\cos 5a \pm 450°$, which is identical to $\cos a \pm 90°$. Odd and

*See also Appendix 4.

Table 7.1 Phase Mapping of PSK Signal Following Multiplication

	Phase Modulation					
Multiplier	0	90°	180°	270°	45°	
2	0	180	0	180	90	
3	0	−90	180	−270	135	
4	0	0	0	0	180	
5	0	90	180	270	225	(minus sign signifies
6	0	0	0	180	270	phase inversion)
7	9	−90	180	−270	−45	
8	0	0	0	0	0	
9	0	90	180	270	45	

even multiplications of any degree produce similar results; that is, for any even multiplication 180° biphase modulation is lost but for odd multiplications it is preserved (inversion is usually of no consequence.)

Table 7.1 is a phase mapping for various modulation shifts, with multiplication by factors of 2 through 9 as an operation. Various amounts of phase modulation are shown to be mapped to zero for some multiplication factors, others are phase inverted, and in some cases the resulting modulation is identical to unmultiplied modulation; for example, quadriphase modulation with multiplication by 4 reduces to a four times carrier with no modulation, whereas multiplication by 5 produces a five times carrier with the original modulation.

Table 7.1 shows that all odd multipliers are usable with biphase or quadriphase signals if an output phase inversion is permissible. When phase inversions are not desirable; then 5, 9, · · ·, $(4n + 1)$ is the only permissible set of multipliers. For multiplies beyond times 9 the table repeats.

The principle of multiplication of modulation also holds for modulation other than in multiplies of 90°, as might be expected. In those cases in which the modulated signal must be multiplied by a factor that would distort or destroy the usual biphase or quadriphase signals this principle has been used to advantage. If, for instance, it is mandatory that a times four multiplier be used but ±90° PSK modulation is desired, we can choose to phase shift ±22.5°. After multiplication by 4 the desired ±90° biphase modulation is produced.

Power Amplification

At the transmission frequency the spread-spectrum-modulated signal is amplified to the required power level and applied to an antenna. Here we are concerned primarily with the characteristics of the power amplifier because many antennas are broadband enough to pass spread spectrum signals satisfactorily. It is necessary, of course, to design a power amplifier with sufficient bandwidth to pass the spread spectrum signal. The question is, "What is sufficient bandwidth?"

For a noncoherent frequency hopping signal it is most important that amplitude linearity of the transmitter be as flat as possible; that is, the transmitted power at each frequency channel should be equal to that transmitted for every other frequency. The reason is that a frequency hopping receiver compares the signal sent in successive channels to make mark/space decisions. Variations in signal power may cause losses in error performance, depending on the receiver's decision criteria. This is especially true when interference is encountered, when the opposite channel from that intended is also occupied (i.e., a "mark" has been sent by transmitting at a given frequency but the frequency channel that would convey a "space" is occupied by interference). Phase linearity is less important in frequency hopping transmitters because the receiver's decisions are usually based only on received power in a channel. If the frequency hopping signal is coherent, however, transmitter phase linearity should be well defined.

Direct sequence modulation is somewhat less critical in its demands on the transmitter, although there are definite advantages in keeping the amplification process as linear as possible. In an earlier paragraph in this chapter Sunde's quoted results show that delay differentials (or phase nonlinearity across the transmitted bandwidth) do not introduce significant losses when phase shift is held to a radian or less at the code clock frequency.

It is interesting to note that direct sequence can be passed through a band-limiting filter with much less effect than a frequency hopping signal can. The frequency hopper, whose power distribution is flat across its occupied band, loses capability immediately when its passband is truncated; that is, if a frequency hopped signal is passed through a bandpass filter whose bandwidth is β percent of the frequency hopping band, the processing gain of the system is reduced to β percent or less of its original capability. Figure 7.12*a* illustrates the power loss for filtering a biphase direct sequency signal. (Quadriphase and double-binary or modified quadriphase power loss as a function of filter bandwidth is shown in Figure 7.12*b*.) Figure 7.12 applies to systems (except for double-binary) in which filtering is done after the final power amplifier. For conventional direct sequence modulation methods filtering before power amplification is impractical because any nonlinearity in a subsequent power amplifier regenerates the $[(\sin x)/x]^2$ spectrum existing ahead of the filter. Double-binary, MSK (minimum shift keying), or some other form of modulation that has its envelope minimally affected by nonlinear processes is mandatory if the signal cannot be filtered after the power amplifier. The curves shown are adapted from C. R. Cahn.[245]

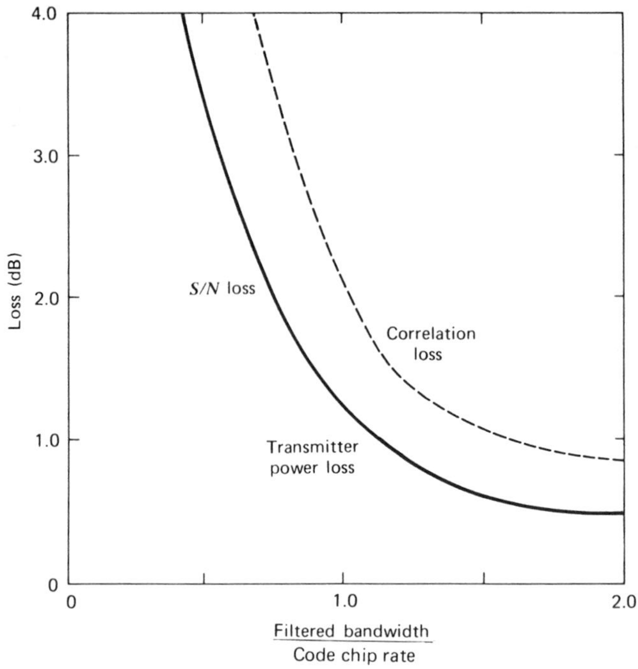

Figure 7.12(*a*) Loss versus bandwidth for conventional biphase signals.

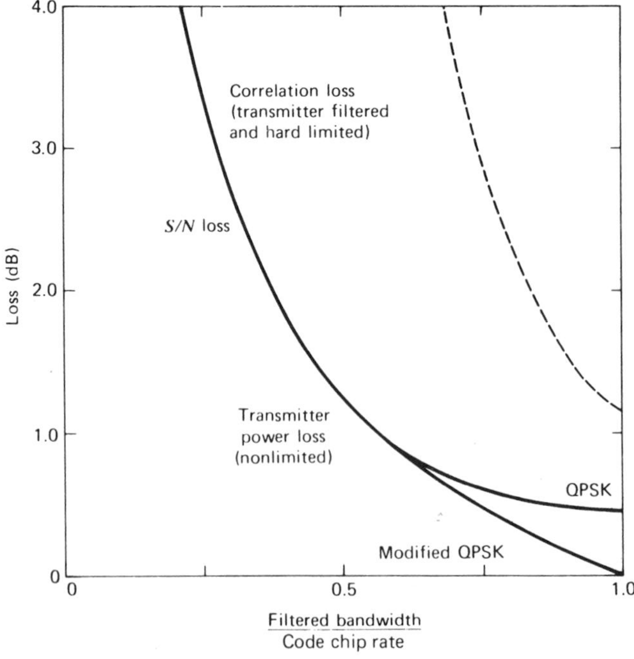

Figure 7.12(*b*) Loss versus bandwidth for conventional and modified quadriphase (double, binary) signals.

VSWR

An area of particular sensitivity for spread spectrum signals, when some care must be taken to ensure proper impedance match, is that of VSWR (voltage standing wave ratio). The broadband nature of spread spectrum signals and the acute frequency dependency of VSWR when a termination is improper combine to work against the spread spectrum system. Here we can say only that in both transmitters and receivers for spread spectrum applications extra care should be taken to ensure that terminations are correct. This includes not only impedance matching but connector mating, cable shield termination, and all the other small things that degrade conventional systems. When there is a VSWR effect in the conventional system, the spread spectrum systems is affected more.

Receiver RF Considerations

Design of the RF section in a spread spectrum receiver is somewhat more critical than transmitter design, mainly because of the presence of interfering signals before the processing stages. Ideally, the receiver would accept the desired signal (along with as little as possible of anything else), amplify it, translate it to the proper intermediate frequency, and deliver it undistorted to the correlator. Concurrently it should also deliver any interfering signal that falls in the direct signal band to the correlator.

Common direct sequence systems can suffer a 6 dB or larger loss in jamming-to-signal ratio when passed through a limiter concurrently with a larger jamming signal. This is the reason for such careful attention to linearity in the receiver RF section. Frequency hopping signals also suffer this same degradation in signal-to-jamming ratio but are unaffected by it from the standpoint of a loss in interference rejection. The frequency hopping receiver does pay for the signal suppression caused by limiting but only in the form of a loss in signal, not in jamming margin. (This can be a significant factor in system design.)

Why does the direct sequence system suffer a loss in interference rejection capability due to limiting but a frequency hopping system does not? First remember that a direct sequence system is most vulnerable to CW jamming and that the frequency of the CW jamming can vary over more than 10% of the signal bandwidth (or more) with nearly equal effect. [Cahn[13] shows that signal-to-noise loss in a limiter is the worst (6 dB) for nonvarying (or CW) signals. He also shows that when noise is used against a limiting spread spectrum receiver the bandwidth of the noise is of little consequence.]

In a frequency hopping receiver the interfering signal is still effective only in the channel or channels it covers, no matter how great its power advantage; that is, even though the limiter may cause the desired signal to be suppressed with respect to the interfering signal, the interferer has no more effect once his signal is ϵ greater than the desired signal. This is quite different from the direct sequence receiver, in which the interference is spread over the entire signal band and a decrease in signal-to-noise ratio due to limiting is effective overall. When it is necessary for a direct sequence receiver to employ limiting (indeed when there is a limiter anywhere in the link), it is customary to employ multiphase (more than two-phase) phase shift keying, which does not suffer the same penalty because of limiting as biphase modulation. Chapter 4 discusses both biphase and quadriphase modulation techniques.

Receiver Signal Handling Capacity

In a receiving system in which linearity must be preserved the linear signal range must be carefully considered. The receiver, from the front end to the correlator, must be designed to handle linearly not only the signal range expected but also the interference. Automatic gain control (AGC) only partly solves the problem in that it can control signal level on a signal, interference, or total power basis. (Because the interference is often much higher than the signal, total power AGC is essentially the same as interference-controlled AGC.)

We are forced to control receiver gain on the basis of correlator output signal level because the level of interference should be ignored as a basis of gain setting. Otherwise a jammer could control signal level or even modulate it by varying his power level. Once the controlled signal level is set each of the receiver stages through which the signal and interference pass together must be able to handle linearly a signal range equal to $(P_{\text{signal}} + M_J)$ dB. For this reason it is often desirable to place the correlator in a spread spectrum receiver as near to the front end as possible so that fewer high-level stages are necessary before the correlator. (This strategy also reduces the local reference signal level requirement.)

Wideband Front Ends

It is often desirable, when a receiver is to cover a broad band of operating frequencies, to use a wideband front end that covers the entire band (or as much of it as possible). The tuning function is then carried out by mixing the received signal with a variable local oscillator, which causes the desired product to be output at a fixed IF frequency. This is a reasonable and even desirable approach in some cases but of little use in a spread spectrum system when interference rejection is a goal. Consider, for instance, that the interferer has only to transmit two signals offset in frequency by an amount equal to the receiver's IF. Then, no matter what the receiver's desired frequency, the first mixer, or correlator, would have an output that falls in the IF. This, in turn could cause the receiver's interference rejection capability to be useless. Similarly, a wideband receiver front end could allow other products, such as images, to cause interference.

In a receiver specifically intended to reject interference it seems singularly inappropriate to allow any interference to enter the correlator other than that for which the system was designed. Therefore wideband amplification (beyond that required by the signal) is not generally recommended.

The Ideal RF Section

An ideal receiver RF section for a spread spectrum system is one in which the desired signal is amplified to the desired level without distortion, along with any interference lying in the desired signal band, and filtered to the desired signal bandwidth. The combined interference and desired signal is then applied to a heterodyne correlator whose output is at an intermediate frequency. This ideal RF section would pass along to the correlator only the section of the signal band that contains the signal; that is, the front end would be tuned and would have a bandwidth approximately twice the code clock rate for a direct-sequence signal or equal to the signal bandwidth for a frequency hopping* signal. Filtering to this bandwidth before the correlator is important to the direct sequence receiver, for interference at the carrier frequency ± clock may be most effective.

A tradeoff may be made between a requirement for a high local reference level and a fixed direct sequence correlator prefilter. This is true because a fixed frequency filter at an IF frequency is attractive, but moving the correlator to an IF requires a larger local reference signal. Alternate configurations are shown in Figure 7.13. Tuned RF amplifiers are shown as the first stage in both configurations. The second, however, which uses a fixed correlator and precorrelation filter, is preferable for direct sequence receivers because the only precorrelation bandpass filtering in the first configuration is in the variable RF stage. Also, carrier suppression in the variable frequency correlator is more difficult to obtain than in a fixed-frequency correlator.

RF sections for frequency hopping receivers need be no more complex than in Figure 7.13*a* (in which a frequency hopping correlator is substituted for the direct sequence correlator shown).

Local oscillators and references for direct sequence or frequency hopping receivers must necessarily be larger than the largest expected signal/jamming combination. This is required to preserve a linear transfer function in mixers and correlators; 3 dB or greater local oscillator and local reference levels (above the largest expected combined signal/

*Frequency hopping signal bandwidth is equal to channel bandwidth times the number of channels.

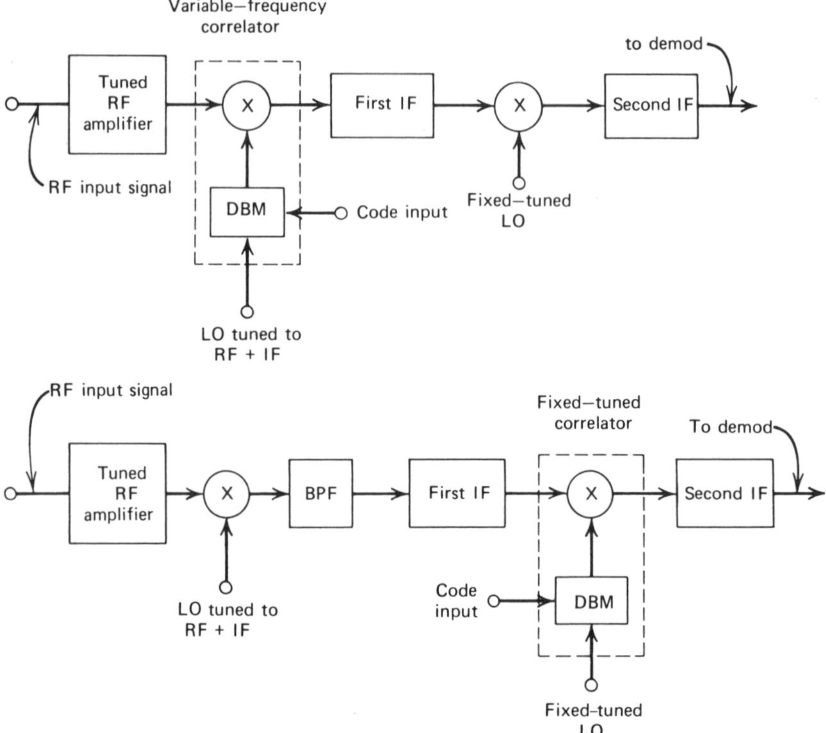

Figure 7.13 Typical direct sequence RF section and correlator configurations for spread spectrum receivers.

jamming level) should be maintained as a minimum. All other stages before the correlator should also maintain total signal handling capability approximately equal to the combined maximum desired and interference signal levels.

A possible alternate to a tuned front end for a spread spectrum receiver is shown in the block diagram in Figure 7.14*a*. In this configuration a bandpass filter covering the expected range of input frequencies precedes the amplifier. Following the wideband amplifier is a frequency converter (mixer) which translates the incoming signal to a fixed IF frequency such that $f_{IF} > 2f_{RF\,max}$. A fixed-tuned IF amplifier and bandpass filter at this frequency then acts to prefilter the signal before application to a correlator. The second multiplier, following the IF, can be the correlator if its locally generated signal is the proper reference. The advantage of this configuration is that RF tuning is completely circumvented. Only a tunable local oscillator input to the first multiplier is necessary. Image rejection (at $2f_{IF} + f_{RF}$) is accomplished by the input bandpass filter's rolloff, and no two frequencies which could give a product equal to the IF frequency can reach the first multiplier. Another advantage is that low-

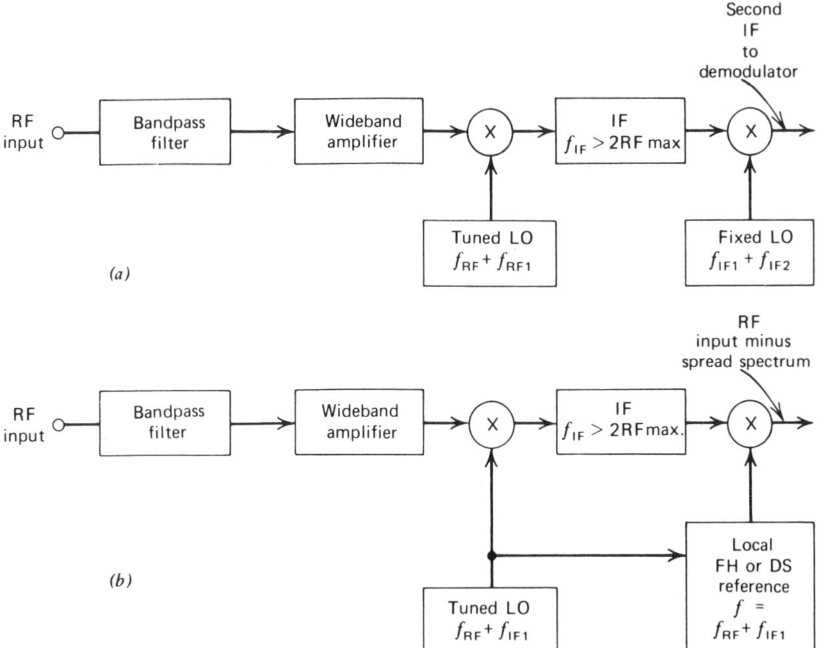

Figure 7.14 Alternative RF configurations with wideband amplification: (*a*) wideband amplifier configuration; (*b*) wideband amplifier configuration adapted to applique use.

noise wideband amplifiers capable of covering wide frequency ranges are readily available.

On the side of disadvantages the wideband amplifier must possess sufficient dynamic range to handle all the signals appearing in the bandwidth of its prefilter. Otherwise limiting and consequent suppression of the desired signal can occur, which would give a jammer control of the signal amplitude, no matter what frequency is transmitted anywhere within the filtered signal band. Of course, AGC can be added to prevent limiting but must also be carefully done.

A modification of Figure 7.14*a* is shown in 7.14*b*. This modification is specifically oriented to use as an applique to existing receivers to allow their conversion to spread spectrum use. Here the same prefiltering, amplification, and conversion to a high IF are used. The same tuned local oscillator, however, also supplies the center frequency for the local spread spectrum reference so that the output signal is at the same frequency as the received signal; only the spread spectrum modulation has been removed, at which point the receiver to which the tuner/correlator has been overlaid will perform its usual functions.

In this discussion we have considered only those cases in which the receiver is a tunable one capable of being set to receive at any one of many frequencies. This is the most difficult case and therefore of most interest.

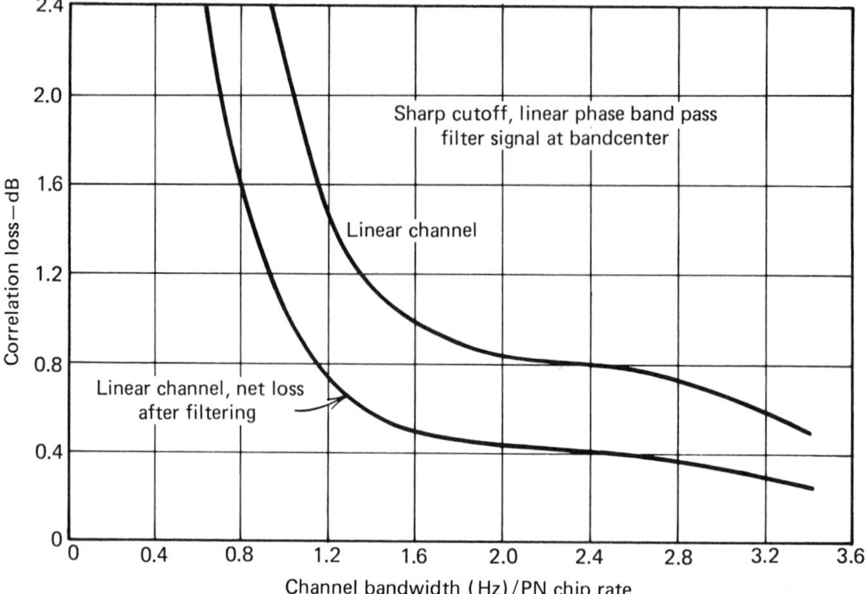

Figure 7.15 Correlation loss due to filtering (from Cahn).

For single-frequency receivers filtering and other design problems are often less severe, although the same principles hold.

Bandwidth and Phase-Shift Effects

A spread spectrum signal can be transmitted or received through a reduced bandwidth channel with reasonably small losses. The direct sequence signals have somewhat better capability for such operation than the frequency hopping signals, however. For frequency hopping signals, a reduction in bandwidth is a direct loss of hopping-frequency channels. Therefore, a 10% loss in bandwidth is equivalent to having ten percent of the channels jammed, since in either case the demodulator loses signal the same amount of time.

In direct sequence systems, a narrowing of bandwidth reduces the harmonic content of the spectrum, which distorts the signal waveform, capability as the loss in bandwidth. That is, a reduction in bandwidth to one-half the spectral width does not cause a 3 dB loss, typically. The loss that occurs is due to a reshaping of the receiver's correlation function from sharply triangular to rounded, as shown in Figure 7.15. In practice, this rounding causes a loss in jamming margin, but is not evident in any other receiver or transmitter operating characteristic.

Phase shifts in RF stages can also cause a loss in jamming margin.

Table 7.2 Effect of Phase-Shift Distortion on Jamming Margin

Filter BW	Loss (dB)
Phase distortion—1.0 radian at first null	
2 R_c	0.96
1.6 R_c	1.20
1.2 R_c	1.51
R_c	2.27
2.0 radians at first null:	
2 R_c	1.2
1.6 R_c	1.25
1.2 R_c	1.61
R_c	2.29
3.0 radians at first null	
2 R_c	1.61
1.6 R_c	1.65
1.2 R_c	1.83
R_c	2.38

Table 7.2 illustrates typical loss expected for various types and amounts of distortion. Typical specifications for minimal loss systems, due to phase shift, call for 0.1 radian or less phase shift between the center frequency and first null of a direct sequence signal.

PROBLEMS

1. Compare the noise power input to a spread spectrum receiver with a 10 MHz bandwidth to a conventional AM receiver with a 50 kHz bandwidth. The spread spectrum receiver has 26 dB process gain. Is the spread spectrum receiver or the conventional AM receiver more sensitive?
2. A spread spectrum receiver is operating in an environment in which the interference present is 20 dB greater than the desired signal. A second interfering signal which is 10 dB greater than the desired also appears. What is the effect of the second interferer?

3. The input noise, plus the contribution of a receiver's front end, causes a desired spread spectrum signal to fall 20 dB below the noise level. If the receiver's jamming margin is 23 dB, how much interference can be tolerated?
4. What is the effect of improper AGC loop damping in a spread spectrum receiver?
5. A frequency hopped signal with 100 MHz center frequency and 10 kHz spacing is to be multiplied by 4. What are the new signal characteristics? A 100 MHz direct sequence, biphase modulated signal is also multiplied by four. What are the characteristics of this new signal?
6. Compare the signal power loss that might be expected when filtering a 5 Mcps code-modulated biphase direct sequence system to a 2.5 MHz RH bandwidth with a quadriphase modulated signal having the same code rate and passing through the same filter.
7. A number of spread spectrum systems are operating in the same frequency band (code division multiplexed) and a large interfering signal comes on the air. Assume that there are 100 simultaneous friendly link users and that the interference is 15 dB above the desired signal at any friendly receiver. What is the effect of the interferer?
8. What noise power might a spread spectrum receiver with a 10 MHz width see? If the system's AJ margin (M_J) is 20 dB, what would the minimum signal be for receiver operation? (Assume a 10 dB noise figure.)
9. What would the effect on minimum signal value be if the above receiver's M_J were increased by increasing its spread spectrum bandwidth to 100 MHz?
10. What would the effect be of multiplying a quadriphase direct-sequence signal by 64?

8

NAVIGATING WITH SPREAD SPECTRUM SYSTEMS

Certainly the best known applications of spread spectrum methods lie in the navigation area. Direct-sequence ranging has been applied in space exploration programs since the early 1960s at least, and the methods employed have been well described in the literature.

The coded modulation characteristic of spread spectrum systems uniquely qualifies them for navigation—from the standpoints of both range measurement and direction finding—enough to allow homing or navigation with respect to any other spread spectrum transmitter. Little use has been made of spread spectrum techniques for direction finding,* but spread spectrum ranging, particularly direct sequence ranging, has been well exploited.

This chapter first describes ranging methods for direct sequence systems and for frequency hoppers. The second part of the chapter is a discussion of direction finding as applied in spread spectrum systems.

8.1 RANGING TECHNIQUES

Any RF signal is subject to a fixed rate of propagation (approximately 6 μsec/mi). The signal reaching a receiver at any given instant left the transmitter that sent it some time before. Because signaling waveforms or modulations are also functions of time, the difference in a signaling waveform. as seen at a receiver, from that present at the transmitter can be related directly to distance between them and used to measure that distance.

*Programs such as GPS (global positioning system) and PLSR (position location reporting system) are now being configured to employ multiple spread spectrum range measurements (multilateration) for position location.

A common type of echo range measurement method is that used in radar systems. Many such systems simply transmit a pulse of RF energy and then wait for the return of a portion of the energy due to its being reflected from objects in the signal path. The radar marks time from the instant of the pulse transmission until its return. The time required for the signal to return is a function of the two-way range to the reflecting object, and because the signal propagation rate is known, the range is easily derived.

Any signal used is subject to the same distance/time relations. An unmodulated carrier, for instance, is delayed precisely the same amount as a spread spectrum signal traveling the same distance. The spread spectrum signal has an advantage, however, in that its phase is easily resolvable. A direct-sequence signal and a simple CW carrier may be likened to a pair of yardsticks, one with feet and inches marked on it and the other blank. With the marked yardstick we can easily resolve distance within about $\frac{1}{8}$ in., but the unmarked yardstick, though just as precise in length, would be of little use for resolving to better than perhaps 1 ft. The marks on a direct sequence signal, of course, are the code sequence modulation, and the basic resolution is one code chip; the higher the chip rate, the better the measurement capability.

Frequency hopping systems do not normally possess high-resolution properties, but this is simply because their hopping rates are not high enough to act as the fine marks on our yardstick. There is no reason, other than that of being less able to construct high-hopping-rate synthesizers, for frequency hopping ranging capability to be less than that of direct sequence.

Let us discuss the operation of spread spectrum ranging systems. In direct sequence ranging code modulation and demodulation are performed in exactly the same manner as in a system used for communications; that is, the code employed phase shift modulates the carrier. (For ranging, whether biphase or quadriphase modulation is used is usually of little consequence. We assume biphase modulation.) The baseband or information channel is unaffected by the ranging operation and may be employed concurrently. It is possible, in fact, to enhance operation by employing a baseband channel in a ranging system.

Figure 8.1 illustrates two simple direct sequence ranging systems. One we call simple, or transmit-then-receive. The other is called duplex, because it transmits and receives simultaneously. In both, the transmitter sends a pseudonoise code-modulated signal. Considering the duplex method, the signal at F_1 is simply translated to a different center frequency F_2 (preserving the code modulation) and retransmitted. This signal then reaches the first transmitter's site with a time delay corresponding to the two-way signal propagation delay between the two units (see Figure 8.2).

A receiver, located at the f_1 transmitter location, then synchronizes to

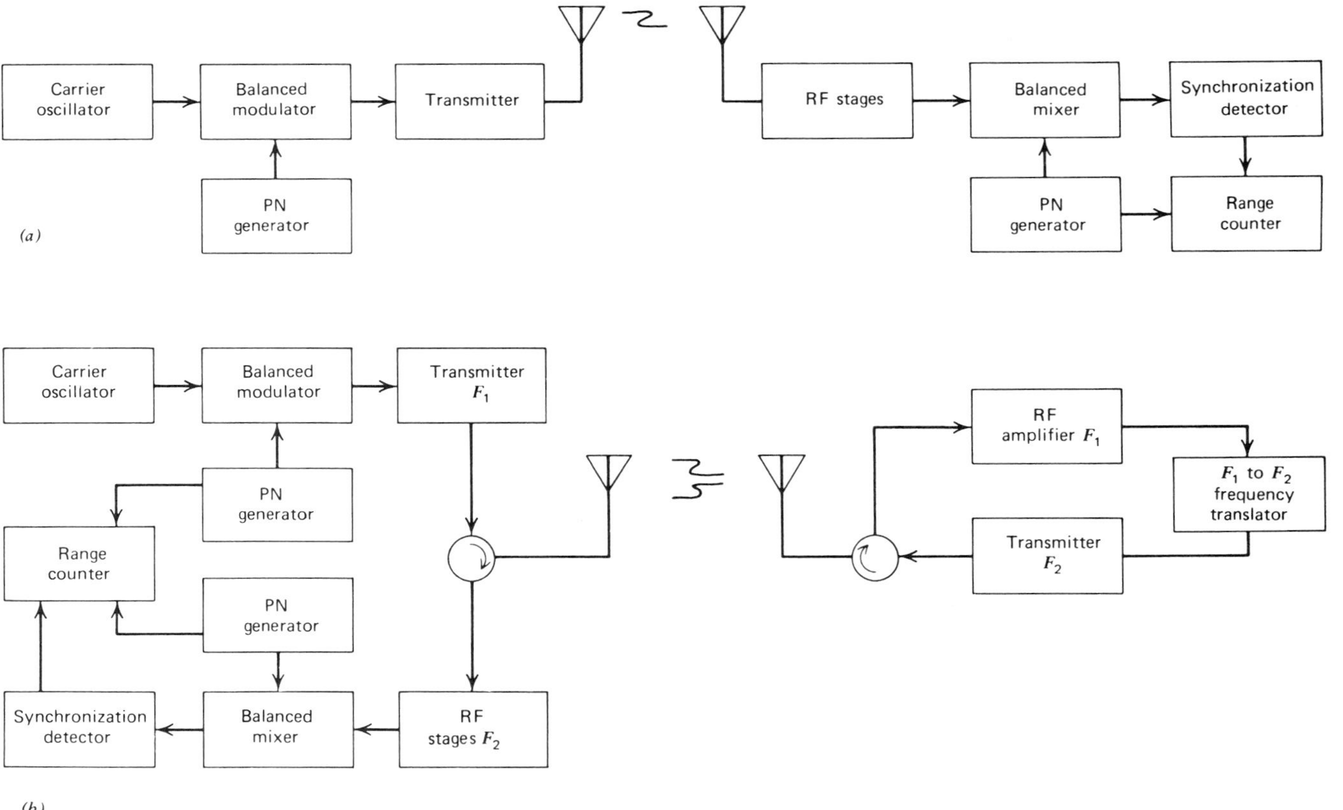

Figure 8.1 Simple direct sequence ranging systems: (*a*) transmit–receive ranging system; (*b*) duplex ranging system.

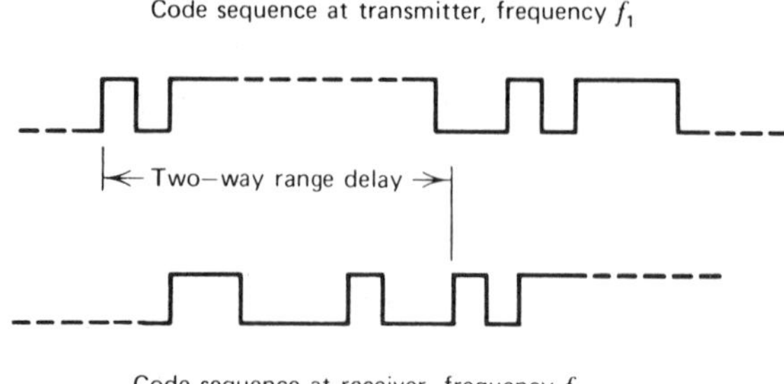

Figure 8.2 Code sequence comparison at transmitter and receiver.

the return signal, and on measuring the number of chips of code delay between the signals being transmitted and received can determine uniquely the range from itself to the repeating station. This type of system requires that the receiving subsystem employ a code sequence generator that is independent of the one used to modulate the transmitter, because the code the receiver must use is by definition out of phase with the transmitted code. The repeating terminal, on the other hand, is reduced in its complexity.

In the simple time-sharing ranging systems transmissions are exactly the same as in the duplex systems, except that transmitters and receivers at the two cooperating terminals do not operate simultaneously and, therefore, only one operating frequency is necessary. First, the terminal initiating the range measurement transmits a code sequence for a period long enough for the desired receiver to which a range measurement is to be made to synchronize itself. It is sufficient to say that the code sequences employed are long enough to support unambiguous range measurement over the distances in question and that an initial synchronization method is provided. The important point is that the receiver matches its internal code reference to the code modulation on the signal it is receiving and that its internal reference is then offset from the internal reference at the transmitter by the distance (propagation time) between them. The interrogator's transmitter, then, is generating the same code sequence as the responder's receiver but is some number of chips ahead of it.

Once initial acquisition has occurred the transmitter (or interrogator) instructs the receiving system to respond to its transmission by switching to the transmit mode at the reception of a given signal. The interrrogating unit then sends a command to the responder to start transmitting, while it goes from the transmission to receiving mode. Both the interrogator and the responder continue to generate the same code without interruption or resetting throughout the range measurement. The responder (the original receiver) is locked to a coded signal, which was offset from the interrogator's transmitted signal by the amount of the range delay, and transmits by using that same code reference to reply. The interrogator must delay its code phase to match the twice-range-delayed responder's signal. The amount of code phase change necessary is again equal to twice the propagation delay from the interrogator (who is measuring the range and initiates the measurement) to the responder. In both cases the range measurement is made by counting chips of offset or fractions of chips and is therefore a discrete measurement, inherently accurate within a 1/2 chip error. In practice measurements are commonly made to within a fraction of a chip period. The highest-resolution spread spectrum systems known can measure range to aproximately 1/1000th of a chip period.

Tone Ranging

A technique for range measurement using multiple, coherent tones that modulate a single carrier is also practical. Figure 8.3 shows the phase relationship for these tones. All tones are started at a single point in time (t_0). At a point distant from the point of origin, but within a range delay less than the peroid of the lowest tone, a receiver that demodulates and compares the phase of the tones can measure range. This is because the three tones have a unique phase relationship that depends on the position of the receiver. Range resolution is limited by the receiver's ability to measure the tone's phases. Thus it is limited in range by signal-to-noise ratio.

The point in discussing tone ranging is that it could be applied to a frequency hopping system, where the tone's separation from the band center would correspond to a particular hop frequency. This capability would depend on having coherence between all of the hopping frequencies used for ranging. Such a method has not been implemented because of the need for coherence, but does offer possible promise for future frequency hopping systems to bring their range resolution more in line with the direct sequence systems.

Figure 8.3 Tone-ranging technique.

Sources of Range Error

Several sources of range error must be considered when employing a spread spectrum ranging system:

Doppler-induced error

Clock-rate offset

Clock-rate error

Bit-count error

The effects of Doppler-induced error, clock-rate offset (due to an error in choice of chip rate), or clock-rate error (due to drift) are very similar, if not the same; that is, whenever the clock rate for a ranging code is offset from the design rate, the number of chips per mile for which the system is calibrated will also be offset. In Chapter 3 it was shown that the clock rate necessary for 1 nautical mile range resolution is 161,875 cps, and for 1 statute mile resolution is 186,333 cps. If a 1% range error is allowable, these chip rates could also be allowed tolerances of $\pm$ 1619 and $\pm$ 1863 cps, respectively.

Doppler-induced errors accumulate for the code clock just as they do at the carrier frequency, that is, at a 1.102 Hz/ MHz·/ Mach rate. Thus for a 10 Mcps clock rate in a system traveling at Mach 2 (twice the speed of sound) the clock rate would be offset by 22.04 cps. Because a 10 Mcps clock rate would allow a basic resolution capability of approximately 100 ft, $(1 \times 10^7)^{-1} = 1 \times 10^{-7}$ sec (which corresponds approximately to a 100 ft range delay) and each chip represents an error to 100 ft, a possible 2204 ft error exists for every second in which the clocks in two terminals are allowed to accumulate respective differences. If the clocks are locked (one tracking the other), however, the accumulation of range error is greatly reduced.

Consider the two ranging methods we have discussed. In the duplex-ranging method measurements are made at one terminal by counting the number of chips difference between the transmitter's code generator and the local receiver's code generator. This measurement is made while the receiver is locked to the return signal, and the receiver's code rate is shifted by twice the one way Doppler shift (see Figure 8.4). Because the measurement occurs while the two codes are locked together, there is little opportunity for accumulation of offset errors. The only error is due to the difference in received code rate compared with the intended code rate. For our 10 Mcps system with 2204 cps (one-way) rate shift the error would be only $4408/10^6 \times 100 = 0.044\%$, or 0.44 mile in 1000. On the other hand, in the simplex system, in which a ranging unit sends an interrogating signal, then switches to the receive mode for the responder's reply, the system is

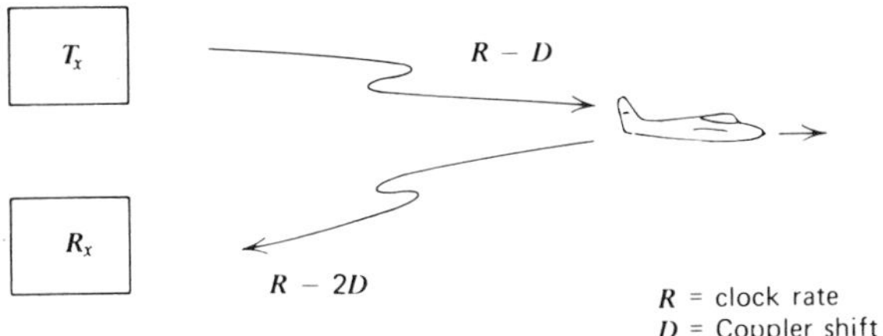

Figure 8.4 Twice-Doppler code-rate-shift accumulation for two-way propagation. (Relative motion closing produces positive frequency shifts.)

out of lock while the range count is being accumulated; that is, the range count actually consists of a count of the number of chips that must be searched to bring the interrogator into synchronization with the responder's transmission. Depending on the number of chips that must be counted, the search process can be quite time consuming—on the order of many seconds. During this time the responder's code clock can drift with respect to interrogator's clock and accumulate range error.

Range count error which accumulates when the two terminal clocks are operating independently accumulates as a chip error; that is, a one-chip cumulative shift during the search time counts as a one-chip error in range. Therefore, if a Doppler clock-rate shift is caused by relative terminal motion during a range measurement in the simplex (transmit-then-receive) ranging systems, range error accumulates at the Doppler-offset rate during the entire time that the interrogator is searching.

One method that could overcome this Doppler and/or clock drift accumulation is to design the responding system in such a way that it shifts its code clock an amount equal to that of the Doppler shift on the interrogating signal it receives but opposite in sense (see Figure 8.5). In other words, if the clock received from the range interrogator is offset by 10 cps in the high direction (signifying that the ranging units are closing on one another), the responder would reply with a code rate offset by 10 cps on the low side. Thus the signal arriving at the interrogator, where the range measurement is made, would be at the nominal code rate and would

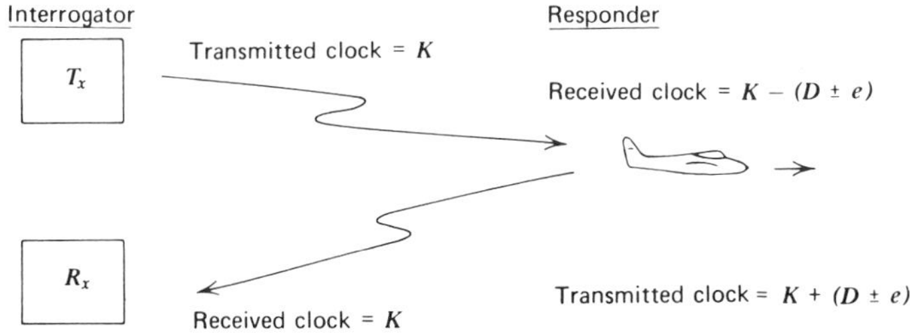

Figure 8.5 Clock-compensation method for range Doppler correction.

not contribute a cumulative error. (Or course, there is no way of telling how much of the offset in the interrogator's clock is due to Doppler shift versus the amount due to simple clock drift.)

One source of error would remain: a possible error in generating the transmitted offset clock which can occur in the responding system. This error is a function of the accuracy of the responder's clock and would not be resolvable by either interrogator or responder. This returns us to the basic need in direct sequence ranging systems—that of maintaining the most accurate possible clock frequency source.

Frequency Hopping Range Measurement

Range measurements can be made with frequency hopping systems just as with direct sequence systems, though they are not often done because of the general lack of resolution available with frequency hopping systems. Remember that frequency hopping systems use code rates in the kilochip to hundred kilochip (per second) range, whereas direct sequence codes start at a low of about 1 Mcps.* This gives the direct sequence-modulated system at least a 10-to-1 resolution advantage in most applications. For this reason frequency hopping is not specifically discussed here as a technique to aid in navigation. It should be realized, however, that the same methods can apply for frequency hopping as for direct sequence systems.

More on the Selection of a Clock Rate†

We have stated that the factors affecting the accuracy of a spread spectrum ranging system are chiefly clock rate, clock drift, and Doppler shift of the clock rate (ignoring possible errors in the digital counting and readout circuits, which should be negligible).

At least two methods of attack are possible in choosing clock rate. In one the clock rate is chosen such that the number of chips per mile is an integer, whereas the other chooses a convenient frequency and corrects for errors.

Assume that we wish to design a ranging unit capable of measuring range within 0.1 nautical mile (1 nm = 1852 m). Therefore, we want to count 10 chips/mile of propagation delay; 0.1 mile corresponds to a chip rate with a wavelength of 185.2 m, or a repetition rate of

$$\frac{c}{R} = \frac{2.997925 \times 10^8}{1.852 \times 10^2} = 1.618750 \text{ Mcps}$$

*The best current frequency hopping systems could not resolve range closer than approximately ± 10,000 ft, based on their chip times.

†See also Chapter 3

where c = speed of light,
R = resolution.

A system that uses a code sequence chip rate of 1.618750 Mcps would therefore have a propagation delay of 10 chips/mile, and a receiver 10 miles away would see a signal delayed by exactly 100 chips; that is, the signal appearing at the receiver was transmitted at a time that, on the transmitter's terms, was 100 chips earlier. Therefore, the receiver's sequence generator must be exactly 100 chips behind the transmitter to achieve maximum correlation and between 99 and 101 chips behind the transmitter to be within the bounds of correlation; 1.618753 MHz is certainly not a standard frequency, but an accurate crystal stabilized source at this frequency is readily obtainable. (As an aside it should be noted that multiples of 0.161875 MHz may be used as clock rates to gain whatever resolution is desired; that is, 1/50th mile would require a clock frequency of $50 \times 0.161875 = 8.093750$ MHz. Of course, the number used for c, the speed of light, is that for light traveling under vacuum, and in any other medium some error will be seen. This is not significant, however, as long as all clocks operate at the same rate; 0.1% error limit due to common frequency offset would allow the clock rate to change up to 1618 bps.) It can be recalled that in both the simple- and duplex-ranging systems the number of code chips of offset due to range is actually equivalent to twice the range being measured. Therefore, in these and similar systems, clock rate required to give a desired resolution is halved.

Once the nominal clock rate is chosen both transmit and receive systems are expected to operate at the chosen frequency (within their error limits) except when a receiving unit is in the "search" mode, looking for correlation with an incoming signal. In the interrogate–respond ranging system this search procedure occurs twice: first, at initial transmission and, second, when the responding unit replies.

Direction of search (forward or backward in code phase) is usually not important to a spread spectrum system, but in transmit–receive ranging systems it is necessary that the unit making the range measurement (the interrogator) search backward; that is, the interrogator's receive clock should run at a rate lower than the responder's transmit clock, so that its code sequence would be slipping backward with respect to the incoming signal. This is important in that the responder's code sequence is retarded with respect to the interrogator's sequence generator at the time of the transmit-to-receive switch, and the interrogator's code sequence must also be retarded a number of chips equal to twice the propagation time in order to recorrelate.

The period during which the interrogating unit is searching for recorrelation with the responding transmitter is the most vulnerable for error entry. During this time any relative clock drift or Doppler shift is accumulated as a range error by the chip search counter in the interrogator.

On the basis of clocks operating near their design frequency, range error for an integral frequency clock system may be expressed by

$$\text{error (in chips)} = (|K_I - K_R| \pm 1.7\ V_R K_R \times 10^{-3})\ T_S$$

where K_I = interrogator clock rate (cps),
K_R = responder clock rate (cps),
V_R = relative velocity (nmph),
T_S = recorrelation search time (sec).

This expresses the accumulated errors due to an interrogator's counting at an assumed clock rate K_I, whereas a responder's reply actually is at a clock rate K_R, offset by the Doppler-frequency shift referred to the responder's clock signal. This is multiplied by search time, for the longer the interrogator's search for correlation, the longer the time that errors are accumulated. Doppler-frequency errors may reduce or add to any errors due to clock offset, depending on the direction of errors and the relative velocity. In some systems it is not practical to use a clock that gives an integral number of clock pulses per mile of delay. These systems may correct for the errors generated by an offset clock frequency. To give an example, with a clock frequency of exactly 2.0 Mcps, the nearest integral multiple of 0.161875 is 1.942500 of Mcps. Therefore, use of a 2.0-Mcps clock would produce an overall error of approximately 3% if 12 chips/mile were used as a measure of range, but because this is a constant and precisely known error rate, it can be corrected readily.

Selecting the Ranging Code

Code sequences used in ranging should be chosen on the basis of consideration for their auto- and cross-correlation properties and in most ranging systems, for sufficient length. Interference between different sequences can nullify selective addressing, or minor correlation in a correct sequence because of false recognition of synchronization can cause an erroneous measurement.

A ranging code used by a system that has no secondary resolving capability must be long enough not to repeat over the maximum distance measured; that is, to measure a range of 1000 miles with a code giving 50 chip/mile resolution would require a code length of at least 50,000 chips. Otherwise, the code sequence would repeat, and the interrogator could recognize synchronization at more than one range. At the 8.093750 Mcps rate required to give 50 chip/mile resolution, the repetition period of the 56,535 ($2^{16} - 1$) chip code required would be 65,355/8,093,750 = 8.1 msec and the bandwidth of the correlation detector should be about 40 Hz. This narrow bandwidth would restrict the search rate severely unless it were possible to accept correlation averaging across less than the complete code length (with the attendant risk of falsely recognizing synchronization at the wrong position of the correct code or even crosscorrelating with a different code).

If a 40 Hz recognition bandwidth is used, searching out a range at the maximum distance could require almost 20 min. Obviously, then, in a simple ranging system high resolution and long distance must be traded off against measurement time. The major problem in direct sequence ranging for aircraft and maneuverable spacecraft has been the requirement for measurement at long range in a reasonably short time. Reduction of ranging time is limited by the maximum search rate a receiving unit is capable of achieving and the length of the code to be searched. Maximum search rate, in turn, is limited by the recongition time of the receiver's correlation detection circuits. (The receiver must be able to recognize acquisition and stop the search process before the point of code synchronization is passed.) To complete the circle the bandwidth of the correlation detectors must be commensurate with the autocorrelation requirements of the codes used.

What techniques are available to reduce ranging time while maintaining accuracy? Much work has been performed in the area of develping special "acquirable" codes which have the required length for long-range measurement but which also have synchronization properties that permit a range to be searched out without traversing the entire code length. Jet Propulsion Laboratories has had great success with code sequences made up of three components codes assembled in such a way that the overall ranging sequence has a length that is the product of the three component code lengths.

Hybrid-Ranging System

Another ranging technique that solves the problem of fast range measurement at long distances is called a hybrid approach, since more than one kind of modulation is used to measure range. In this technique a *PN* code, a few thousand chips long, and a digital "range message" (whose bit rate is the repetition rate of the *PN* sequence) simultaneously modulate the transmitter. The *PN* code does not determine maximum range and therefore its length can be short to reduce time. The only bound on reducing code length is that system cross-correlation requirements must be met.

The ambiguity in range caused by using a short *PN* code is resolved by modulating with a "slow" digital pattern whose length is such that its repetition period is longer than the propagation delay of the longest range to be measured. No separate search process is needed to synchronize this and is therefore synchronized by the initial search process. All that is needed is a measurement of the relative phase of received range messages, which is as simple as counting the number of *PN* code repetitions between range message markers.

The range message itself is the simplest kind of digital signal, made up of a number of square waves whose half-wave period is the code repetition rate, inverted in phase after a time greater than the maximum expected propagation delay. This phase inversion is the marker that resolves the range ambiguity. Modulation is such that the *PN* code biphase modulates the transmitter carrier at its basic rate, permitting high resolution. The range message, however, is sent as a baseband signal at a comparatively low bit rate (the repetition rate of the code), frequency shift keying the same carrier that is modulated by the *PN* code. Composite transmitted output then is a $[(\sin x)/x]^2$ spectrum which shifts frequency at the range message rate.

Figure 8.6 shows simplified block diagrams of a hybrid ranging system and illustrates the timing of transmissions in interrogating and responding systems. The interrogator and responder units may actually be identical; range may be measured by either one. One range cycle is described as follows (see timing diagrams):

1. The interrogating unit transmits a *PN* modulated signal for a period long enough to ensure that the intended responder has synchronized.
2. The range message is sent consisting of a series of one–zero bits (a square wave) at the *PN* code repetition rate, and is followed by a phase inversion of the message. This information FSK modulates the *PN* modulated carrier.
3. At the *PN* code repetition marker following the phase inversion several things occur; (a) the responder starts generating an identical range message to that sent by the interrogator (i.e., a given number of square wave periods, an inversion, repeating), (b) the interrogator switches to receive, and (c) the responder switches to transmit. (Search processing for resynchronization is delayed in the interrogator for a period long enough to exceed the *T-R* switching time. In this way switching time in the two units is removed as a source of range error.)
4. The responder now transmits a *PN* (only) modulated signal for a synchronization period and the interogator counts the number of chips searched to regain synchronization, storing the chip count.
5. Responder then FSK-modulates the *PN*-modulated signal with the range message. (Remember that it has been generating the range message since the *T-R* switch.)
6. Interrogator receives the range message and counts the number of message bits between the first phase inversion in the received message and the next occurring in its own range message (which has also been generated since the *T-R* switch).
7. Each message bit counted is weighted, added to the *PN* chip count already entered, and read out as a range.

Selection of parameters used in hybrid systems is illustrated by the following example.

Suppose (again) that ranging is to be performed at distances up to 1000 miles and with resolution of 0.1 miles but only 1 sec is allowed for the measurement. We wish to complete a range measurement in 1 sec or less, which means that we must

transmit a code sequence to allow the intended responder to synchronize,

add FSK information to the transmitted signal, telling the responder to transmit,

switch to receive and count the number of bits searched over,

receive and accumulate the range message information.

Because the two synchronization search processes consume most of the overall measurement time, let us allow 0.45 sec for each synchronization. A 2047-chip *PN* code would have to be searched at a rate of approximately 4600 chip/sec to ensure recognition in this time.

A 5600-cps search rate would require synchronization recognition in less than 1/2300 sec (because the correlation function is 2 chips wide). Therefore, the recognition bandwidth should be 805 Hz or greater. Because

$$\frac{\text{code length}}{\text{chip rate}} = \text{repetition period} = \frac{0.35}{\text{recognition BW}}$$

the chip rate should be

$$\frac{1}{\tau} = \frac{\mathrm{BW}_R \times L}{0.35}$$

where τ = chip period,
BW_R = recognition bandwidth,
L = code length,
$0.35 = \mathrm{BW} \times T_R$ for the synchronization recognition circuit.

Code chip rate, using a 2047-chip code, searching 4600 chips/sec, and using a synchronization recognition circuit 806 Hz wide, would be at least

$$\frac{805 \times 2047}{0.35} = 4.7 \text{ Mcps}$$

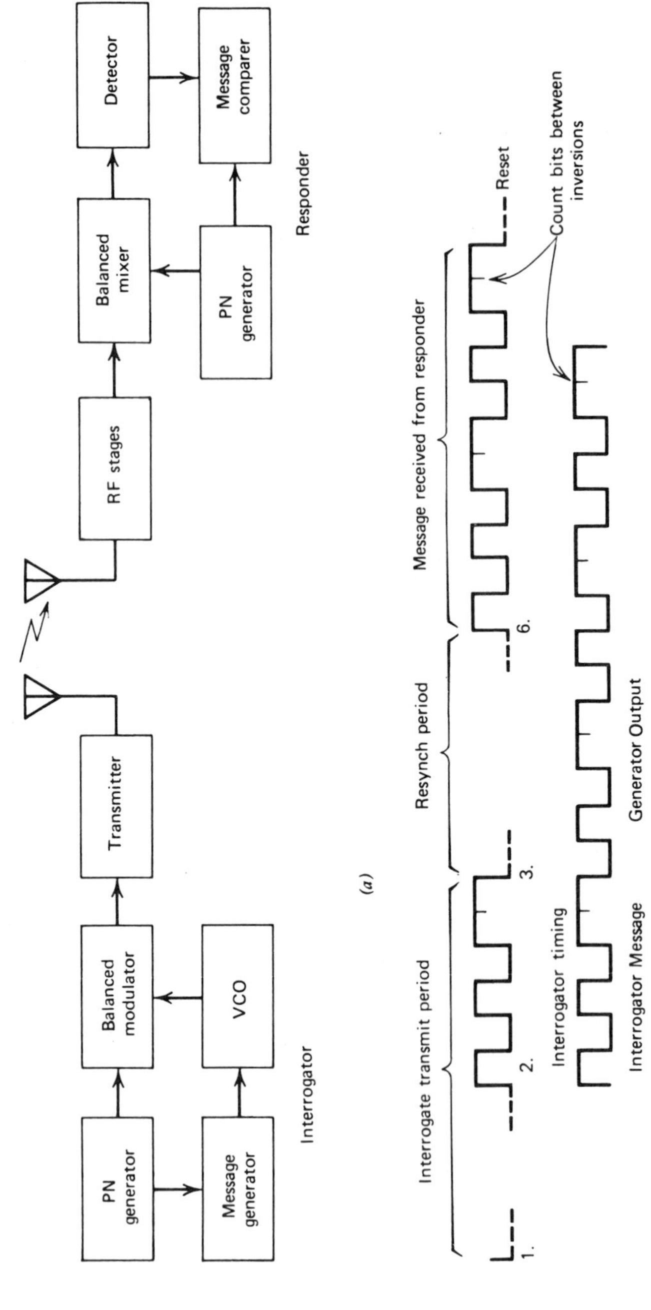
PN generator
Message generator
Balanced modulator
VCO
Transmitter
Interrogator
RF stages
Balanced mixer
PN generator
Detector
Message comparer
Responder
(a)
Interrogate transmit period
Resynch period
Message received from responder
1.
2.
3.
6.
Reset
Count bits between inversions
Interrogator timing
Interrogator Message
Generator Output

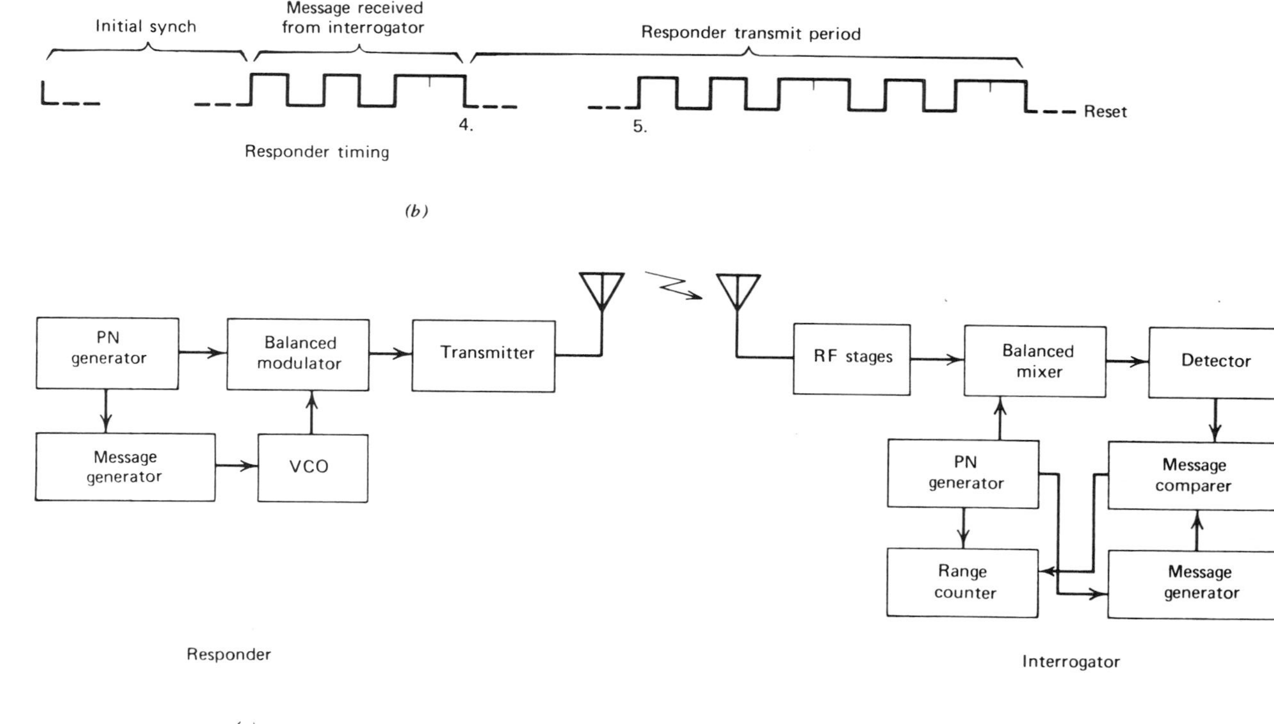

Figure 8.6 Hybrid ranging system; (*a*) interrogate mode; (*b*) message timing diagram; (*c*) respond mode.

The nearest multiple of 0.161875 is 30 × 0.161875, or 4.856250 Mcps. (Resolution capability would be 1/60th of a mile.)

Next we construct the range message, which must have a period greater than the greatest propagation delay expected (more than 6.3 msec). The repetition period of the code sequence is 0.42 msec; therefore, the code repeats almost 15 times in this range.

Making use of the property of maximal code sequence generators that the all-one's vector is repeated once and only once in each code sequence, we use this repetition to generate a word made up of 15 or more bits and a bit period equal to the *PN* sequence repetition rate. This word is the "range message" and consists of a square wave (one-zero sequence) 15 bits long, a phase reversal, 15 bits, and a phase reversal.

In measurements that use the range message, the phase reversal marks the beginning of a sequence and each bit is weighted 2047/30 = 68.23 miles. Therefore, for each bit of offset in the range message the interrogator adds 68.23 miles to the range. Overall range measurement time, then, using the parameters described, would be approximately 950 msec. Range resolution would be 1/60th of a mile, and maximum range would be 1091 miles. (Adding bits in the range message would increase maximum range capability by 68.23 mi./bit, and measurement time 0.42 msec/bit. Length of the range message does not affect resolution.)

Accuracy of the hybrid approach is excellent in that resynchronization search time is short and clock drift is minimized. Any error in counting range message bit difference, however, will produce errors in range at the weighted bit rate.

8.2 DIRECTION FINDING

For navigation purposes ranging capability is, of course, an extremely valuable tool. But what of the companion problem—knowing the direction to the system supplying the range response? Let us examine some of the ways in which a spread spectrum system can derive a bearing on, as well as range to, a desired responder, so that a user can home in on or navigate with respect to another spread spectrum system.

Special Antennas

Antennas designed for direction are available for all types of use. Here we consider only one example to describe typical use with a spread spectrum receiver. Figure 8.7 shows an antenna pattern typical of a direction finder designed to operate with a separate receiver. (In fact, it is a composite of two antennas with their cardioid-shaped patterns superimposed.) This type of direction finder operates by switching alternately between a pair of opposed antennas, detecting (usually by envelope detection) the signal,

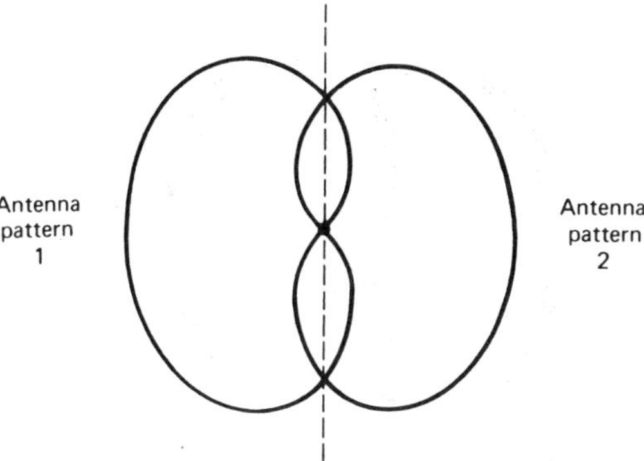

Figure 8.7 Dual cardioid pattern set for typical direction-finding antenna.

and comparing the levels from the two antennas. The entire antenna assembly is then rotated until the transmitting station being received lies on the centerline between the two antenna patterns. At this point the signal received by both antennas is equal because both lie at the same angle with respect to the received signal.

This type of direction-finding approach works as well with spread spectrum systems as with any other because the antenna switching operation effectively amplitude-modulates any incoming signal, whether spread spectrum or CW. Because envelope detection might not function quite so well on a spread spectrum signal as on others, the AGC signal internal to the spread spectrum receiver is normally taken as an antenna control signal. Two difficulties exist with this two-pattern switched-antenna direction finder:

1. Poor sensitivity, for the antenna directs its lowest response point toward the desired receiver.
2. Two separate null points exist so that the antenna could point toward the transmitter or in the exact opposite direction. This ambiguity is usually corrected by spoiling the composite pattern; the sharply concave section then exists on only one side.

When code sequence rates are high enough, a different two-antenna approach is feasible; code phase is used as the criterion for measurement rather than comparative amplitude. For code-phase direction finding the antennas used are conventional communications types. No special provision is necessary. A block diagram of a two-antenna, code phase difference measuring direction finder is shown in Figure 8.8. It is effectively two receivers both locked to the same received signal; a measurement of phase difference in their local code reference generators determines direction. Alternatives to this configuration are possible and are discussed shortly, but first let us describe the operation of this one.

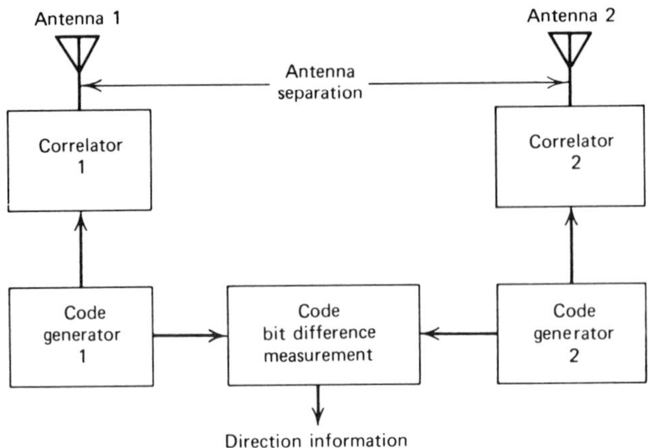

Figure 8.8 Simple two-receiver direct sequence direction finder.

A signal appearing at antenna 1 causes the code sequence at antenna 1 to be synchronized with the signal's code modulation. The same signal at antenna 2, which is some known distance s away, also causes the second code sequence to be synchronized. The two code sequences are then synchronized with the incoming signal but not with each other, because the signal phase is different at each antenna location. The amount of difference is dependent on the antenna separation and the angle of signal approach; that is, if the antennas are separated by a distance corresponding to two code chips and the phase difference measured is two code chips, the signal can approach only along a common axis passing through both antennas. (Figure 8.9 further clarifies this point.)

When the signal approaches the two antennas from an angle θ, however, the angle measurement is somewhat more complicated.* Because the distance between antennas is accurately known and the distance to each antenna from the signal source is measured, the three sides of an extremely elongated triangle are defined and so are its included angles. Therefore, the angular direction to the transmitter is known to be one of two nearly supplementary angles θ and (assuming that the distance to the transmitter is many times greater than the antenna to antenna distance). The direction finder must then determine which of the two angles is correct. (See Figure 8.10 for illustration.) This angle determination can be made by observing which of the distances measured is shortest, for the shortest distance must be opposite the desired angle.

A second angle estimation method simply makes use of the observation that the difference in distance measurement at the two antennas,

*Except, of course, when $\theta = 90°$, for which condition the range measured at both antennas would be identical.

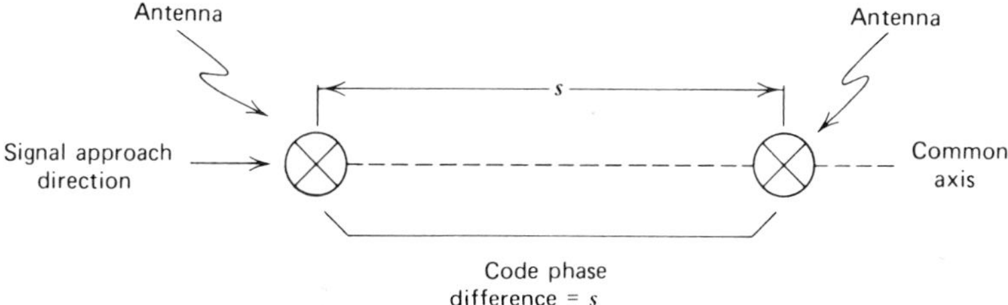

Figure 8.9 Signal approach along a common two-antenna axis ($\theta = 0$).

compared with the distance between the antennas, varies approximately as the cosine of the angle to the signal source (for small antenna separation compared with the distances measured).

Referring to Figure 8.10*b*, in which the angle being measured is the angle enclosed by lines *a* and *c*,

$$\text{distance } a_1 = \frac{c_1}{\cos \theta_1}$$

$$\text{distance } a_2 = \frac{c_2}{\cos \theta_2}$$

$$a_1 - a_2 = \frac{c_1}{\cos \theta_1} - \frac{c_2}{\cos \theta_2}$$

but

$$\theta_1 \sim \theta_2 = \theta$$

(see the geometry of Figure 8.9) and

$$a_1 - a_2 \sim \frac{c_1 - c_2}{\cos \theta}$$

$$\cos \theta = \frac{c_1 - c_2}{a_1 - a_2}$$

The distance $c_1 - c_2$ is just the antenna separation, and $a_1 - a_2$ is the difference measured by the two antennas. This angle estimate is simple to make and requires little hardware, compared with the previous described method, which requires an estimate based on knowing the three sides of a triangle.

In either of these two-antenna methods an ambiguity exists in that the receiving system cannot tell if a transmitter is in the top or bottom

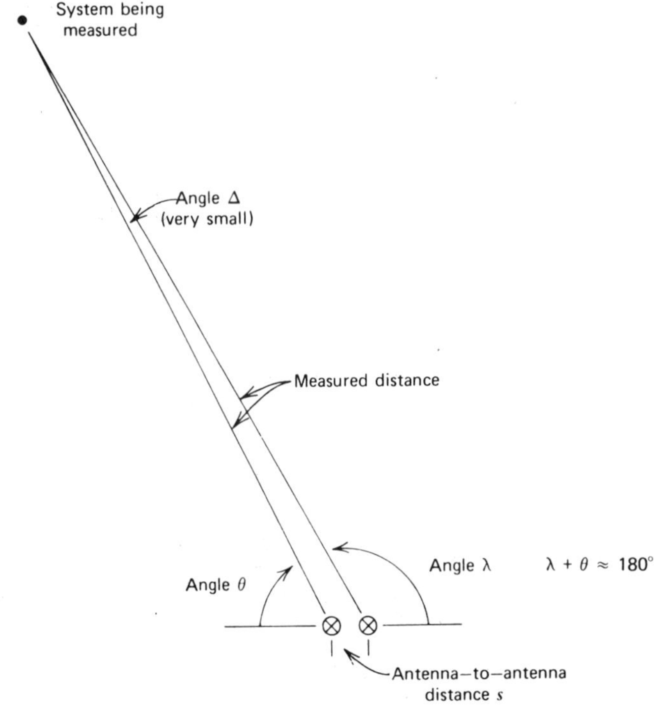

Figure 8.10*a* Illustration of measurement geometry, $s \ll D$.

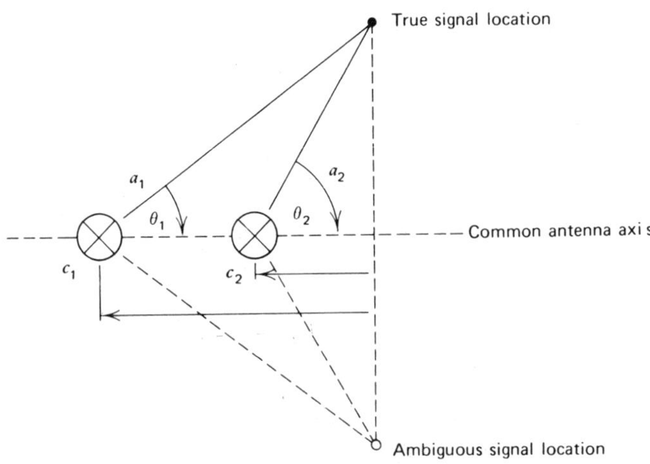

Figure 8.10*b* Illustration of distance difference measurement geometry, showing possible ambiguity in transmitter location.

hemisphere, as separated by the common antenna axis. When such ambiguity can cause difficulty, three antennas are needed to resolve it. The third antenna is required to show only whether the signal time of arrival has been advanced or retarded with respect to the signal arriving at the other two antennas.

The accuracy of direction measurement when a two- or three-antenna array is used is determined by two parameters:

1. The distance between antennas used for time of arrival comparisons (assuming the distance is accurately known, of course.)
2. The degree to which distance can be resolved as received by the processors attached to the two (or more) antennas.

Many direct-sequence systems can resolve range to within one-tenth clock chip, and some can do as well as one-thousandth chip. Consider a system which has one-tenth chip capability, say a system whose code rate is 10 Mcps. For 10 chip antenna separation, the distance between antennas would be approximately 1000 ft—not compatible with aircraft mounting. Yet, the maximum distance difference measurable would be ± 10 chips, for 200 resolvable increments (100 increments per 90° quadrant). Direction measurements would be accurate to approximately 0.9°. Of course, a higher code rate or larger antenna separation would improve the measurement capability. What about the receiving system which cannot afford the 100 ft separation we mentioned above? Many aircraft installations cannot affort a 10 ft antenna separation, to say nothing of 1000 ft, and 10 ft separation allows for only 1 chip, even when code rate is 100 Mcps. It is true that a 100 Mcps code, coupled with 1/100th chip resolution, and 10 ft antenna separation could give 0.9° resolvability, but such systems are in general too advanced to consider for airborne use, at least until 100 Mcps processing systems are proven.

A possible alternative is to make a pair of measurements, separated in time, so that on a moving aircraft or other vehicle the effect is that of having a pair of antennas separated by the distance desired. In this way a 1 Mcps code could be used, with an effective 10,000 ft separation to give the same resolution as higher rate codes with fixed antennas. Also, a single pair of antennas would be able to resolve the hemispherical direction ambiguity, previously discussed, because measurements would be made at two locations.

The same principles are applicable to frequency hopping systems but with the general penalties imposed by the slower frequency hopping code rates. Spread spectrum signals can, in general, perform as navigational signals, comparing very favorably with other signal formats. Although acting as distance or direction-finding signals, none of the interference rejection or other properties is impaired. Therefore, spread spectrum signals offer perhaps a new dimension to navigational signals. Viewed in another way, spread spectrum communicating signals offer a new dimension for communications systems—ranging and direction-finding capability.

It is worthy of note that a spread spectrum system can provide range measurements over any range at which the receiver will synchronize. Furthermore, the range resolution is never worse than a small fraction of a code chip period. This exceeds the performance of many other ranging methods, since they tend to degrade with distance because of their greater dependence on a good signal to noise ratio to measure signal rise time or phase.

PROBLEMS

1. What basic range resolution may be expected from a direct sequence system with a 10 Mcps code rate?
2. What sequence generator length is the minimum that should be employed for unambiguous ranging at 350 mi if the code employed is at 20 Mcps?
3. If exactly 25 chip / mi range resolution is desired, what code chip rate is required?
4. Two range measurements are 191.03 and 192.05 mi, made by systems whose antennas are separated by 0.1 mi. What is the direction to the system being ranged on with respect to the common line between receiving antennas?
5. What directional resolution can be expected from a receiving system for 50 Mcps coded signals if its effective antenna separation is 0.5 m?
6. A 100 khop / sec frequency hopping system uses its hopping interval to measure range. Give an estimate of its range resolution capability.
7. In a direct sequence system what is the effect on a range measurement of a carrier frequency offset; and a clock rate offset?
8. What effect has distance inaccuracy on angle measurement when distance measurement is used for direction finding?

9

APPLICATIONS OF SPREAD SPECTRUM METHODS

This chapter is intended to show a few of the ways in which spread spectrum techniques have been and are being used in the hope that those shown will spur new ideas and further applications. It is also hoped that the reader will be encouraged by the knowledge that spread spectrum methods have provided practical, real-world solutions to some tough, real-world problems.

As far as general communications are concerned, applications of spread spectrum techniques are relatively obvious. Certainly the usual voice, data, and other common information transfer uses are practical, though in some cases restrictions like frequency allocation might prevent the use of a wideband signal. Some areas in which spread spectrum methods (or at least system components) have already been put to good use are given in Table 9.1, and in the remainder of this chapter we describe a few of the specific ways in which these methods are employed.

Space Systems

In space systems, especially communications satellites, which may be stationary and therefore continuously accessible to interference, spread spectum methods have proved effective. In general, however, satellites do not employ processing on board because the added complexity is significant and the number of possible users may be sharply reduced.

In most applications of satellites for communications there is little advantage to placing a spread spectrum processor on board. A simple repeating satellite is more economical and does almost as well. In those systems in which nonprocessing satellite repeaters are installed all spread spectrum modulation and demodulation must be performed on the ground, as in Figure 9.1.

Table 9.1

Area	Application	Type of System	Primary User
Space systems	Communications, ranging, multiple accessing, jamming protection	Direct sequence, frequency hopping	Military
Avionics systems	Communications, position location, discrete addressing, low detectability, jamming protection, radar, collision avoidance	Direct sequence, frequency hopping, hybrid direct sequence/frequency hopping, chirp	Military
Test systems and equipment	Bit-error detection, noninterfering in-service testing, signal correlation, privacy, (pseudo-) random selection, number generation	Direct sequence	Commercial
Signal protection (nonmilitary)	Interference rejection, voice and data scrambling	Direct sequence	Commercial
Position location	Collision avoidance, direction finding, ranging	Direct sequence	Many

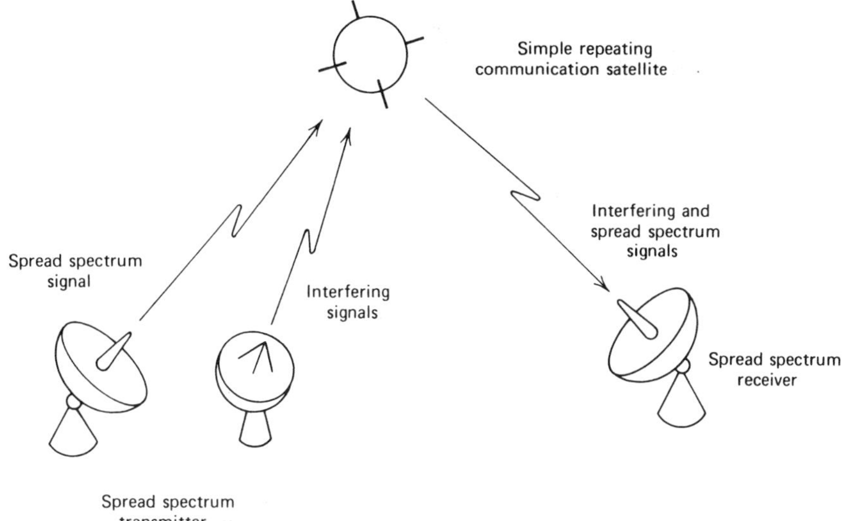

Figure 9.1 Satellite relay configuration for a simple repeater.

Basic advantages and disadvantages to on-board processing are:

The complexity of a spread spectrum demodulator is avoided on the spacecraft without on-board processing.

The satellite is forced to retransmit an uplink interference signal that reduces the spacecraft transmitter power available for sending the desired signal without on-board processing.

Every would-be receiver must have a spread spectrum demodulator. Large numbers of downlink users are therefore precluded, in general, because of the cost of the equipment with on-board processing.

Typical spread spectrum equipments for operation through a repeating type satellite are:

1. URC-55
   (a) 9.6 mcps direct sequence (QPSK) modulation.
   (b) 1, 2, or 4 analog voice or TTY signals; one digital data channel, 75 bps to 4.9 kbps; one FSK orderwire.
   (c) Long code synch acquisition with real time start.
   (d) Time transfer capability.
2. URC-61
   (a) Compatible with URC-55.
   (b) 9.6 Mcps direct sequence (QPSK) modulation.

- (c) One analog voice or TTY signal; one digital data channel, 75 bps to 4.9 kbps.
- (d) One FSK orderwire.
- (e) Long code acquisition plus frequency-hopping acquisition capability.
- (f) Range measurement.
- (g) Time transfer capability.

3. USC-28
   - (a) 40 Mcps direct sequence (QPSK) modulation.
   - (b) Multiple multiplexed users; total data rate up to 5.0 Mbps; net control orderwire.
   - (c) Software implementation of communication and fault location functions.
   - (d) Multiple acquisition modes.
   - (e) Doppler and range offset compensation.
   - (f) Time transfer capability.

Figures 9.2, 9.3, and 9.4 show three of these systems. The most complex is the USC-28, which spreads baseband-multiplexed user signals (total data rate $\leqslant$ 5.0 Mbps) to a bandwidth of 35 MHz using quadriphase modulation. This system also automatically adjusts its baseband rate as a function of channel performance, which allows optimization of system capacity under varying conditions. Its processing gain varies between approximately

$$10 \log \frac{3.5 \times 10^7}{75} = 56.7 \text{ dB}$$

and

$$10 \log \frac{3.5 \times 10^7}{5.0 \times 10^6} = 8.5 \text{ dB}$$

as the necessity arises. Many of the demodulation and tracking functions of the USC-28 are implemented in software.

A commercial use of spread spectrum modulation has been accomplished in a direct sequence system developed by Equatorial Communications, which operates in the 3.9–4.2 GHz satellite band. This system is used for transmission of business information. It allows use of satellite receiving antennas as small as 2 ft in diameter.

For comparison purposes Figure 9.5 shows the configuration of a spread spectrum system that employs on-board processing. What are the advantages of on-board processing? When on-board processing is used, the satellite receives the signal on the uplink, demodulates it, and uses the

Figure 9.2 URC-55 modem (courtesy Magnavox Co.).

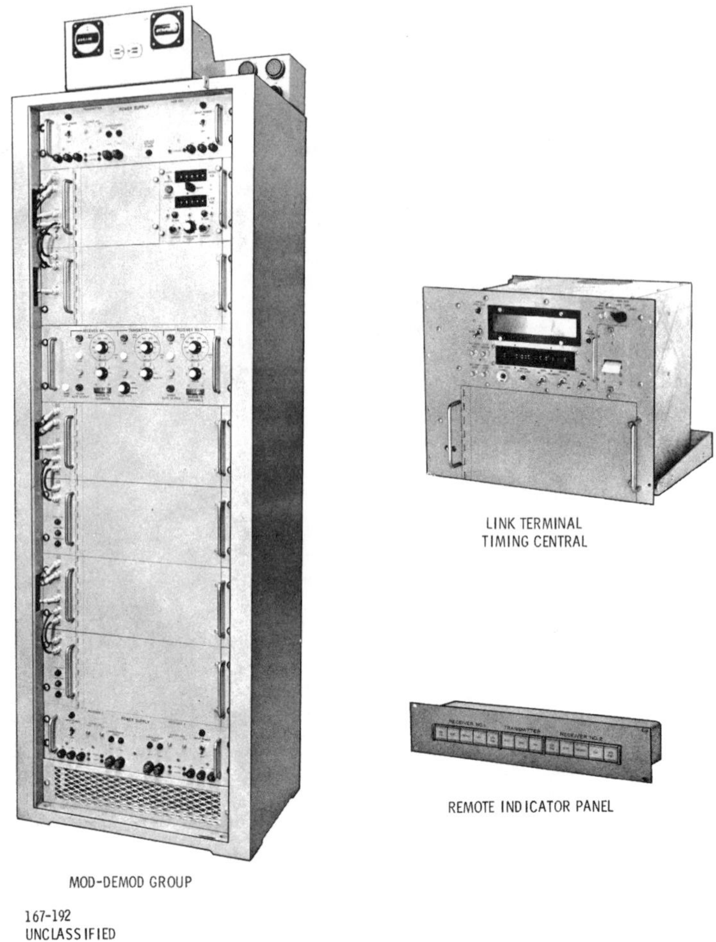

Figure 9.3 URC-61 modem (courtesy Magnavox Co.).

information to modulate the downlink. In the demodulation process any uplink interference is suppressed by an amount approximately equal to receiver process gain. The downlink transmitter is not required to retransmit signals other than the desired signal. Therefore, when on-board spread spectrum processing is employed, effective transmitter power requirements can be reduced because it is not necessary to retransmit an interfering signal.

Of course, the downlink can also be spread spectrum modulated, but this is not so useful as on-board processing with conventional (AM, FM, etc.) modulation on the downlink. Little advantage is gained in having a spread spectrum downlink because the distances involved are such that a spread spectrum signal could not protect the receiver from interference

Figure 9.4 USC-28 modem (courtesy Magnavox Co.).

that is close by when the satellite downlink transmitter is often 20,000 miles away. A great deal is gained, however, by transmitting a spread spectrum uplink signal to take advantage of the interference protection afforded while sending the information to the receivers on an easily accessed downlink.

One space-oriented system that makes use of spread spectrum

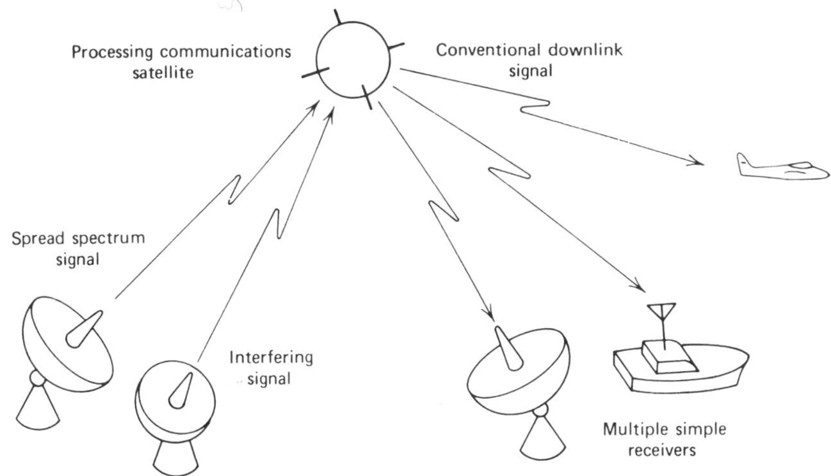

Figure 9.5 Satellite relay configuration using on-board processing with conventional narrowband downlink.

modulation to advantage is the TDRSS (Tracking and Data Relay Satellite System) system. This system is one of the few that make use of spread spectrum modulation for some reason other than antijamming. The reason for its use in TDRSS is to reduce the power density of the signals as seen at the earth's surface. Signal levels from space are limited to a density of 145 dBm in a 4 kHz bandwidth, in a square meter aperture, and TDRSS accomplishes this by transmitting a 3.08 Mcps direct-sequence modulated signal.

Two separate spread spectrum signals are transmitted simultaneously in TDRSS. One employs a data modulated 1023-chip Gold code biphase modulating carrier. The second signal is also biphase modulated, but with a linear maximal code whose length is $256 \times 1023 - 261{,}888$, truncated from a normal $2^{18} - 1$ code length (262,143).

Figure 9.6 shows a block diagram of the TDRSS modulation and demodulation subsystems. One of the most interesting characteristics of the TDRSS spread spectrum subsystem is its synch acquisition technique. The two transmitted signals are sent simultaneously at the same frequency, with the long-code ($2^{18} - 1$) modulated carrier orthogonal to the Gold code modulated carrier, and attenuated some 10 dB below it. The receiver then treats the two signals as independent biphase signals. (Alternately, they could be considered to be an unbalanced quadriphase signal. See Figure 9.7.)

Code synch acquisition is greatly accelerated in TDRSS by synchronizing the start and end of the long code with the repetition period of the 1023 chip Gold code. Figure 9.8 illustrates this relationship. In the transmitter, the short and long codes are started at the same time, and the long code is truncated and started again after 256 repetitions of the short code. In the receiver, short-code synch is acquired by a search through the total uncertainty (1023 chips maximum). Then, the receiver resets its long-code generator each time the short code repeats and integrates over the short-code period. If it recognizes long-code synch between short-code repetitions, it continues the long code without resetting. If long-code synch has not been recognized by the time the short code repeats, it resets and tries again. In this way, the long-code synch occurs in not more than 256 repetitions of the short code, which is approximately $1023/(3 \times 10^6) \times 256 = 88$ msec maximum.

Another unusual feature of the TDRSS system is that the data transmitted are added to the short code and not to the long code (see Figure 9.6). The long code is used primarily for range ambiguity resolution.

The TDRSS system will provide relay of signals from orbiting spacecraft, where 100% coverage is provided by only two TDRSS spacecraft for all users whose orbits are at least 1200 km above the earth's surface. Figure 9.9 illustrates the system's geometry.

Another satellite-type spread spectrum system is the Global Posi-

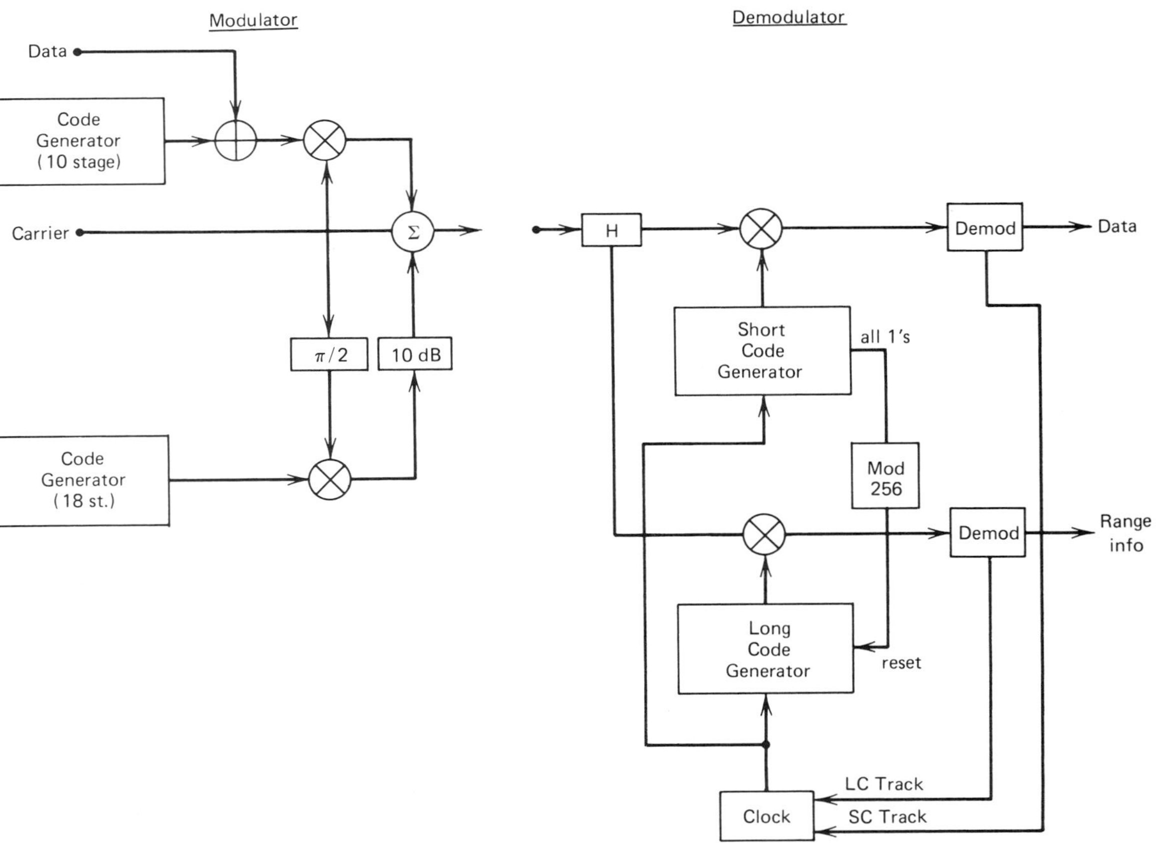

Figure 9.6 TDRSS modulation and demodulation subsystems.

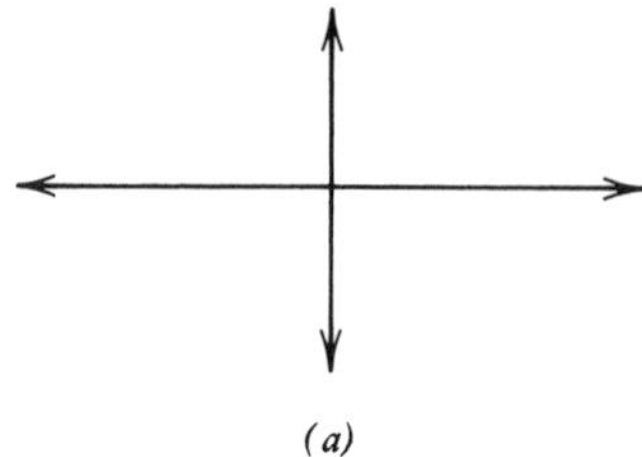

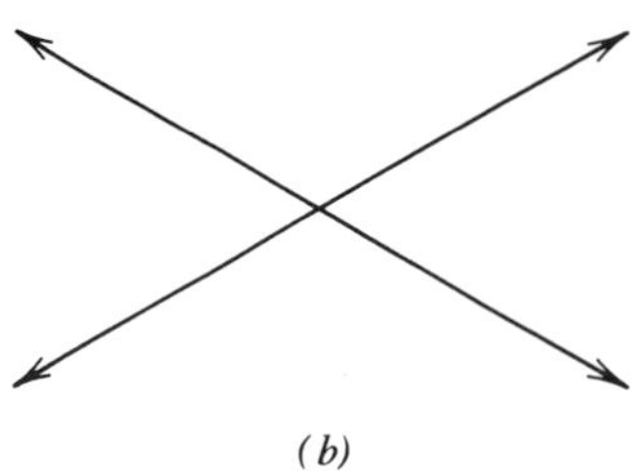

Figure 9.7 TDRSS-type signals as (*a*) independent biphase signals, and (*b*) composite unbalanced quadriphase.

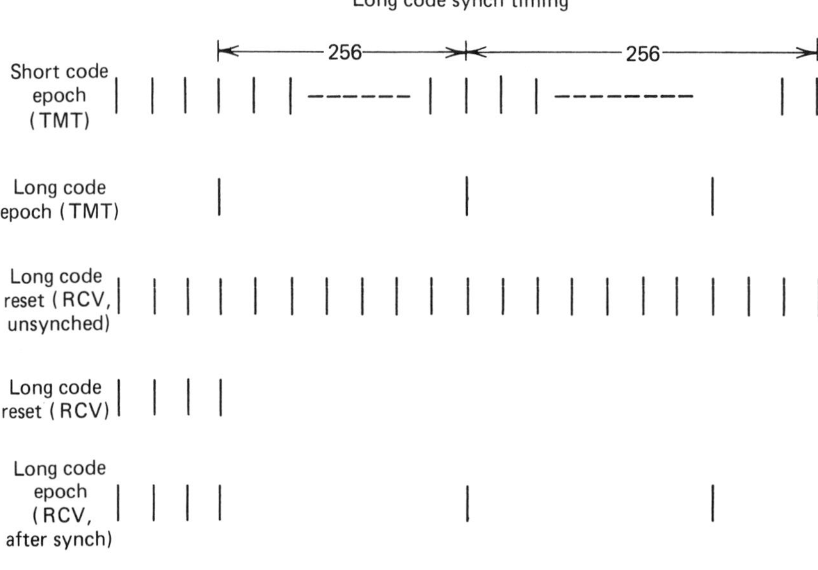

Figure 9.8 TDRSS long-code synchronization timing.

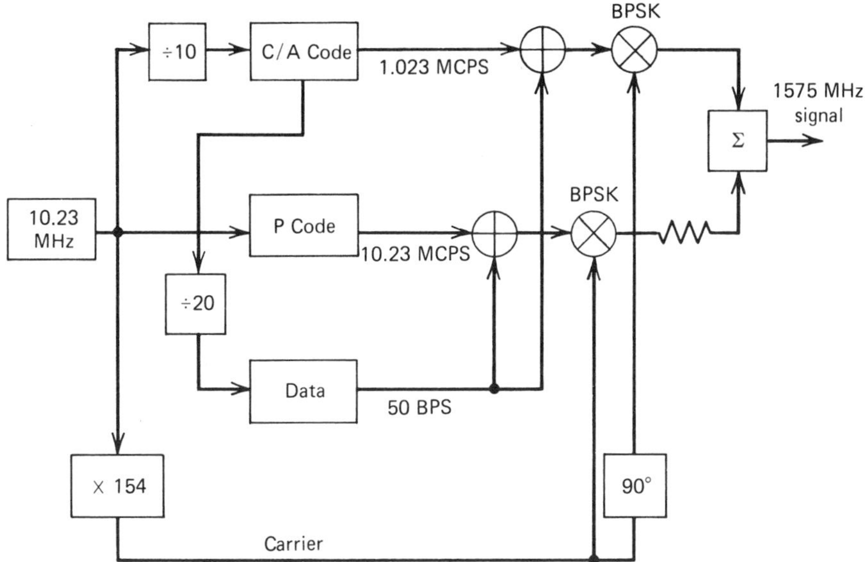

Figure 9.9 GPS modulator configuration. (L_1 signal)

tioning System, or GPS. This system will make use of a constellation of 18 satellites, each in a 12-hr orbit, with up to six satellites in view at any time over a large portion of the earth's surface. Each of the satellites transmits three signals:

1. A long code (one week repetition period) biphase modulated 1227 MHz signal.
2. A long code (same code) biphase modulated signal at 1575 MHz.
3. A short code (1023 chip Gold code) biphase modulated signal on an orthogonal carrier at 1575 MHz.

The 1023 chip Gold code (called a *C/A* code) transmitted by each satellite is different from all others. All satellites use the same long code (called a P code), however. Thus, initial acquisition is accomplished using the 1023 chip short code, which is modulated with 50 bps data that provide information to allow the receiver to synchronize to the long code. Figure 9.9 shows the basic modulator configuration used in GPs. Note that the GPS modulator adds the same 50 bps modulation to both long and short codes, so either long code or short code receivers can demodulate the same information.

The GPS signal structure is similar in some ways to the TDRSS signal structure; that is, both employ a pair of quadrature biphase modulated carriers, with different codes modulating them. The similarity does not end there, as TDRSS and GPS both employ 1023 chip Gold codes for initial acquisition. The long codes used in the two systems are quite different, however. A comparison of the two can be summarized as:

GPS	TDRSS
Short code (clear access or C/A):	Short code:
1.023 Mcps	3.08 Mcps
1023 chips long	1023 chips long
Gold codes	Gold codes
period, 1.0 msec	period, 0.33 msec
Long code (protected or P):	Long code:
10.23 Mcps	3.08 Mcps
one week long	$256 \times 1023 = 261{,}888$ chips long
6.19×10^{12} chips	linear maximal code
mildly nonlinear code	period, 85 msec
Modulation:	Modulation:
1575 MHz	I carrier, short code
I carrier, C/A code	Q carrier, long code
Q carrier, P code (suppressed 3 dB)	(suppressed 10 dB)
1227 MHz	Several frequencies
P code only	

The idea behind GPS is to transmit spread spectrum signals that allow a range measurement, from a known satellite location (the transmitter on board the satellite tells the receiver the satellite location). Then, with a knowledge of the transmitter location and the distance to the satellite, the receiver can locate itself on a sphere whose radius is the distance measured. After receiving signals and making range measurements on other satellites, the receiver can calculate its position based on the intersection of several spheres.

It is interesting to note that position location can be accomplished using either the short or long code modulated GPS signals. Figure 9.10 illustrates one typical GPS receiver configuration. Receivers are of two basic designs, one of which synchronizes to up to four satellite signals at the same time (satellite signals are differentiated by having different Gold codes). The alternate, lower-cost GPS receiver configuration uses only one receiver, which must sequentially acquire and make range measurements on at least three satellite signals. The second receiver approach is slower and less accurate than the first, but offers more commercial applicability because of less complexity and cost.

In other space applications spread spectrum signals have been employed for ranging at interplanetary distances.

Figure 9.10 Four-channel GPS receiver (courtesy Magnavox).

Avionics Systems

Avionics systems can employ spread spectrum methods in any of the ways available to satellite and other systems. Ranging, direction finding, discrete addressing, and communicating are all within the realm of practicality. The major unique consideration for avionics applications is the large and unpredictable range of Doppler frequency offsets. A pair of supersonic aircraft present Doppler shifts corresponding to a worst case of as much as relative Mach 10 (plus or minus, because the two may be flying directly toward or away from one another) or any amount up to that approximate maximum. The worst part is that, in general, the amount and sign of the shift is not predictable and any equipment used in such an application must be capable of covering the entire range of frequency uncertainty. Although satellite systems can generate higher Doppler shifts, both position and velocity of satellite systems are normally well defined and frequency uncertainties encountered may be compensated out as a demodulator design consideration.

Avionics systems have received heavy spread spectrum emphasis. Several systems have been developed for avionics use, or are currently under development. Some of these are:

ARC-50. The oldest of the avionics systems. First produced in the early 1960s. This is a direct-sequence modem for voice communications. Its code rate is approximately 5 Mcps and its modulation format is biphase. The ARC-50 is also capable of ranging, and reads on range to $\frac{1}{10}$ mile at ranges of up to 300 miles. Figure 9.11 illustrates the ARC-50 system, whose spread spectrum modem originally contained some 650

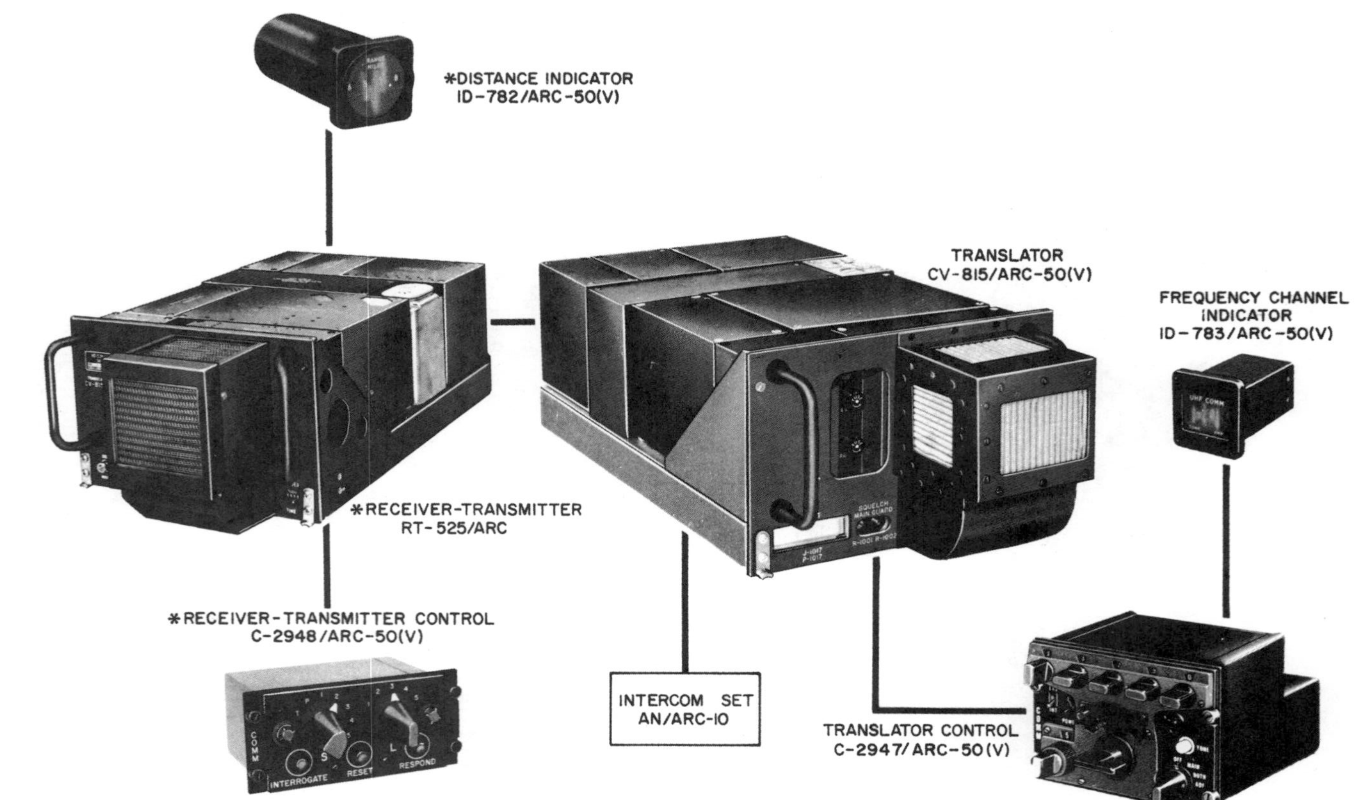

Figure 9.11 ARC-50 system (courtesy Magnavox Co.).

Figure 9.12 ARC-148 spread spectrum modem (part of ARC-148 system).

discrete transistors. The system is also capable of output power control over a wide range. The spread spectrum modem for a later version of the ARC-50 (the ARC-148) is shown in Figure 9.12 for comparison.

Have Quick. A frequency hopping UHF AM radio adapted from the ARC-164 radio. Have Quick will be the first widely used spread spectrum radio. The Have Quick configuration will be a standard for at least the next 10 years. (See Figure 9.13.)

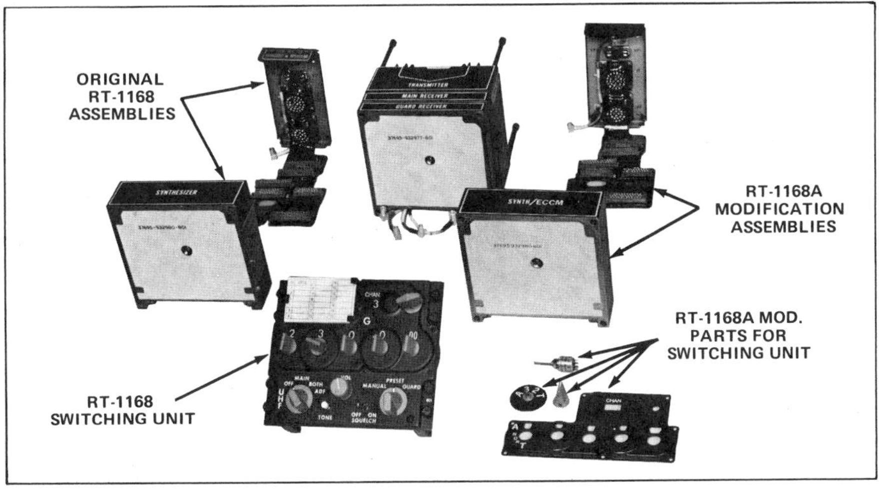

Figure 9.13 ARC-164 modification to Have Quick configuration.

JTIDS. Joint Tactical Information Distribution System. Although really only a data transmission system, JTIDS has been touted for every function in modern aircraft except radar. The system is the heir to many previous programs (previous names were PLRACTA, SEEK BUS, CNI, UCNI, MYSTIC LNIK, ITACS, and many others). Operating in the 960–1215 MHz band, JTIDS must provide compatibility with IFF and TACAN systems, which share the band.

The signal format of JTIDS employs frequency hopped direct sequence modulation; the direct sequence modulation is MSK at 5 Mcps. This direct sequence signal is then frequency hopped over a 52 frequency set, with separation of 3 MHz between frequency choices.

Three classes of JTIDS systems are under development. Their capabilities are summarized as follows:

Characteristic	Class 2	Class 1A	Class 1
Data rate, kbps			
transmit	80	160	280
receive	140	140	280
transmit and receive	160	160	300
Open message start rates, Hz	50/100	50/100	100/200
Number function channels	64	64	128
Number voice channels	2	4	10
Number conferencing voice channels	1 × 3	1 × 3	2 × 3
TACAN	Yes	Yes	N/A
Number metachannels	6	8	16
Quaternary function channel	2	2	4
Number of users	512	512	512
Antennas	2	1	2
RF power, watts	200	200/400* 1600†	200/400* 3200†
Range, n.mi.	300	300	300

*Military power mode.
†Burn-through power mode.

Note that the voice channel capability of JTIDS is very limited, even though an overall JTIDS network is designed to handle in excess of 96,000 data users. Figure 9.14 illustrates one of several JTIDS formats.

Figure 9.15 shows the three classes of JTIDS equipment as configured by the current equipment developers.

Another system currently under development, though not strictly for

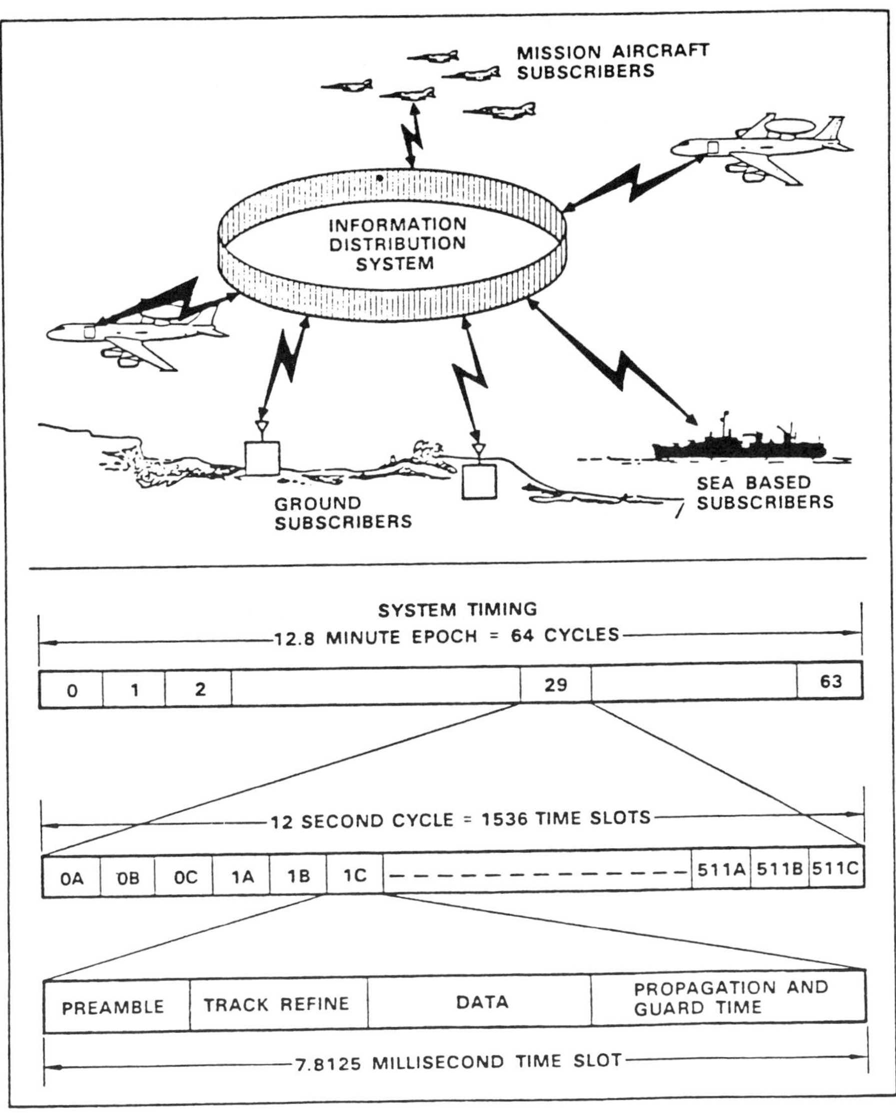

(*a*)

Figure 9.14 JTIDS format (*a*) Information distribution system;

avionics use, is SINCGARS (Single Channel Ground-Airborne Radio System). SINCGARS is a frequency hopping system that hops over the 30–88 MHz band, with 25 kHz frequency spacing. Though only two candidate systems are under contract, several frequency hopping alternates for the same band have already been developed. A summary of these 30–88 MHz band frequency hopping systems is:

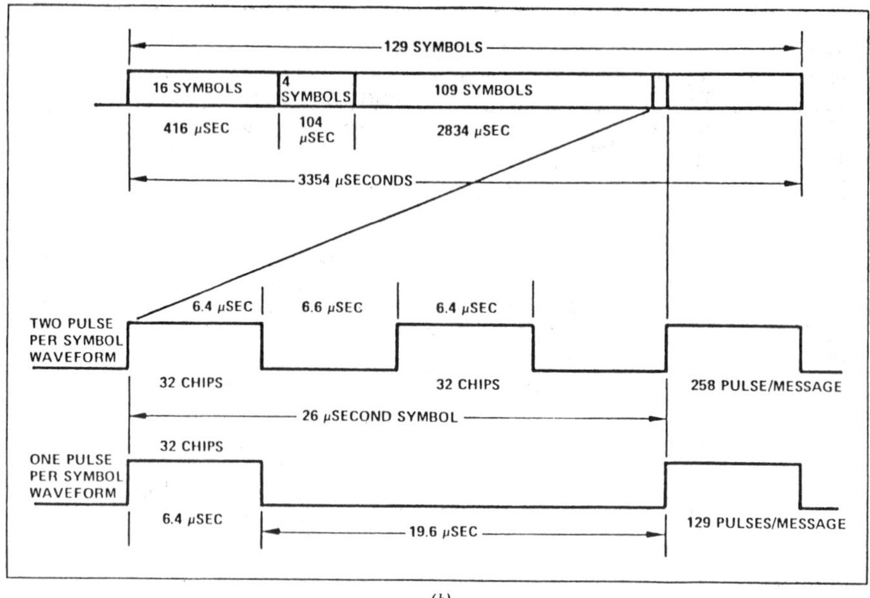

(*b*)

Figure 9.14 (*b*) waveform modulation.

Unit	Developer
SINCGARS	Cincinnati Electronics
SINCGARS	ITT
JAGUAR	Racal
SCIMITAR	Marconi
PRC-117	Harris
PRC-118	Tadiran
PRC-77	Rockwell

All of these units have 25 kHz frequency spacing, and are generally intended for voice or data. Their hopping rate is in the range of a few hundred hops per second. Some have the ability to hop over any of the 2320 channels available, and some hop over a set of 256 of the 2320 available.

Figure 9.16 shows the ITT SINCGARS radio.

Test Systems and Equipment

Spread spectrum techniques (or at least techniques common to spread spectrum systems) have recently become popular in test systems. References 216, 217, 235, and 237–240 describe a few of the spread spectrum techniques or subsystems. Transmission test sets[217,237,240] (actu-

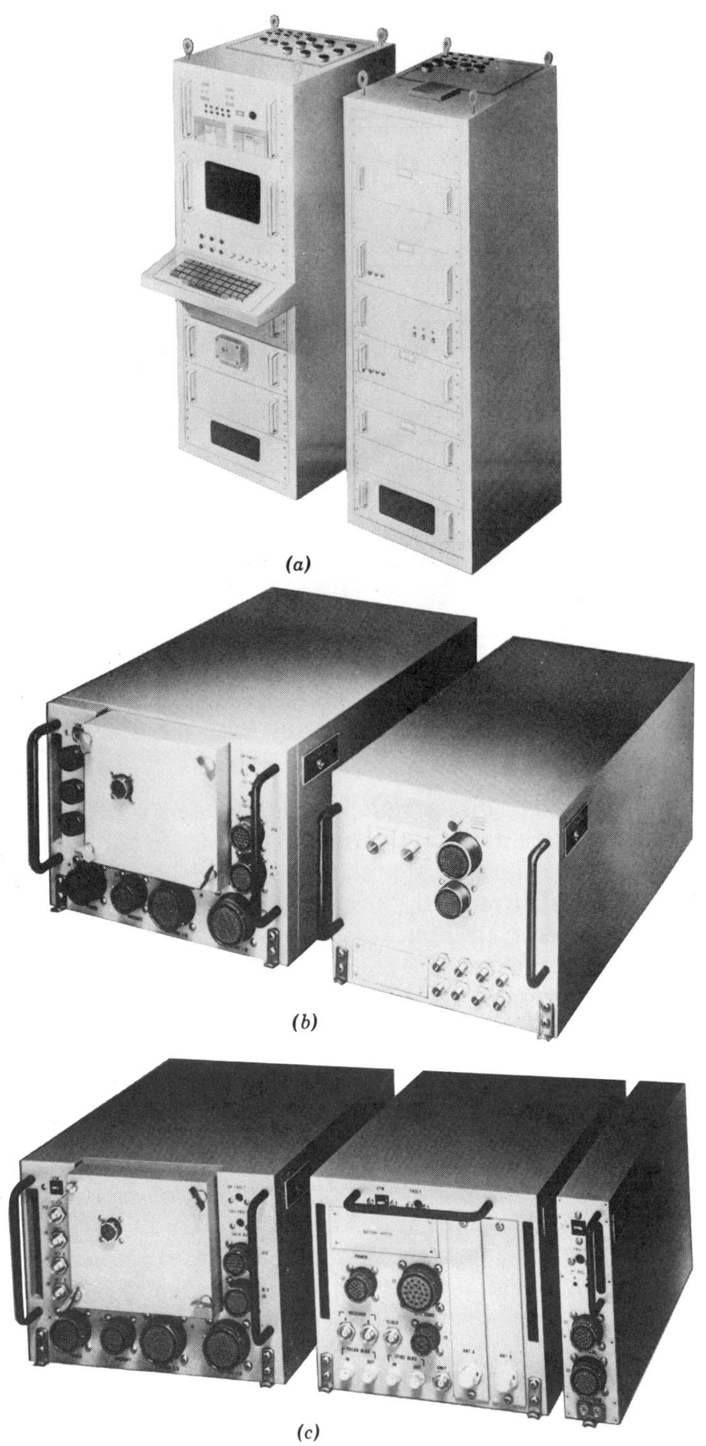

Figure 9.15 Three JTIDS configurations: (*a*) class 1; (*b*) class 1A; (*c*) class 2.

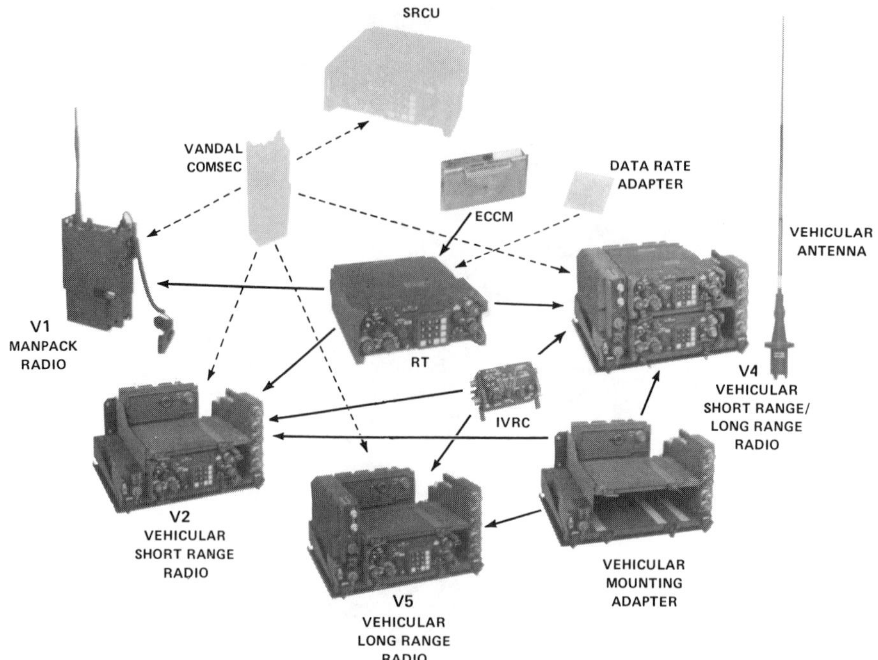

Figure 9.16 ITT SINCGARS radio.

ally almost complete direct sequence modulated transmitters and receivers) characterize data transmission systems. In such applications a pseudonoise code sequence modulates the system under test. The coded signal then passes through the transmission path to the test set, which locks on to the received code sequence and compares it with its local reference for differences. Any chip difference between the received and locally generated signals is recorded as an error.

Another type of transmission test set allows the transmission system to be under test at the same time it is used for normal operation. This application makes use of the low power density of direct sequence modulated signals by transmitting the direct sequence signal simultaneously with the desired information-bearing signal. Mutual interference is low because the wideband test signal can be transmitted well below the desired signal and the desired narrowband signal looks like just another noncoherent jamming signal to the spread spectrum receiver. Noise gating[239] and similar methods are related techniques. Figure 9.17 shows block diagrams of typical test setups employing direct sequence methods.

The pseudorandom codes used in spread spectrum systems have also been applied alone for "random" number generation, sources of "noise"

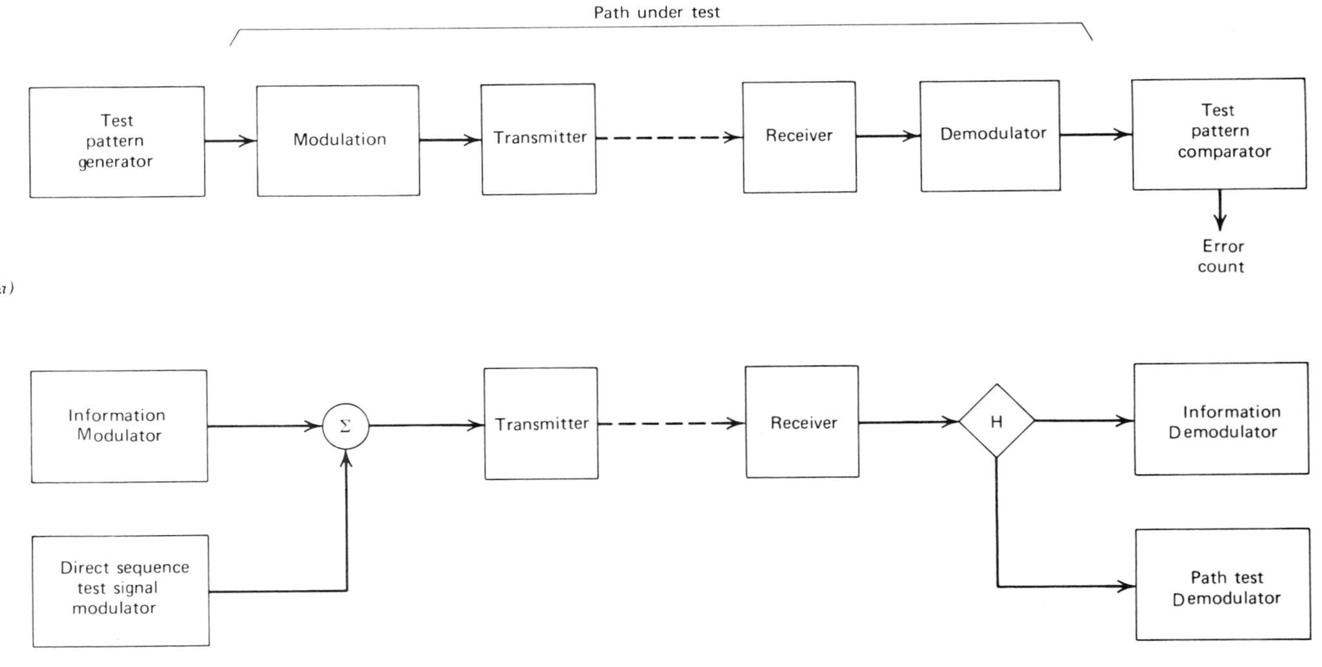

Figure 9.17 Typical transmission path-test setups by direct sequence techniques: (*a*) code pattern transmission test; (*b*) suppressed signal transmission test.

modulation, data scrambling, and in most other areas in which noise or random sequences can be put to use. Commercial code sequence generators[216] have been introduced for these uses.

Message Protection

Most of the newer spread spectrum systems provide for fully secure, crypto-type codes to be used for generating their signal spreading functions. Thus, the systems are capable of resisting code breakage and spoofing to high a level as required.

Position Location

The advantages of position-location methods are obvious, whether the system is a spread spectrum system or not. Collision avoidance, homing, ranging, surveying, navigation, and many others depend on position location. Spread spectrum methods are not always best for every application, but they are often capable of providing precision where other techniques fail. One spread spectrum system, for instance, is capable of

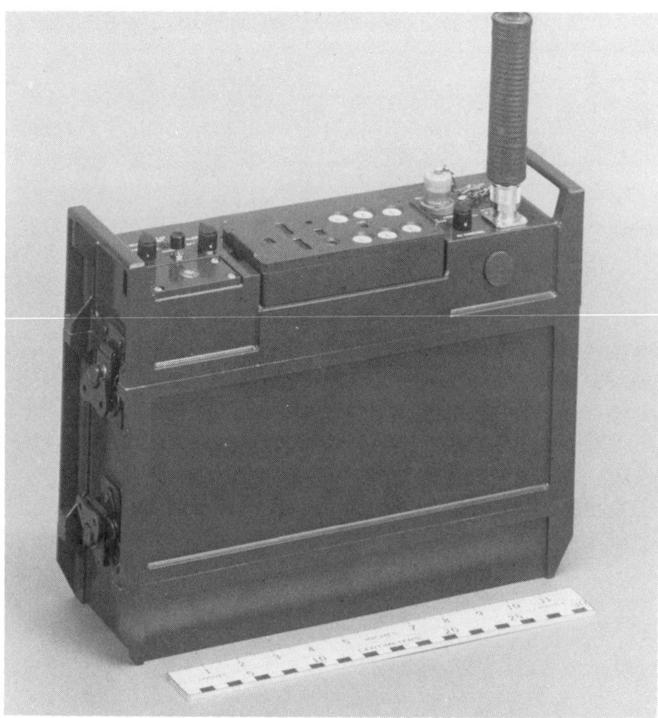

Figure 9.18 PLRS user unit (Courtesy Hughes Aircraft Co.).

resolution of range to 3 in. This capability is rivaled only by laser rangefinders, which cannot operate in inclement weather.

Spread spectrum methods certainly must be considered wherever position measurement is to be carried out.

Real Systems

We have already mentioned GPS, which is an important example of a spread spectrum position location system. Another spread spectrum position location system currently under development is PLRS (Position Location Reporting System). This system employs a format very similar to that of JTIDS, but operates in a different frequency band and with a different number of hopping channels. The number of hopping channels is eight, with direct sequence modulation at 5 Mcps using minimum shift keying. A PLRS user unit is illustrated in Figure 9.18.

A Final Word

It is hoped that this and the preceding chapters have encouraged those who have read them to employ spread spectrum methods wherever they are advantageous. As with any other technological area, spread spectrum systems cannot solve every problem, but they are a different approach toward some old problems. Those who read this book will undoubtedly conceive new applications and better techniques. It is hoped that what has been said here will aid in these conceptions.

A few of the systems developed in recent years have been described. These only represent the overall spread spectrum community, however, and show that such systems are practical. With practicality established, it is sure that spread spectrum systems will proliferate in the future.

APPENDIX 1
GLOSSARY OF TERMS

This glossary is intended to define the terms used in this book in ways that are meaningful here but do not necessarily carry the same impact as they do elsewhere. Also defined are a number of terms commonly abbreviated without explanation. This is done as much as for any other reason to ensure that what is said here does have some relation to the real world.

AGC. Automatic gain control.

AJ. Antijamming.

AJ Margin. Antijamming margin; the worst case jamming-to-signal ratio at which a receiver will deliver rated performance.

AM. Amplitude modulation.

Antijamming. Capability of withstanding jamming or interference.

Autocorrelation. Multiplication of a signal with a time-delayed replica of itself, with product integration.

$$\psi_{\text{auto}} = \int f(t)\, f(t - \tau)\, dt$$

Baseband. The basic information channel for a communication system. For voice systems approximately 3 kHz low-pass baseband bandwidths are usually required.

Biphase. A term used to signify two-phase ($\pm$ 90°) phase-shift keying.

Carrier. A term used to identify the center frequency of a signal, whether modulated or not.

Chip. A signal frequency output from a frequency hopping signal generator. Chip time signifies the time occupied by that single frequency. Also the period of a code clock.

Chirp. Term applied to designate pulsed-FM modulated signals.

Clock. The frequency or chip-rate reference used to set the rate of code generation in a spread spectrum system.

Codes. *Composite;* A code made up of two or more distinct codes, usually generated by modulo-2 addition. *Gold:* A particular type of composite code. Oriented to multiple access. *JPL:* A particular type of composite code. Oriented to ranging and rapid synch acquisition. *Linear:* A class of codes in which only linear operations are allowed for generation (e.g., for binary codes only modulo-2 addition or subtraction is allowed). *Maximal:* Codes whose length is $2^N - 1$, where N is the generator length. *Nonlinear:* A class of codes in which nonlinear operations are performed. All codes other than linear. These codes encompass the truly secure codes. *Nonmaximal:* All codes other than maximal. *Quasiorthogonal:* A term used to describe codes with good mutual cross-correlation properties. *Synchopated:* Composite codes generated with non-synchronous or nonsimultaneous chip timing.

Coherent. A synchronized, phase-matched condition between a receiver's reference and the desired signal.

Correlation. The degree of agreement between a pair of signals.

Correlator. That section of a spread spectrum system in which a received signal and the local reference are compared for agreement. The desired, synchronized signal is despread and undesired signals are spread.

Costas Demodulator. A special compound phase-lock loop intended for demodulation of suppressed carrier, double-sideband signals.

Cross-correlation. The degree of agreement between two unlike signals; that is,

$$\psi_{\text{cross}} = \int f(t)\, g(t - \tau)\, dt$$

CW. Abbreviation for continuous wave, signifying an unmodulated carrier signal.

Dehop. The process (dehopping) by which a frequency hopping receiver removes or remaps the spread spectrum frequency hopping modulation from an incoming signal into a single frequency.

Delay Lock. A form of clock frequency tracking loop which employs two correlators whose references are offset by some fraction of a chip. Used with direct sequence systems.

Direct Sequence. A form of modulation wherein a code sequence is used to directly modulate a carrier, usually by phase-shift keying.

Direct Spread. Same as direct sequence.

Direct Synthesis. A frequency synthesis method in which all output frequencies are developed by multiplication or division from one or more frequency sources. No phase-lock circuits are employed.

Dither. A form of clock frequency tracking loop in which the clock train supplied to the receiver's code reference is deliberately jittered a controlled amount—usually a fraction of a chip—generating an error signal for clock control.

DS. Direct sequence.

Feedback Tap. A point of feedback output used in a sequence generator for generating code sequences.

Frequency Hopper. A type of spread spectrum system in which the wideband signal is generated by hopping from one frequency to another over a large number of frequency choices. The frequencies used are chosen (pseudo-) randomly by a code similar to those used in direct sequence systems.

Frequency Shift Keying. A form of modulation similar to frequency hopping, except that normally two frequencies are employed and the choice of frequency is made on the basis of a "one" or "zero" being sent.

Frequency Synthesis. Generation of multiple frequencies from one or a few sources. See direct and indirect synthesis categories.

FH. Frequency hopping.

FH/DS. Combined frequency hopping, direct sequence modulation.

FM. Frequency modulation.

FSK. Frequency shift keying.

G_p. See Process gain.

Heterodyne. A process in which a signal is mixed with another or multiplied with it, usually for the purpose of frequency translation.

Hybrid Ranging. A ranging method in which short code sequences are combined with a synchronous tone to give faster acquisition than is available with longer sequences but longer range than available with short sequences alone.

Incremental Phase Modulation. A phase modulation technique used to shift one code sequence with respect to another, usually for the purpose of a synchronization search.

Indirect Synthesis. A frequency synthesis method in which output frequencies are developed by the use of phase-lock methods.

Integrate and Dump. A form of synchronous detector that integrates over an information bit period and is discharged (dumped) at the end of that period. The integrator is sampled just before discharge to determine its state. When synchronized with a received signal, it is a matched filter.

Interference. Any signal that tends to hamper the normal reception of a desired signal. Equivalent to jamming except considered nonhostile in origin.

IPM. Incremental phase modulation.

Jammer. A source of deliberate interference, usually hostile.

Jamming. The signal produced by a jammer, and intended to interfere with reception of a desired signal.

Jamming Margin (M_j). The amount of interference a system is able to withstand while producing the required output signal-to-noise or bit-error rate.

Limiter. A nonlinear device used to prevent the passage of signals above some chosen limiting level. Below this level signals are passed unimpaired.

Limiting. The action performed by a limiter. Limiting action, because of its nonlinear character, produces signal products similar to those produced by multiplication or modulation.

***m*-ary.** Used to designate signal formats other than binary; multilevel signal formats such as ternary (3), and quaternary (4).

Matched Filter. The ideal filter, designed for reception of a specific signal; passing the signal with minimum loss while passing minimum noise.

Maximal Sequence. A type of code sequence which is the longest that can be generated by a feedback code generator. For binary code generation its length is $2^N - 1$, where N is the generator length in chips.

Mersenne Prime. A particular type of maximal code sequence in which the code length ($2^N - 1$) in chips is a prime number. Important when cross-correlation is a major factor.

M_j. See jamming margin.

Modem. Abbreviation for modulator–demodulator.

Modular Shift Register Generator. A particular type of code sequence generator in which each delay element is followed by a modulo-2 adder.

Modulo-2 Adder. Arithmetic logic element used in binary code generators to form feedback. Sometimes called an exclusive-or gate, or half-adder, because it performs the addition function but does not include a shift output. Logically its output is $1 = \overline{A}B + A\overline{B}$.

MQPSK. Modified quadriphase shift keying. Four-phase PSK modulation modified by shifting chip clocks so that phase transitions larger than 90° do not occur. Same as OQPSK and SQPSK.

MSK. Minimum shift keying. A form of modulation closely allied to FSK and staggered QPSK.

MSRG. Abbreviation for modular shift register generator.

Noise Figure. The amount of noise added (in dB) by a signal-handling stage to that existing at its input. Usually applied to a first RF stage to signify the amount of noise output over the noise input.

OQPSK. Offset quadriphase shift keying. Same as MQPSK and SQPSK.

Orthogonal. Used to signify signals that are mutually transparent or non-interfering. Frequency and amplitude modulation are ideally considered orthogonal.

PDM. Pulse duration modulation.

Phase-lock Loop. A feedback loop designed to cause an oscillator internal to the loop to track an incoming carrier's frequency. Generally used for angle demodulation.

Preamble. A special code sequence used for acquisition of synchronization in spread spectrum systems. Usually short compared with codes used in communicating.

Process Gain (G_p). The gain or S/N improvement enjoyed by a spread spectrum system due to coherent band spreading and remapping of the desired signal. Expressed by the equation $BW_{RF}/R_{info} = G_p = TW$.

Propagation. The process by which a signal proceeds from the transmitter to a receiver. A general term; for example, "propagation path."

Pseudonoise. A term used to signify any of a group of code sequences that exhibit noiselike properties. Also sometimes used as a name for systems that employ pseudonoise code modulation.

PSK. Abbreviation for phase-shift keying.

Pulsed FM. Also called "chirp." A signal format in which a pulsed carrier is transmitted and is swept in frequency during the transmission. This format permits a special type of matched filter (chirp) detector to be used for signal reception.

Quadriphase. Same as QPSK.

QPSK. Four-phase, phase-shift keyed signal format. Phase shifts occur at 0°, ± 90°, and 180°, depending on the data being transmitted.

Remapping. The process of correlation in which a spread spectrum desired signal is transformed into a coherent narrowband signal and undesired signals are transformed into wide bandwidths.

SCPDM. Abbreviation for suppressed clock PDM, a signal format similar to common PDM, except that the bandwidth required for its passage is reduced by removing clock information.

Scrambler. One of a group of signal processors that cause an input signal to be incoherent to anyone not having the proper key for descrambling.

Search. A process by which spread spectrum systems acquire synchronization.

Secure. A term used to signify that a signal is protected from both inadvertent and deliberate attempts at decoding the signal. For all practical purposes a secure signal can be decoded only by one possessing the proper key.

Sequence. A code or train of chips.

Settling Time. The time required for a frequency hopping signal to change from one frequency to a new one.

Shift Register. A sequentially connected group of delay elements used to generate code sequences in spread spectrum systems.

Shift Register Generator. The combination of a shift register and modulo-2 addition for code sequence generation.

Simple Shift Register Generator. A form of shift register sequence generator in which the minimum number of modulo-2 addition stages is used to generate the particular code desired.

Spread Spectrum. Any of a group of modulation formats in which an RF bandwidth much wider than necessary is used to transmit an information signal so that a signal-to-noise improvement may be gained in the process.

SQPSK. Staggered quadriphase shift keying. Same as OQPSK and MQPSK.

SRG. Abbreviation for shift register generator.

SSRG. Abbreviation for simple shift register generator.

Subcarrier. A carrier, modulated with information separate from carrier modulation, which in turn modulates a carrier.

Suppressed Carrier. A particular type of modulation format in which the carrier signal itself is suppressed or held to a level below the modulation components.

Synchronization. Timing agreement or the act of gaining timing agreement between a spread spectrum transmitter and its receiving system.

Tau-jitter. Same as dither.

TH. Abbreviation for time hopping.

Time/Frequency Hopping. A combination of time-hopping and frequency hopping modulations for the purpose of overcoming near/far problems.

Time Hopping. Similar to simple pulse modulation, except that the pulse pattern may be pseudorandom.

***TW* (*TW* Product).** The equivalent of G_p, or process gain, where T is the period of the information sent, and W is the bandwidth employed to send it; that is, $T = 1/R_{\text{info}}$.

APPENDIX 2

BIBLIOGRAPHY ON SPREAD SPECTRUM SYSTEMS AND RELATED TECHNOLOGY

The following bibliography is a listing of articles, memoranda, and technical papers that pertain either directly to spread spectrum systems or to the technology drawn no in implementing them. The papers listed are generally obtainable from the sources given or in many cases from any reasonably good technical library. The range of technical depth should satisfy either those that are just curious or those who are in need of rigorous theoretical and mathematical treatment.

INDEX OF CATEGORIES

1. ANTIJAMMING (INTERFERENCE REJECTION)

I1. J. M. Aein, "On the Output Power Division in a Captured Hard-Limiting Repeater," *IEEE Trans. Comm. Tech.*, June 1966.

I2. S. A. Bohrer, "Example of Frequency Hopping Performance Analysis: Magnavox Research Laboratories, MX-TM-6755-3006-66, December 1966.

I3. C. R. Cahn, "A Note on Signal-to-Noise Ratio in Band-Pass Limiters," *IRE Trans. Info. Th.*, January 1961.

I4. C. R. Cahn, "Increasing Frequency Hopping Rate of Digital Synthesizers," MRL STN-14, December 1964.

I5. C. R. Cahn, "Data Coding to Reduce Vulnerability to Pulse Jamming," MRL STN-13, November 1964.

I6. C. R. Cahn, "Comparison of Frequency Hopping and PN for AJ Data Transmission," MRL STN-15, December 1964.

I7. A. Corneretto, "Spread Spectrum Com. System Uses Modified PPM," *Electronic Design*, June 21, 1961.

I8. J. P. Costas, "Poisson, Shannon and the Radio Amateur," *IRE Proc.*, December 1959.

I9. H. J. Friedman, "Jamming Susceptibility," *IEEE Trans. Aerosp. Elect. Syst.*, July 1968.

I10. K. Gilmore, "The Secret Keepers," *Popular Electronics,* August 1962.

I11. L. M. Goodman and S. B. Russell, "The TATS Master—A Net Controller for Tactical Satellite Communications."

I12. I. Jacobs, "The Effects of Video Clipping on the Performance of an Active Satellite PSK Communication System," *IEEE Trans. Comm. Tech.*, June 1965.

I13. I. Kadar and H. H. Schreiber, "Performance of Spread Spectrum Systems in a Multipath Environment," WESCON, 1970.

I14. S. L. Levine, "Anti-Jam Communications Systems," Astia Document AD445897, April 1964.

I15. J. F. Mason, "Air Force Plans Integrated Avionics Network," *Electronic Design,* August 15, 1968.

I16. T. J. Mattis, "Tactical Satellite Communications Requirements," *Signal,* September 1969.

I17. A. J. Miadich, "On-Board vs. Ground A. J. Processing," TRW Correspondence 7323-2-257, June 1968.

I18. W. C. Morchin, "Radar Range in a Jamming Environment," *Microwave Journal,* June 1968.

I19. A. Pfanstiehl, "Intelligent Noise," *Analog Science Fiction,* 1961 (approx.).

I20. J. L. Sevy, "The Effect of Limiting a Biphase or Quadriphase Signal Plus Interference," *IEEE Trans. Aeros. Elect. Sys.*, May 1969.

I21. D. P. Sullivan, "Future Trends in Military Communications Satellite Repeaters," TRW, 1968.

I22. R. Thorensen, "On the Theory of Jam-Resistant Communication Systems," MRL Report R-502, April 1962.

123. A. J. Viterbi, "Maximum Problems in Coding for Jammed Channels," MRL STM-27, June 1966.

124. A. J. Viterbi, "Bandspreading Combats Multipath and RFI in Tactical Satellite Net," *Communications Designer's Digest,* December 1969.

125. D. Shklarsky, P. Das, and L. B. Milstein, "Adaptive narrow-band interference suppression," in *Proc. Natl. Telec. Conf.*, 1979, pp. 15.2.1–15.2.3.

126. S. L. Johnston, "Radar ECCM history," in *Proc. NAECON,* May 1980, pp. 1210–1214.

127. R. H. Pettit, "Error probability for NCFSK with linear FM jamming," *IEEE Trans. Aerosp. Electron. Syst.*, vol. AES-8, pp. 609–614, September 1972.

128. H. D. Lewis, "Array Radars Solve Communication Jams" *Microwaves,* April 1982.

129. M. K. Simon and A. Polydoros, "Coherent Detection of Frequency-Hopped Quadrature Modulations in the Presence of Jamming—Parts I and II," *IEEE Trans. Comm.*, November 1982.

130. L. B. Milstein and D. Schilling, "Performance of a Spread Spectrum Communication System Operating over a Frequency-Selective Fading Channel in the Presence of Tone Interference," *IEEE Trans. Comm.*, January 1982.

131. M. J. Bouvier, Jr., "The Rejection of Large CW Interferers in Spread Spectrum Systems," *IEEE Trans. Comm.*, vol. COM-26, pp. 254–256, February 1978.

132. D. Shklarsky, P. Das, and L. B. Milstein, "Adaptive Narrow-band Interference Suppression," *Proc. Natl. Telecommun. Conf.*, 1979, pp. 15.2.1–15.2.3.

133. G. R. Johnson, "Jamming Low Power Spread Spectrum Radar," *Electron. Warfare,* pp. 103–112, Sept.–Oct. 1977.

134. M. Matsumoto and G. R. Cooper, "Multiple Narrow-band Interferers in an FH-DPSK Spread-Spectrum Communications System," *IEEE Trans. Veh. Technol.*, vol. Vt-30, pp. 37–42, February 1981.

135. S. W. Houston, "Modulation Techniques for Communications—Part I: Tone and Noise Jamming Performance of Spread Spectrum M-ary FSK and 2, 4-ary DPSK Waveforms." *Proc. IEEE 1975 Natl. Aerosp. Electron. Conf.*, June 10–12, 1975, pp. 51–58.

136. J. K. Omura, B. Levitt, and R. Stokey, "FH/MFSK Performance in a Partial Band Jamming Environment," *IEEE Trans. Commun.* (to be published).

137. D. Avidor, "Anti-jam Analysis of Frequency Hopping M-ary FSK Communication Systems in HF Rayleigh Fading Channels," Doctoral Dissertation, School Eng. Appl. Sci., University of California, Los Angeles, 1981.

138. L. B. Milstein, S. Davidovici, and D. L. Schilling, "The Effect of Multiple-tone Interfering Signals on a Direct Sequence Spread Spectrum Communication System." *IEEE Trans. Commun.*, vol. COM-30, pp. 436–446, March 1982.

139. S. W. Houston, "Tone and Noise Jamming Performance of a Spread Spectrum M-ary FSK and 2, 4-ary DPSK Waveforms." *Proc. Natl. Aerosp. Electron. Conf.*, June 1975, pp. 51–58.

140. R. H. Pettit, "A Susceptibility Analysis of Frequency Hopped M-ary NCPSK—Partial-band Noise on CW tone Jamming." presented at the Symp. Syst. Theory, May 1979.

141. M. Matsumoto and G. R. Cooper, "Multiple narrow-band interferers in an FH-DPSK spread-spectrum communication system," *IEEE Trans. Veh. Technol.*, vol. VT-30, pp. 37–42, February 1981.

142. P. A. Kullstam, "Spread Spectrum Performance Analysis in Arbitrary Interference" *IEEE Trans. Comm.*, August 1977.

143. J. W. Ketchum and J. G. Proakis, "Adaptive Algorithms for Estimating and Suppressing Narrow-Band Interference in PN Spread Spectrum Systems," *IEEE Trans. Comm.*, August 1977.

144. L. M. Li and L. B. Milstein, "Rejection of Narrow-Band Interference in PN Spread-Spectrum Systems Using Transversal Filters," *IEEE Trans. Comm.*, August 1977.

145. M. Matsumoto and G. R. Cooper, "Performance of a Nonlinear FH-DPSK Spread Spectrum Receiver with Multiple Narrow-Band Interfering Signals," *IEEE Trans. Comm.*, August 1977.

146. J. S. Lee and L. E. Miller, "Error Performance Analyses of Differential Phase-Shift-Keyed/Frequency-Hopping Spread-Spectrum Communication System in the Partial-Band Jamming Environments," *IEEE Trans. Comm.*, August 1977.

147. M. K. Simon, "The Performance of M-ary FH-DPSK in the Presence of Partial-Band Multitone Jamming," *IEEE Trans. Comm.*, August 1977.

148. H. P. Baer, "Interference Effects of Hard Limiting in PN Spread-Spectrum Systems," *IEEE Trans. Comm.*, August 1977.

149. A. B. Glenn, "Low Probability of Intercept (LPI) Performance in Jamming and Nonjamming Environments for SHF and EHF Satellite Communication Systems," *IEEE Milcom. Conf. Proc.*, October 1981.

150. G. K. Huth and C. L. Weber, "The Performance of Direct Sequence Systems in the Presence of Jammers," *IEEE Milcom. Conf. Proc.*, October 1981.

151. R. J. McEliece and W. E. Stark, "The Optimal Code Rate vs. A Partial Band Jammer," *IEEE Milcom. Conf. Proc.*, October 1981.

152. J. E. Blanchard, "A Slow Frequency Hopping Technique That is Robust to Repeat Jamming," *IEEE Milcom. Conf. Proc.*, October 1981.

153. W. E. Stark, "Coding for Coherent Frequency-Hopped Spread-Spectrum Communication in the Presence of Jamming," *IEEE Milcom. Conf. Proc.*, October 1981.

154. T. C. Huang and L. Yen, "Error Probability Performance of Noncoherent MFSK/FH Receivers in the Presence of Interference and Gaussian Noise," *IEEE Milcom. Conf. Proc.*, October 1981.

155. M. Spellman, "A Comparison Between Frequency Hopping and Direct Spread PN as Antijam Techniques," *IEEE Milcom. Conf. Proc.*, October 1981.

156. I. R. Smith, "Adaptation of Spread Spectrum Processing Gain and Voice Processing Algorithms in a Jamming Environment," *IEEE Milcom. Conf. Proc.*, October 1981.

157. M. K. Simon and C. Wang, "Limiter Discriminator Detection of Narrowband FM (CPDSK) in Jamming and Multipath," *IEEE Milcom. Conf. Proc.*, October 1981.

158. B. Levitt, "Comprehensive Analysis of FH/MFSK with Multi-tone Jamming," *IEEE Milcom. Conf. Proc.*, October 1981.

159. J. Omura, "Anti-Jam Performance of FH/MFSK in Rayleigh Fading," *IEEE Milcom. Conf. Proc.*, October 1981.

160. R. P. Rafuse, "Interference Rejection Properties of Pseudonoise and Frequency-Hopping Systems—A Review," *Int. Conf. Comm. Proc.*, June 1980.

161. R. A. Scholtz, "Centered CW Interference Rejection Using Spread Spectrum Techniques," *Int. Conf. Comm. Proc.*, June 1980.

162. M. K. Ward and H. M. Gibbons, "The Performance of Frequency Hopped Noncoherent M-ary FSK Modulation in the Presence of Repeat Back Jamming," *Int. Conf. Comm. Proc.*, June 1980.

163. R. H. Lang and R. L. Pickholtz, "Transient Analysis of an Adaptive Array in the Presence of a Step Jammer," *Natl. Telec. Conf. Proc.*, Nov. 1980.

164. S. Mahood, "Performance of an FHSS Signal in White Gaussian Noise Using a CCD/C2T Implementation," *Natl. Telec. Conf. Proc.*, November 1980.

165. G. R. Cooper, "Narrowband Interference in an FH-PPSK Spread Spectrum Communication System," *Natl. Telec. Conf. Proc.*, November 1980.

166. L. B. Milstein and D. L. Schilling, "The Effect of a Comb Jammer on a Direct Sequence Spread Spectrum System," *Natl. Telec. Conf. Proc.*, November 1980.

167. R. S. Orr, "Performance of PN/FH Hybrid Spread Spectrum Systems Against Partial-Band Noise Jamming," *Natl. Telec. Conf. Proc.*, November 1980.

168. S. Lee and J. H. Park, "Spreading Nulling Tradeoffs in an Interference Environment," *Natl. Telec. Conf. Proc.*, November 1980.

169. M. J. Bouvier, Jr., "The Rejection of Large CW Interferers in Spread Spectrum Systems," *IEEE Trans. Comm.*, February 1978.

170. F. M. Hsu and A. A. Giordano, "Digital Whitening Techniques for Improving Spread Spectrum Communications Performance in the Presence of Narrowband Jamming and Interference," *IEEE Trans. Comm.*, February 1978.

171. M. K. Simon and K. T. Woo, "The Performance of Suppressed Carrier Receivers in a Pulsed RFI Environment," *IEEE Trans. Comm.*, May 1980.

172. L. B. Milstein and D. L. Schilling, "Performance of a Spread Spectrum Communication System Operating Over a Frequency-Selective Fading Channel in the Presence of Tone Interference," *IEEE Trans. Comm.*, January 1982.

173. S. L. Johnston, "Guided Missile ECM/ECCM," *Microwave Journal,* September 1978.

174. R. L. Buskirk and E. J. Nossen, "ECM/ECCM Effects on Voice Transmission," *ITC Conf. Proc.*, 1978.

175. L. Li and L. B. Milstein, "Rejection of Pulsed CW Interference in PN Spread Spectrum Systems Using Complex Adaptive Filters," *IEEE Trans. Comm.*, January 1983.

176. O. A. H. Shabsigh, "On the Effects of CW Interference on MSK Signal Reception," *IEEE Trans. Comm.*, August 1982.

177. C. E. Cook, "Anti-Intercept Margins of Relay-Augmented Data Links," *IEEE Trans. Comm.*, July 1981.

2. APPLICATIONS OF SPREAD SPECTRUM METHODS

21. L. R. Allain, "First Steps to Radar-Waveform Design," *Microwaves,* December 1965.

22. W. B. Allen and E. C. Westerfield, "Digital Compressed Time Correlators and Matched Filters for Active Sonar," *J. Acoust. Soc. Amer.*, January 1964.

23. G. C. Anderson and M. A. Perry, "A Calibrated Real-Time Correlator/Averager/Probability Analyzer," *HPJ.*, November 1969.

24. B. B. Barrow, L. G. Abraham, Jr., S. Stein, and D. Bitzer, "Tropospheric-Scatter Propagation Tests Using a Rake Receiver," *IEEE Comm. Conv.*, June 7, 1965.

25. D. R. Bitzer, D. A. Chesler, R. Ivers, and S. Stein, "A Rake System for Tropospheric Scatter," *IEEE Trans. Comm. Tech.*, August 1966.

26. H. Blasbalg, D. Freeman, and R. Keeler, "Random Access Communications Using Frequency Shifted PN (Pseudo-Noise) Signals," IBM Corporation.

27. H. Blasbalg, H. Najjar, R. D'Antonio, and R. Haddad, "Air-Ground, Ground Air Communications Using Pseudo-Noise Through a Satellite," *IEEE Trans. Aerosp. Elect. Sys.*, September 1968.

28. W. T. Brandon, A. G. Camera, and Q. C. Wilson, "A New Approach to Tactical Aircraft-Satellite Communications," ESD-TR-68-296, Mitre Corp.

29. D. Chesler, "Performance of a Multiple Access RADA System," *IEEE Trans. Comm. Tech.*, August 1966.

210. C. E. Cook, "Pulse Compression—Key to More Efficient Radar Transmission," *IRE Proc.*, March 1960.

211. A Corneretto, "Better Radars Designed Using Pulse Compression," *Electronic Design,* August 30, 1963.

212. R. H. Cushman, "Make the Most of Noise: Correlate It," EDM, March 1, April 15, 1971.

213. D. S. Dayton, "Coming to Grips with Multipath Ghosts," *Electronics,* November 27, 1967.

214. P. M. Diamond, "Satellite Systems for Integrated Communications, Navigation and Identification," AF Report SAMSO-TR-70-160, Aerospace Corp.

215. R. C. Dixon, "A Spread Spectrum Ranging Technique for Aerospace Vehicles," SW IEEE CON, April 1968.

216. E. S. Donn, "Manipulating Digital Patterns with a New Binary Sequence Generator, *HPJ,* April 1971.

217. E. S. Donn, "Measuring Digital Error Rate with Pseudorandom Signals," *Telecommunications,* November 1971.

218. P. R. Drouilhet, Jr. and S. L. Bernstein, "TATS—A Bandspread Modulation-Demodulation System for Multiple Access Tactical Satellite Communication," EASCON, 1969 record.

219. S. W. Golomb, et al., *Digital Communications with Space Applications,* Prentice-Hall, Englewood Cliffs, New Jersey, 1964.

220. R. Y. Huang and P. Hooten, "Communication Satellite Processing Repeaters," *IEEE, Proc.*, February 1971.

221. R. M. Hultberg, F. H. Jean, and M. C. Jones, "Time Division Access for Military Communications Satellites," *IEEE Trans. Aerosp. Elect. Sys.*, December 1965.

222. W. J. Judge, "Multiplexing Using Quasiorthogonal Functions," *AIEEE Winter General Mtg.*, January 1962.

223. J. R. Klauder, A. C. Price, S. Darlington, and W. J. Albersheim, "The Theory and Design of Chirp Radars," *BSTJ*, July 1960.

224. R. Price and P. E. Green, Jr., "A Communication Technique for Multipath Channels," *IRE Proc.*, March 1958.

225. R. Price, "Wideband Analog Transmission Through Multipath by Means of a Pseudo-Noise Carrier," *URSI Spring Mtg.*, April 30, 1962.

226. R. L. Rex and G. T. Roberts, "Correlation, Signal Averaging, and Probability Analysis," *HPJ*, November 1969.

227. L. G. Roberts, "Picture Coding Using Pseudo-Random Noise," *IRE Trans. Info. Th.*, February 1962.

228. S. G. Shepherd and A. E. Seman, "Design with Integrated Circuits, Part 2," *Electronic Design*, July 19, 1965.

229. J. C. Springett, "Pseudo-Random Coding for Bit and Word Synchronization of PSK Data Transmission Systems," *Jet Propulsion Laboratories*.

230. W. Spoonemore, "UHF Propagation Experiment Using a Pseudonoise MODEM," AFAL Report TOD NR. 0510 PROJ. NR. 7662-04, January 1972.

231. A. J. Talamini, Jr. and E. C. Farnett, "New Target for Radar: Sharper Vision with Optics," *Electronics*, December 27, 1965.

232. R. B. Ward, "Digital Communications on a Pseudo-Noise Tracking Link Using Sequence Inversion Modulation," *IEEE Trans. Comm. Tech.*, February 1967.

233. R. B. Ward and F. L. Strubel, "Remove Clock Synchronization Experiment Progress 1968," LMSC/A946603 Lockheed MASC, December 1968.

234. W. J. Wester and H. G. Carter, "Spacecraft Sequence Generator Utilizes Integrated Circuitry," *EDN*, December 1964.

235. "Model 3721A Correlator," Hewlett-Packard, August 1969.

236. "Unified-Carrier Link Proposed for Lunar Mission," *Electronic Design*, September 27, 1963.

237. "Range Rider Pseudo-Noise Transmission Test Sets," Data Sheets 1000, 1100, International Data Sciences, September 1970.

238. "PN Carrier Modem Brings Privacy to Commercial Mobile Radio Networks," *Communications Designers' Digest*, January-February 1971.

239. "New Methods for Continuous, In-Service Monitoring of Microwave Systems," International Microwave Corp., Cos Cob, Connecticut.

240. "System 5000/5100 Digital Communication Test Set," Tau-Tron Inc., June 1971.

241. "MX230a Voice Modem," Magnavox Research Laboratories Report R-2081A.

242. J. F. Crush and B. C. Grossman, "Applications of Spread Spectrum Technology to a Low Cost Aircraft Traffic Control System," *Proc. Symp. Sp. Spec. Comm.*, March 1973.

243. C. L. Cuccia, "Communication by Satellite: A Status Report," *Microwave Systems News,* April/May 1974.

244. L. E. Hoff, "Design Consideration of Spread Spectrum Communication Systems for the HF Band," *Proc. Symp. Sp. Spec. Comm.*, March 1973.

245. C. R. Cahn, Paper No. 5, AGARD-NATO Lecture Series No. 58 on "Spread Spectrum Communications," May 1973.

246. C. L. Cuccia, "Spread Spectrum Systems Serve Nearly all C^3 Aspects," *MSN,* April 1982.

247. P. M. Goggans, D. A. Jelinek, and W. F. Nielsen, "SECURICOM: An Antijam, Secure-Message, Security Force Radio Communication System, Technical Summary," Sandia Laboratories SAND80-0347, March 1981.

248. R. C. Dixon, "SECURICOM: An Anti-Jam, Secure-Message, Security Force Radio Communication System. Design Feasibility Study," Sandia Laboratories SAND80-7049, March 1981.

249. M. Spellman, "Spread-Spectrum Radios Thwart Hostile Jammers," *Microwaves,* September 1981.

250. K. Custance, "Hop, Skip, and Jump. Who Wins?" *Telecommunications,* August 1981.

251. E. Ribchester, "Frequency-Hopping Radios Outwit 'Smart' Jammers," *Microwaves,* November 1979.

252. W. C. Melton, "Time-of-Arrival Measurements Possible with Use of Global Positioning System," *Defense Electronics,* December 1981.

253. N. E. Bolen, "JTIDS Coordinates Tactical Elements," *Defense Electronics,* January 1982.

254. G. R. Cooper and R. W. Nettleton, "A Spread-Spectrum Technique for High Capacity Mobile Communications," *IEEE Trans. Veh. Techn.*, vol. VT-27, pp. 264–275, November 1978.

255. D. J. Goodman, P. S. Henry, and V. K. Prabhu, "Frequency-Hopped Multilevel FSK for Mobile Radio," *Bell Syst. Tech. J.*, vol. 59, pp. 1257–1275, September 1980.

256. B. Goldberg, "Applications of Statistical Communications Theory," presented at the Army Sci. Conf. West Point, NY, June 20–22, 1962 (AD 332048): republished in *IEEE Commun. Mag.*, vol. 19, pp. 26–33, July 1981.

257. P. F. Sass, "Army Spread Spectrum—Evolution or Revolution?" *IEEE Milcom. Conf. Proc.*, October 1981.

258. J. S. Slechta and G. G. Willman, "Design Tradeoffs for Frequency Hopping VHF Net Radio," *IEEE Milcom. Conf. Proc.*, October 1981.

259. B. M. Leiner, "Performance of an Advanced ECCM Packet Radio," *IEEE Milcom. Conf. Proc.*, October 1981.

260. H. B. Goldman, "A Comparison of JTIDS and Packet Multiple Access Techniques," *IEEE Milcom. Conf. Proc.*, October 1981.

261. W. C. Scales, "Maritime Uses of Spread Spectrum," *IEEE Milcom. Conf. Proc.*, October 1981.

262. D. Sant, "Spread Spectrum and Satellite Systems," *IEEE Milcom. Conf. Proc.*, October 1981.

263. W. E. Howell, "Civil Aeronautical Applications," *Natl. Telec. Conf. Proc.*, November 1980.

264. J. Lane, "Geophysical Applications," *Natl. Telec. Conf. Proc.*, November 1980.

265. J. M. Ligon, "The GPS Z-Set Goes to Sea," *Natl. Telec. Conf. Proc.*, November 1980.

266. J. H. Painter, V. Rhyne, and P. Noe, "Z-Set Results in the Gulf of Mexico," *Natl. Telec. Conf. Proc.*, November 1980.

267. K. P. Yiu, F. Lee, and R. Eschenback, "Land Navigation of a Low Cost GPS Receiver," *Natl. Telec. Conf. Proc.*, November 1980.

268. N. E. Bolen, "JTIDS—Status and Issues," *Natl. Telec. Conf. Proc.*, November 1980.

269. P. Freret, "Application of Spread Spectrum Radio to Wireless Terminal Communications," *Natl. Telec. Conf. Proc.*, November 1980.

270. A. Girodano, H. Sunkneberg, H. dePadro, P. Stynes, D. Brown, and S. C. Lee, "A Spread-Spectrum Simulcast MF Radio Network," *IEEE Trans. Comm.*, May 1982.

271. P. Freret, R. Eschenbach, D. Crawford, and P. Braisted, "Applications of Spread-Spectrum Radio to Wireless Terminal Communications," *IEEE Natl. Telec. Conf. Proc.*, 1980.

272. G. L. Cariolaro, S. Cucchi, and F. Molo, "Transmultiplexer Via Spread-Spectrum Modulation," *IEEE Trans. Comm.*, July 1982.

273. D. J. Goodman and U. Timor, "Spread-Spectrum Mobile Radio with Variable-Bit-Rate Speech Transmission," *IEEE Trans. Comm.*, March 1982.

274. R. C. Dixon and F. P. Kaiser, "Applications of Anti-Jamming Techniques in Voice Communications Systems," NRL Workshop on Technology for Communications ECCM, February 1976.

275. W. B. Davenport, Jr., "NOMAC data transmission systems," presented at the RDB Symp. Inform. Theory Appl. Guided Missile.

276. P. E. Green, Jr., "Correlation detection using stored signals," M.I.T. Lincoln Lab., Tech. Rep. 33, August 4, 1953 (AD 020524).

277. B. M. Eisenstadt, P. L. Fleck, Jr., O. G. Selfridge, and C. A. Wagner, "Jamming tests on NOMAC systems," M.I.T. Lincoln Lab., Tech. Rep. 41. September 25, 1953 (AD 020419).

278. P. E. Green, Jr., "The Lincoln F9C radioteletype system," M.I.T. Lincoln Lab., Tech. Memo. 61, May 14, 1954 (not available from M.I.T.).

279. P. E. Green, Jr., R. S. Berg, C. W. Bergman, and W. B. Smith, "Performance of the Lincoln F9C radioteletype system," M.I.T. Lincoln Lab., Tech. Rep. 88, October 28, 1955 (AD 080345).

280. J. W. Craig, Jr., "An experimental NOMAC voice communication system," M.I.T. Lincoln Lab., Rep. 34G-0007, August 29, 1960 (AD 319610).

281. D. J. Gray, "A new method of teletype modulation," M.I.T. Lincoln Lab., Tech. Rep. 9, September 22, 1952 (AD 000928).

282. J. P. Costas and L. C. Widmann, "Data transmission system," U.S. Patent 3 337 803, August 22, 1967 (filed Jan. 9, 1962). "Reliable tactical communications," General Electric Res. Lab., Schenectady, NY, Final Rep., Contr. DA-36-039sc-42693, March 2, 1954 (AD 30344).

283. E. R. Baldwin, Jr., "SINCGARS: The Army's New Frequency Hopping Jam Resistant Combat Net Radio," *Signal Magazine,* November 1982.

284. E. B. Parker, "Micro Earth Stations Make Business Satcom Affordable," *MSN Magazine,* November 1982.

285. R. P. Denaro, "Navstar: The All Purpose Satellite" *Spectrum,* May 1981.

3. CODING

31. D. R. Anderson, "A New Class of Cyclic Codes," *SIAM J. Appl. Math.*, **16,** No. 1, 1968.

32. D. R. Anderson, "Periodic and Partial Correlation Properties of Sequences," TRW I.C. 7353. 1-01, July 1969.

33. L. Baumert, M. Easterling, S. W. Golomb, and A. Viterbi, "Coding Theory and Its Applications to Communications Systems," JPL Report 32-67, March 1961.

34. T. G. Birdsall and M. P. Ristenbatt, "Introduction to Linear Shift-Register Generated Sequences," Tech. Report 90, University of Michigan Research Institute, October 1958.

35. R. B. Blizard, "Quantizing for Correlation Decoding," *IEEE Trans. Comm. Tech.*, August 1967.

36. L. J. Bluestein, "Interleaving of Pseudo-Random Sequences for Synchronization," *IEEE Trans. Aerosp. Elect. Sys.*, July 1968.

37. R. H. Braasch, "The Distribution of (n-m) Terms for Maximal Length Linear Pseudo-Random Sequences," *IEEE Trans. Comm. Info. Th.*, July 1968.

38. R. C. Curry, "A Method of Obtaining All Phases of a Pseudo-Random Sequence," Naecon 1967.

39. R. C. Dixon, "100 MHz PN Code Generator," TRW 10C 7325. 3–19, July 1968.

310. G. D. Forney, Jr., "Coding and Its Application in Space Communications," *IEEE Spectrum,* June 1970.

311. J. B. Freymodsson, "A Note on the Run-Length Distribution of Ones and Zeros in Maximal-Linear Shift Register Sequences," unpublished memo dated June 11, 1963.

312. R. Gold, "Study of Correlation Properties of Binary Sequences," Magnavox Research Laboratories Report AFAL TR-66-234, August 1966.

313. R. Gold, "Characteristic Linear Sequences and Their Coset Functions," *J. SIAM Appl. Math.*, **14,** No. 5, September 1966.

314. R. Gold, "Optimal Binary Sequences for Spread Spectrum Multiplexing," *IEEE Trans. Info. Th.*, October 1967.

315. S. W. Golomb, "Sequences With Randomness Properties," Glenn L. Martin Co., June '55.

316. S. W. Golomb, *Shift Register Sequences,* Holden-Day, San Francisco, 1967.

317. S. C. Gupta and J. H. Painter, "Correlation Analyses of Linearly Processed Pseudo-Random Sequences," *IEEE Trans. Comm. Tech.*, December 1966.

318. H. Haberle, "Modulation, Synchronization and Coding," *Intl. Conf. Digital Satellite Comm.*, London, November 1969.

319. J. C. Hansen, "Modern Algebra for Coding," *Electro-Technology,* April 1965.

320. W. P. Horton, "Shift Counters," Computer Control Corporation.

321. D. Kahn, "Modern Cryptology," *Scientific American,* July 1966.

322. J. H. Lindholm, "An Analysis of the Pseudo-Randomness Properties of Subsequences of Long *m*-Sequences," *IEEE Trans. Info. Th.*, July 1968.

323. W. C. Lindsey, "Coded Noncoherent Communications," *IEEE Trans. Space Elect. Telem.*, March 1965.

324. G. D. O'Clock, Jr., G. L. Grasse, and D. A. Gandolfo, "Switchable Acoustic Surface Wave Sequence Generator," *IEEE Proc.*, October 1972.

325. J. H. Painter, "Designing Pseudo-Random Coded Ranging Systems," *IEEE Trans. Aerosp. Elect. Sys.*, January 1967.

326. W. W. Peterson, "Error Correcting Codes," MIT Press, Wiley, New York.

327. C. E. Shannon, "Communication Theory of Secrecy Systems," Bell Labs.

328. D. A. Swick, "Wideband Ambiguity Function of Pseudo-Random Sequences: An Open Problem," *IEEE Trans. Info. Th.*, July 1968.

329. R. C. Tausworthe, "Random Numbers Generated by Linear Recurrence Modulo-Two," *Math. Computation,* **19,** April 1965.

330. R. C. Titsworth, "Optimal Ranging Codes," *IEEE Trans. Space Elec. Telem.*, March 1963.

331. K. E. Toerper, "Biphase Barker-Coded Data Transmission," *IEEE Trans. Aerosp. Elect. Syst.*, March 1968.

332. A. J. Viterbi, "On Coded Phase-Coherent Communications," JPL Report 32-25, August 1960.

333. S. Wainberg and J. K. Wolf, "Subsequences of Pseudo-Random Sequences," *IEEE Trans. Comm. Tech.*, October 1970.

334. E. J. Watson, "Primitive Polynomials (Mod 2)," Math. Comp. 1962.

335. R. C. White, Jr., "Experiments with Digital Computer Simulation of Pseudo-Random Noise Generators," *IEEE Trans. Elect. Comp.*, June 1967.

336. J. K. Wolf, "On the Application of Some Digital Sequences to Communication," *IEEE Trans. Comm. Syst.*, December 1963.

337. N. Zierler, "Several Binary-Sequence Generators," Tech. Report 95 MIT, Lincoln Laboratories, September 1955.

338. N. Zierler, "Linear Recurring Sequences," *J. Soc. Indust. Appl. Math,* **7,** 1959.

339. "Surface Acoustic Waves Open New Avenues in Signal Processing," *Microwaves,* July 1971.

340. "Microwave Acoustics; Waves of the Future," *Microwave Systems News,* December/January 1974.

341. P. A. De Vito, P. H. Carr, W. J. Kearns, and J. H. Silva, "Encoding and Decoding with Elastic Surface Waves at 10 Megabits Per Second," *IEEE Proc.*, October 1971.

342. I. M. Jacobs, "Practical Applications of Coding," *IEEE Trans. Info. Th.*, May 1974.

343. J. Lee and D. R. Smith, "Families of Shift-Register Sequences with Impulsive Correlation Properties," *IEEE Trans. Info. Th.*, March 1974.

344. G. Solomon, "Optimal Frequency Hopping Sequences for Multiple Access," *Proc. Symp. Sp. Spec. Comm.*, March 1973.

345. A. R. Eckler, "The construction of missile guidance codes resistant to random interference," *Bell Syst. Tech. J.*, vol. 39, pp. 973–994, July 1960.

346. R. Gold, "Maximal recursive sequences with 3-valued cross-correlation functions," *IEEE Trans. Info. Th.*, vol. IT-14, pp. 154–156, January 1968.

347. R. A. Marolf, "200 Mbit/s pseudo random sequence generator for very wide band secure communication systems," in *Proc. NEC,* Chicago, IL. 1963, vol. 19, pp. 183–187.

348. R. C. Mathes, "Secret telephony," U.S. Patent 3 967 066, June 29, 1976 (filed September 24, 1941).

349. R. K. Potter, "Secret telephony," U.S. Patent 3 967 067, June 29, 1976 (filed September 24, 1941).

350. S. V. Jones, "After 35 years, secrecy lifted on encoded calls," *New York Times,* p. 27, July 3, 1976.

351. A. J. Busch, "Signalling circuit," U.S. Patent 3 968 454, July 6, 1976 (filed September 27, 1944).

352. A. E. Joel, Jr., "Pulse producing system for secrecy transmissions," U.S. Patent 4 156 108, May 22, 1979 (filed January 21, 1947).

353. W. W. Hansen, "Secret Communication," U.S. Patent 2 418 119 Apr. 1, 1947 (filed April 10, 1943).

354. L. A. deRosa and M. Rogoff, "Secure single sideband communication system using modulated noise subcarrier," U.S. Patent 4 176 316, November 27, 1979 (filed March 20, 1953); reissue appl. filed September 4, 1981, Re. Ser. No. 299 469.

355. E. Chaffee and E. Purington, "Method and means for secret radiosignalling," U.S. Patent 1 690 719, November 6, 1928 (filed March 31, 1922).

356. C. E. Shannon, "A mathematical theory of cryptography," Bell Tel. Lab., memo., September 1, 1945; later published in expurgated form as "Communication theory of secrecy systems," *Bell Syst. Tech. J.*, vol. 28, pp. 656–715, October 1949.

357. A. J. Viterbi and I. M. Jacobs, "Advances in coding and modulation for noncoherent channels affected by fading, partial band, and multiple access interference," in *Advances in Communication Systems,* vol. 4, Academic, New York, 1975.

358. G. D. O'Clock, Jr., C. L. Grasse, and D. A. Grandolfo, "Switchable Acoustic Surface Wave Sequence Generator," *IEEE Proc.*, October 1971.

359. P. Schobi, "Memory Structure Speeds Generation of Pseudonoise Sequences," *Computer Design,* March 1981.

360. M. B. Pursley and D. V. Sarwate, "Performance Evaluation for Phase-Coded Spread Spectrum Multiple Access Communication—Part II, Code Sequence Analysis," *IEEE Trans. Comm.*, vol. COM-25, pp. 800–803, August 1977.

361. M. B. Pursley and H. F. A. Roefs, "Numerical Evaluation of Correlation

Parameters for Optimal Phases of Binary Shift-Register Sequences," *IEEE Trans. Comm.*, vol. COM-25, pp. 1597–1604, August 1977.

362. G. Solomon, "Optimal Frequency Hopping Sequences for Multiple Access," in *Proc. 1973 Symp. Sp. Spec. Comm.*, vol. 1, AD915852, pp. 33–35.

363. D. V. Sarwate and M. B. Pursley, "Hopping Patterns for Frequency-Hopped Multiple-Access Communication," in *Proc. 1978 IEEE Int. Conf. Comm.*, vol. 1, pp. 7.4.1–7.4.3.

364. F. D. Garber and M. B. Pursley, "Optimal Phases of Maximal-length Sequences for Asynchronous Spread-Spectrum Multiplexing," *Electron. Lett.*, vol. 16, pp. 746–747, September 1980.

365. M. B. Pursley and H. F. A. Roets, "Numerical Evaluation of Correlation Parameters for Optimal Phases of Binary Shift Register Sequences," *IEEE Trans. Comm.*, vol. COM-27, pp. 1597–1604, October 1979.

366. D. V. Sarwate and M. B. Pursley, "Crosscorrelation Properties of Pseudorandom and Related Sequences," *Proc. IEEE,* vol. 68, pp. 593–619, May 1980.

367. G. R. Cooper, "Covert Communication with a Purely Random Spreading Function," *IEEE Milcom. Conf. Proc.*, October 1981.

368. F. S. Gutlebert, "Spread Spectrum Multiplexed Noise Codes," *IEEE Milcom. Conf. Proc.*, October 1981.

369. S. Pupolin and C. Tomasi, "Moments of the Weights of Pseudo-Noise Subsequences," *IEEE Milcom. Conf. Proc.*, October 1981.

370. Beeler, Bolt, Beranek & Newman "Finding Good Signalling Codes with Custom VLSI" *IEEE Milcom. Conf. Proc.*, October 1981.

371. A. K. Elhakeem, "Optimal Codes and Filters for Hard-Limited OQPSK/DS Multiple Access Systems," *IEEE Milcom. Conf. Proc.*, October 1981.

372. S. E. El-Khamy, "Matched Spread Spectrum Techniques," *IEEE Milcom. Conf. Proc.*, October 1981.

373. J. Dupuis, "Hardware and Software Tradeoffs in Spread Spectrum Code Generation," *ITC Conf. Proc.*, October 1981.

374. P. Schobi, "Memory Structure Speeds Generation of Pseudonoise Sequences," *Computer Design,* March 1981.

375. R. C. Titsworth, "Correlation properties of cyclic sequences." Ph.D. dissertation, California Institute of Technology, Pasadena, 1962.

376. R. S. Lunayach, "Performance of Direct Sequence Spread Spectrum System with Long and Short Period Code Sequences," *IEEE Trans. Comm.*, March 1983.

377. N. B. Chakrabarti and M. Tomlinson, "Design of Sequences with Specified Autocorrelation and Cross Correlation," *IEEE Trans. Comm.*, November 1976.

378. L. B. Milstein and R. R. Ragonetti, "Combination Sequences for Spread Spectrum Communications," *IEEE Trans. Comm.*, July 1977.

379. S. G. Glisic, "Power Density Spectrum of the Product of Two Time Displaced Versions of a Maximum Length Binary Pseudonoise Signal," *IEEE Trans. Comm.*, February 1983.

380. M. B. Pursley, "The Role of Coding in Multiple-Access Satellite Communication Systems," University of Illinois, Report R-724 April 1976.

381. W. E. Stark, "Performance of Concatenated Codes on Channels With Jamming," *IEEE I.C.C. Proc.*, 1982.

4. MATCHED FILTERS/CONVOLVERS

41. D. B. Armstrong, "Solid State Surface Acoustic Delay Lines," Litton Industries TN70-1 30, January 1970.

42. J. H. Collins and P. J. Hagon, "Applying Surface Wave Acoustics," *Electronics,* November 10, 1969.

43. C. E. Cook, "Pulse Compression—Key to More Efficient Radar Transmission," *IRE Proc.*, March 1960.

44. G. A. Coquin and R. Tsu, "Theory and Performance of Perpendicular Diffraction Delay Lines," *IEEE Proc.*, June 1965.

45. E. de Atley, "Surface Acoustic Waves Offer Cheap Signal Processing," *Electronic Design,* December 6, 1970.

46. G. A. Coquin and R. Tsu, "Theory and Performance of Perpendicular Diffraction Delay Lines," *IEEE Proc.*, June 1965.

47. J. H. Eveleth, "A Survey of Ultrasonic Delay Lines Operating Below 100 MC/S," *IEEE Proc.*, October 1965.

48. J. H. Eveleth, "Dispersive and Nondispersive Ultrasonic Delay Lines," Anderson Laboratories Tech. Note., undated.

49. P. Franklin, "Ultrasonics, Microwaves Join Forces," *Microwaves,* November 1968.

410. P. Franklin, "Echo Processor Offers 12 μs Delay with 50 dB Gain," *Microwaves,* October 1969.

411. C. L. Grasse and D. A. Gandolfo, "400 MHz Acoustic Surface-Wave Pulse Expansion and Compression Filter," *IEEE Trans. Micro. Th. Tech.*, June 1971.

412. R. C. Haavind, "Praetersonics: Microwaves of the Future?" *Electronic Design,* July 18, 1968.

413. J. R. Klauder, A. C. Price, S. Darlington, and W. J. Albersheim, "The Theory and Design of Chirp Radars," *BSTJ,* July 1960.

414. E. G. Lean and A. N. Broers, "Microwave Acoustic Delay Lines," *The Microwave Journal,* March 1970.

415. P. D. Newhouse, "Simplify EMC Design," *Microwaves,* May 1970.

416. F. A. Olson, "Today's Microwave Acoustic (Bulk Wave) Delay Lines," *The Microwave Journal,* March 1970.

417. M. J. Riezerman, "Lazer-Acoustic Line Reverses Time Functions," *Electronic Design,* June 7, 1969.

418. H. M. Gerard, W. R. Smith, W. R. Jones, and J. B. Harrington, "The Design and Applications of Highly Dispersive Acoustic Surface-Wave Filters," *IEEE Trans. MTT,* April 1973.

419. T. A. Martin, "The IMCON Pulse Compression Filter and Its Applications," *IEEE Trans. MTT,* April 1973.

420. G. S. Kino, S. Ludvik, H. J. Shaw, W. R. Shreve, J. M. White, and D. K. Winslow, "Signal Processing by Parametric Interactions in Delay-Line Devices," *IEEE Trans. MTT,* April 1973.

421. A. G. Bert, B. Epsztein, and B. Kantrowicz, "Signal Processing by Electron-Beam Interaction with Piezoelectric Surface Waves," *IEEE Trans. MTT,* April 1973.

422. D. T. Bell, Jr., J. D. Holmes, and R. V. Ridings, "Application of Acoustic Surface-Wave Technology to Spread Spectrum Communications," *IEEE Trans. MTT,* April 1973.

423. J. Burnsweig and J. Wooldridge, "Ranging and Data Transmission Using Digital Encoded FM-"Chirp" Surface Acoustic Wave Filters," *IEEE Trans. MTT,* April 1973.

424. P. M. Grant, J. H. Collins, B. J. Darby, and D. P. Morgan, "Potential Applications of Acoustic Matched Filters to Air-Traffic Control Systems," *IEEE Trans. MTT,* April 1973.

425. P. J. Hagon, F. B. Micheletti, R. N. Seymour, and C. Y. Wrigley, "A Programmable Surface Acoustic Wave Matched Filter for Phase-Coded Spread Spectrum Waveforms," *IEEE Trans. MTT,* April 1973.

426. D. P. Morgan and J. G. Sutherland, "Generation of Pseudonoise Sequences Using Surface Acoustic Waves," *IEEE Trans. Micro. Th. Tech.*, April 1973.

427. A. J. Berni and W. D. Gregg, "On the Utility of Chirp Modulation for Digital Signalling," *IEEE Trans. Comm.*, July 1973.

428. L. B. Milstein and P. K. Das, "A SAW Implemented Wideband Interference Rejection Spread Spectrum System," *IEEE I.C.C. Proc.*, 1982.

5. DIGITAL TRANSMISSION/EDAC

51. J. N. Birch, "Design Considerations for PCM and Delta Mod Systems," Magnavox ASAO Report TP 68-2174.

52. C. R. Cahn, "Performance of Digital Phase-Modulation Communication Systems," *IRE Trans. Comm. Syst.*, May 1959.

53. C. R. Cahn, "Application of Wagner Code to Digital Data Transmission System," Magnavox Research Laboratories STN-24, December 1965.

54. C. R. Cahn, "On Transmitting Digital Data Over an FM System," Magnavox Research Laboratories MX-TM-3091-70, November 1970.

55. C. J. Creveling, "Comparison of the Performance of PCM and PFM Telemetry Systems," NASA GSFC 1965.

56. H. J. Landon, "Sampling, Data Transmission and the Nyquist Rate," *IEEE Proc.*, October 1967.

57. A Lenden, "Correlative Digital Communication Techniques," *IEEE Trans. Comm. Tech.*, December 1964.

58. R. Lugannani, "Intersymbol Interference and Probability of Error in Digital Systems," *IEEE Trans. Info. Th.*, November 1969.

59. B. M. Oliver, J. R. Pierce, and C. E. Shannon, "The Philosophy of PCM," *IRE Proc.*, November 1948.

510. J. Salz and V. G. Koll, "Experimental Digital Multilevel FM Modem," *IEEE Trans. Comm. Tech.,* June 1966.

511. R. W. Sanders, "Communication Efficiency Comparison of Several Communication Systems," *IRE Proc.,* April 1960.

512. T. T. Tjhung, "Band Occupancy of Digital FM Signals," *IEEE Trans. Comm. Tech.*, December 1964.

513. M. E. Wall and T. G. Kuhn, "Wideband Data Transmission: Which Medium Is Best?" *Microwaves*, April 1966.

514. J. W. Whelan, "Analog-FM vs. Digital-PSK Transmission," *IEEE Trans. Comm. Tech*., June 1966.

515. P. A. Wintz and R. E. Totty, "Principles of Digital Communications," *Electro-Technology*, February 1967.

516. "Hybrid PCM Improves Noise Performance When Bandwidth is Limited," *Communications Designer's Digest*, November 1969.

517. "Frame Sync Study Points Up Optimum Parameters for PCM Transmission," *Communications Designer's Digest*, February 1970.

518. A. J. Viterbi, and I. M. Jacobs. "Advances in Coding and Modulation for Noncoherent Channels Affected by Fading Partial Band, and Multiple Access Interference," in *Advances in Communication Systems*, vol. 4, Academic, New York, 1975.

519. K. Yao, "E or Probability of Asynchronous Spread-Spectrum Multiple-Access Communication—Part I: System Analysis," *IEEE Trans. Comm*., vol. COM-25, pp. 803–809, August 1977.

520. K. T. Wu, "Average Error Probability for DS-SSMA Communications: The Gram-Charlier Expansion Approach," in *Proc. 19th Annu. Allerton Conf. Comm. Contr. Comput*., September 1981, pp. 237–246.

521. P. D. Shaft, "Low-Rate Convolutional Code Applications in Spread-Spectrum Communications," *IEEE Trans. Comm*., August 1977.

522. J. K. Omura, and B. K. Levitt, "Coded Error Probability Evaluation for Antijam Communication System," *IEEE Trans. Comm*., May 1982.

523. E. A. Geraniotis, "Coding for PSK and QPSK SFH/SSMA Communications via Fading Channels," *IEEE Milcom. Conf. Proc*., October 1981.

524. E. W. Chandler and G. R. Cooper, "Error Correction Coding Performance Bounds for Spread Spectrum Systems," *IEEE Milcom. Conf. Proc*., October 1981.

525. N. Shacham, "Performance of ARQ with Sequential Decoding Over One-Hop and Two-Hop Radio Links," *IEEE Milcom. Conf. Proc*., October 1981.

526. R. C. Hanlon and C. S. Gardner, "Error Performance of Direct Sequence Spread Spectrum Systems on Non-Selective Fading Channels," *IEEE Trans. Comm*., November 1979.

527. P. M. Hopkins, and R. S. Simpson, "Probability of Error in Pseudonoise (PN)-Modulated Spread Spectrum Binary Communications Systems," *IEEE Trans. Comm*., April 1975.

528. E. H. German, Jr., "A Comment on "Probability of Error in PN-Modulated Spread Spectrum Binary Communications Systems," *IEEE Trans. Comm*., June 1978.

529. W. Hirt, and S. Pasupathy, "Continuous Phase Chirp (CPC) Signals for Binary Data Communication," *IEEE Trans. Comm.*, June 1981.

6. DIRECT SEQUENCE

61. C. R. Cahn, "Spectrum Reduction of Biphase Modulated (2-PSK) Carrier," Magnavox Research Laboratories MX-TM-3103-71.

62. C. E. Gilchreist, "Pseudonoise System Lock-In," JPL Research Summary No. 36-9.

63. P. W. Nilsen, "PN Receiver Carrier and Code Tracking Performance," Magnavox Research Laboratories MX-TM-8-674-3043-68.

64. G. F. Sage, "Serial Synchronization of Pseudonoise Systems," *IEEE Trans. Comm. Tech.*, December 1964.

65. R. B. Ward, "Acquisition of Pseudonoise Signals by Sequential Estimation," *IEEE Trans. Comm. Tech.*, December 1965.

66. L. E. Zegers, "Common Bandwidth Transmission of Information Signals and Pseudonoise Synchronization Waveforms," *IEEE Trans. Comm. Tech.*, December 1968.

67. M. Lewis, "PLLs Upconvert Chirp Radar Signals," *Microwaves*, June 1981.

68. O.-C. Yue, "Performance of Frequency-Hopping Multiple-Access Multilevel FSK Systems with Hard-Limited and Linear Combining," *IEEE Trans. Comm.*, November 1981.

69. D. L. Schilling, L. B. Milstein, R. L. Pickholtz, and R. W. Brown, "Optimization of the Processing Gain of an M-ary Direct Sequence Spread Spectrum Communication System," *IEEE Trans. Comm.*, August 1980.

610. M. B. Pursley, "Performance Evaluation for Phase-Coded Spread-Spectrum Multiple-Access Communications—Part I: System Analysis," *IEEE Trans. Comm.*, vol. COM-25, pp. 795–799, August 1977.

611. M. Mizuno, "Randomization Effect of Errors by Means of Frequency Hopping Techniques in a Fading Channel" *IEEE Trans. Comm.*, May 1982.

612. F. D. Garber, "Analysis of Generalized Quadriphase Spread-Spectrum Communication," *Natl. Telec. Conf. Proc.*, November 1980.

613. M. B. Pursley and D. V. Sarwate, "New Results on Frequency-Hop, Spread-Spectrum, Multiple-Access Communications," *Natl. Telec. Conf. Proc.*, November 1980.

614. I. M. Jacobs, "Dama-Frequency Hopping and Pre-Correction for a Processing Satellite," *Int. Conf. Comm. Proc.*, June 1980.

615. W. C. Lindsey, L. Beiderman, and R. P. Sherwin, "Coding and Modulation Tradeoffs for Frequency-Hopped Channels," *Int. Conf. Comm. Proc.*, June 1980.

616. L. B. Milstein, R. L. Pickholz, D. L. Schilling, and R. Brown, "Optimization of the Processing Gain of an M-ARY Direct Sequency Spread Spectrum Communication System," *Int. Conf. Comm. Proc.*, June 1980.

617. R. C. Dixon, "Frequency Hopping Synthesizers Employing Conventional, Commercially-Available Integrated Circuits," *ITC Conf. Proc.*, October 1981.

618. B. G. Haskell, "Computer Simulation Results on Frequency Hopped MFSK Mobile Radio-Noiseless Case," *Natl. Telec. Conf. Proc.*, November 1980.

619. B. K. Levitt, "On Direct Sequence Spread Spectrum Systems," *Natl. Telec. Conf. Proc.*, November 1980.

620. J. Low and S. M. Waldstein, "A Direct Sequence Spread-Spectrum Modem for Wideband HF Channels," *IEEE Milcom. Proc. Conf.*, October 1981.

621. D. L. Schilling, L. B. Milstein, R. L. Pickholtz, and R. W. Brown, "Optimization of the Processing Gain of an M-ary Direct Sequence Spread Spectrum Communication System," *IEEE Trans. Comm.*, August 1980.

See also 16, 17, 110, 113, 114, 117, 119, 120, 121, 122, 26, 27, 28, 214, 215, 219, 220, 224, 225, 226, 229, 231, 232, 235, 236, 237, 238, 239, 33, 310, 318, 325, 331, 51, 58, 59.

7. FREQUENCY HOPPING

71. N. Abramson, "Bandwidth and Spectra of Phase- and Frequency-Modulated Waves," *IEEE Trans. Comm. Syst.*, December 1963.

72. C. R. Cahn, "Noncoherent Frequency Hop Sync Mode Performance," Magnavox Research Laboratories, STN-12, March 1964.

73. O. H. George, "Performance of Noncoherent M-ary FSK Systems With Diversity Under the Influence of Rician Fading," *IEEE Int. Conf. Comm.*, June 1968.

74. G. K. Huth, "Detailed Frequency-Hopper Analysis," Magnavox Research Laboratories, STN-29, August 1966.

75. A. Kaplan, "Detection and Analysis of Frequency Hopping Radar Signals," *Sylvania Elect. Syst.* W.D.L. Mountain View, California.

76. H. H. Schreiber, "Self-Noise of Frequency Hopping Signals," *IEEE Trans. Comm. Tech.*, October 1969.

77. F. G. Splitt, "Combined Frequency and Time-Shift Keyed Transmission Systems," *IEEE Trans. Comm. Syst.*, December 1963.

78. C. M. Thomas, "A Matched Filter Concept for Frequency Hopping Synchronization," TRW 10C 7353.6-05.

79. R. Malm, and K. Schreder, "Fast Frequency Hopping Techniques," *Proc. Symp. Spr. Spec. Comm.*, March 1973.

710. E. J. Nossen, "Fast Frequency Hopping Synthesizer," *Proc. Symp. Sp. Spec. Comm.*, March 1973.

711. P. S. Henry, "Spectrum efficiency of a frequency-hopped-DPSK spread spectrum mobile radio system," *IEEE Trans. Veh. Tech.*, vol. VT-28, November 1979.

712. J. D. Edell, "Wideband, noncoherent, frequency-hopped waveforms and their hybrids in low probability-of-intercept communications," Naval Res. Lab., Washington, DC, NRL Rep. 8025, November 8, 1976.

713. D. J. Goodman, P. S. Henry, and V. K. Prabhu, "Frequency-hopped multilevel FSK for mobile radio," *Bell Syst. Tech. J.*, vol. 59, pp. 1257–1275, September 1980.

714. R. F. Pawula and R. F. Mathis, "A Spread Spectrum System with Frequency Hopping and Sequentially Balanced Modulation—Parts One and Two," *IEEE Trans. Comm.*, May 1980.

715. M. K. Simon, G. K. Huth, and A. Polydoros, "Differentially Coherent Detection of QASK for Frequency-Hopping Systems, Parts I and II," *IEEE Trans. Comm.*, January 1982.

716. O.-C. Yue, "Performance of Frequency-Hopping Multiple-Access Multilevel FSK Systems with Hard-Limited and Linear Combining," *IEEE Trans. Comm.*, November 1981.

717. M. K. Simon, and A. Polydoros, "Coherent Detection of Frequency-Hopped Quadrature Modulations in the Presence of Jamming—Parts I and II," *IEEE Trans. Comm.*, November 1981.

718. L. B. Milstein, R. L. Pickholtz, D. L. Schilling, "Optimization of the processing gain of an FSK-FH system," *IEEE Trans. Comm.*, vol. COM-28, pp. 1062–1079, July 1980.

719. J. K. Omura, B. Levitt, and R. Stokey, "FH/MFSK performance in a partial band jamming environment," *IEEE Trans. Comm.*, to be published.

720. D. Avidor, "Anti-jam analysis of frequency hopping *M*-ary FSK communication systems in HF Rayleigh fading channels," Doctoral dissertation, School Eng. Appl. Sci., University of California, Los Angeles, 1981.

721. D. V. Sarwate and M. B. Pursley, "Hopping patterns for frequency-hopped multiple-access communication," in *Proc. 1978 IEEE Int. Conf. Comm.*, vol. 1, pp. 7.4.1–7.4.3.

722. P. S. Henry, "Spectrum efficiency of a frequency-hopped-DPSK spread spectrum mobile radio system," *IEEE Trans. Veh. Tech.*, vol. VT-28, pp. 327–329, November 1979.

723. O. C. Yue, "Hard-limited versus linear combining for frequency-hopping multiple-access systems in a Rayleigh fading environment," *IEEE Trans. Veh. Tech.*, vol. VT-30, pp. 10–14, February 1981.

724. R. W. Nettleton and G. R. Cooper, "Performance of a frequency-hopped differentially modulated spread-spectrum receiver in a Rayleigh fading channel," *IEEE Trans. Veh. Tech.*, vol. VT-30, pp. 14–29, February 1981.

725. E. A. Geraniotis and M. B. Pursley, "Error probability bounds for slow frequency-hopped spread-spectrum multiple access communications over fading channels," in *Proc. 1981 IEEE Int. Conf. Comm.*

726. A. J. Budreau, A. J. Slobodnick, Jr., and P. H. Carr, "Fast Frequency Hopping Achieved With SAW Synthesizers," *Microwave Journal*, February 1982.

727. E. Ribchester, "Frequency Hopping Techniques Vary With Frequency," *Microwaves & RF*, March 1983.

728. S. M. Sussman and P. Kotiveeriah, "Partial Processing Satellite Relays for Frequency-Hop Antijam Communications," *IEEE Trans. Comm.*, August 1982.

729. A. K. Elhakeem, "Overall SNR Optimization of a FH/MFSK Pulse Code and Adaptive Data Modulation Systems in Mixed Jamming," *IEEE I.C.C. Proc.*, 1982.

730. S. M. Elnoubi, "Error Rate Performance of Frequency Hopped MSK Spread Spectrum Mobile Radio System With Differential Detection," *I.C.C. Proc.*, 1982.

731. J. E. Blanchard, "Performance of M-Ary FSK/FH Against Optimum Multitone Jamming," *I.C.C. Proc.*, 1982.

732. C. Niyonizeye, M. Lecours, and H. T. Huynh, "Address Assignment in a Multiple Access FH-FSK System," *I.C.C. Proc.*, 1982.

733. R. Muammar and S. C. Gupta, "Performance of a Frequency-Hopped Multilevel FSK Spread Spectrum Receiver in a Rayleigh Fading and Log-Normal Shadowing Channel," *I.C.C. Proc.*, 1982.

See also 12, 14, 15, 16, 18, 110, 111, 113, 114, 117, 124, 29, 216, 217, 218, 219, 220, 226, 33, 310, 315, 323, 325, 332, 336, 510, 512, and Frequency Synthesis.

8. FREQUENCY SYNTHESIS

81. J. L. Barnum, "A Multioctave Microwave Synthesizer," *The Microwave Journal*, October 1970.

82. R. W. Burnell, "Phase-Lock Frequency Synthesizer," TRW 10C 7325.2-99, July 1968.

83. L. F. Blachowicz, "Dial Any Channel to 500 MHz," *Electronics*, May 2, 1966.

84. M. G. Davis, Jr., "Phase-Lock Frequency Synthesizer Equivalent to Lincoln Labs' Frequency Synthesizer," TRW 10C 7325-18, April 22, 1968.

85. J. Delaune, "MTTL and MECL Avionics Digital Frequency Synthesizers," Motorola AN-532, July 1970.

86. N. C. Hekemian, "Digital Frequency Synthesizers," *Frequency*, July/August 1967.

87. L. Illingworth, "Digital Methods Synthesize Frequency," *Electronics Design*, May 23, 1968.

88. R. S. Kahn, "Frequency Synthesizer Survey," Magnavox Research Laboratories MX-TM-8-672-3009-67, February 1967.

89. B. Koeper, "Shift Frequency Automatically Without Transients," *EDN*, November 11, 1968.

810. D. R. Lohrmann and A. R. Sills, "Cut Synthesizer Current Consumption," *Electronic Design*, December 5, 1968.

811. R. A. Maag, "Spurious Output and Acquisition Time of an Indirect Frequency Synthesizer," TRW 10C 7352.1-20, February 1969.

812. J. Noordanus, "Frequency Synthesizers—A Survey of Techniques," *IEEE Trans. Comm. Tech.*, April 1969.

813. E. Renshler and B. Welling, "An Integrated Circuit Phase-Locked Loop Digital Frequency Synthesizer," Motorola AN-463, March 1963.

814. R. R. Stone, Jr. and H. F. Hastings, "A Novel Approach to Frequency Synthesis," *Frequency*, September/October 1963.

815. "Modern Methods of Frequency Synthesis," *Antekna Tech. Bull.* 20-02.

See also Frequency Hopping.

9. INFORMATION TRANSMISSION

91. T. Berger, "Optimum PAM Compared with Information-Theoretic Bounds," *NEREM Record*, 1966.

92. C. R. Cahn, "Theoretical Comparison of Analog and Digital Modulations for Voice," Magnavox Research Laboratories STN-23, January 1966.

93. W. Cohen, "Signal-to-Noise Ratio in Modulated Systems," Parts 1 and 2, *Electro-Technology*, October 1965, January 1966.

94. J. A. Develet, "Coherent FDM/FM Telephone Communication," *IEEE Proc.*, September 1958.

95. C. C. Diaz and B. R. Norvell, "Pulse Modulation Intelligibility," *Electro-Technology*, March 1966.

96. L. W. Gardenhire, "Selecting Sample Rates," *ISA Journal*, April 1964.

97. J. J. Hupert, "Frequency-Modulation Techniques," *Electro-Technology*, February 1965.

98. J. J. Hupert, "Distortion in Angle-Modulation Networks," *Electro-Technology*, September 1965.

99. H. J. Landau, "Sampling, Data Acquisition, and the Nyquist Rate," *IEEE Proc.*, October 1967.

910. G. A. Miller and J. C. R. Licklider, "The Intelligibility of Interrupted Speech," *J. Acous. Soc. Amer.*, March 1950.

911. J. A. Porter, "Prediction of Speech Intelligibility Over Voice Communications Systems," NASA EB-65-R2001.

912. L. Schuchman, "Dither Signals and Their Effect on Quantization Noise," *IEEE Trans. Comm. Tech.*, December 1964.

913. C. P. Walsh, "Frequency-Modulation Principles," *Electro-Technology*, January 1965.

914. D. D. Weiner and B. J. Leon, "The Quasi-Stationary Response of Linear Systems to Modulated Waveforms," *IEEE Proc.*, June 1965.

915. J. A. Young, "Design Parameters Associated with Multilevel Transmission Systems," *NEREM Record*, 1966.

916. N. S. Jayant, "Characteristics of a Delta Modulator," *IEEE Proc.*, March 1971.

917. W. Hirt, and S. Pasupathy, "Continuous Phase Chirp (CPC) Signals for Binary Data Communication—Parts I and II," *IEEE Trans. Comm.*, June 1981.

918. O. Shimbo and M. Celebiler, "The Probability of Error Due to the Intersymbol Interference and Gaussian Noise in Digital Communication Systems," *IEEE Trans. Comm.*, vol. COM-19, pp. 113–119, April 1971.

919. P. H. Anderson, F. M. Hsu, and M. M. Sandler, "A New Adaptive Modem for Long Haul HF at Data Rates >1BPS/Hz," *IEEE Milcom. Conf. Proc.*, October 1981.

920. S. Crozier, R. Lyons, and K. Tiedmann, "Wideband Adaptive Serial Data Modem for HF Communications," *IEEE Milcom. Conf. Proc.*, October 1981.

921. P. Das and D. Shklarsky, "The Use of the Hilbert Transform to Double the Information Rate of a SAW Implemented Direct Sequence Spread Spectrum System," *Natl. Telec. Conf. Proc.*, November 1980.

922. S. Dhar and D. Perry, "Equalized Megahertz-Bandwidth HF Channels for Spread Spectrum Communications," *IEEE Milcom. Conf. Proc.*, October 1981.

923. J. Otto, "Chirping RPV Data Links for ECM Protection," *Microwaves Magazine*, December 1974.

924. A. Hammer and D. J. Schaefer, "Performance Analysis of M-ary Code Shift Keying in Code-Division Multiple Access Systems," *IEEE I.C.C. Proc.*, 1982.

10. MODULATION/DEMODULATION

101. J. G. Adashko, "Design of Phase Sensitive Demodulators," *Electronic Design*, April 12, 1962.

102. E. J. Baghdady, "On the Noise Threshold of Conventional FM and PM Demodulators," *IEEE Proc.*, September 1963.

103. C. R. Cahn, "Phase Detector Characteristic's Effect on Signal-to-Noise Ratio, General Theory," STL 10C, April 1962.

104. C. R. Cahn, "Analysis of Correlation System with Receiver Filter," Magnavox Research Laboratories, STN-20, September 1965.

105. C. R. Cahn, "Error Bound for Non Optimally Demodulated Channel," Addendum to Magnavox Research Laboratories, STN-27, June 1966.

106. J. P. Chandler, "IC's Fill Need for Low-Drift Phase-Shift Keyed Detector," *EDN* January 15, 1969.

107. E. Channell, "The Semiconductor Ring Modulator," *EDN*, January 1964.

108. P. W. Cooper, "Correlation Functions for the Random Binary Wave," *IEEE Trans. Comm. Syst.*, December 1963.

109. J. P. Costas, "Synchronous Communications," *IRE Proc.*, December 1956.

1010. J. A. Datillo, "Incremental Phase Modulator," Magnavox Research Laboratories MX-TR-8-765-2002, March 1967.

1011. J. A. Develet, Jr., "An Analytic Approximation of Phase-Lock Receiver Threshold." *TRW* 9332.6-2, April 1962.

1012. J. L. Eckstrom, "Coherent Matched Filter Detection of Quadratically Phase-Distorted Carrier-Band Pulses, with Application to Transionospheric Signalling," *IEEE Trans. Space Elect. Telem.*, March 1964.

1013. J. T. Frankle, "Threshold Performance of Analog FM Demodulators," *RCA Review*, December 1964.

1014. T. A. O. Gross, "Increasing the Dynamic Range of AM Detectors," *EEE*, November 1963.

1015. D. T. Hess, "Equivalence of FM Threshold Extension Receivers," *IEEE Trans. Comm. Tech.*, October 1968.

1016. S. R. Hirsch, "Convolution—A Graphical Interpretation," *EDN*, February 1967.

1017. W. J. Judge, "A Passive Correlator of Arbitrarily Large T-W Product," Magnavox Research Laboratories R-502, April 1962.

1018. E. H. Katz and H. H. Schreiber, "Design of Phase Discriminators," *Microwaves*, August 1965.

1019. C. Kurth, "Analysis and Synthesis of Diode Modulators," *Frequency,* January/February 1966.

1020. I. M. Langenthal, "Correlation and Probability Analysis," Signal Analysis Industries Corp., TB14.

1021. R. I. Levine, "Correlation—Theory and Practice," *Electronic Products*, November 1963.

1022. R. B. Mouw and S. M. Fukuchi, "Broadband Double Balanced Mixer/Modulators," Parts I and II, *The Microwave Journal*, March/May 1969.

1023. C. W. Ogar, "Putting Diode Modulators to Work," *Electronic Industries*, July 1961.

1024. J. H. Polson, "Wideband Digital Data Link Quadriphase Demodulator," TRW 7322.05-202, December 1968.

1025. A. W. Stoll, "The Electronic Correlator," *Electronic Industries*, August 1965.

1026. M. S. Stone, "Analytical Considerations for Multiphase Modulation and Demodulation," TRW 7323.2-245, April 1968.

1027. R. B. Stone and G. M. White, "Correlation Detector Detects Voice Fundamental," *Electronics*, November 22, 1963.

1028. W. G. Urbach, Jr., "Produce Fast and Clean RF Pulse," *Electronic Design*, September 1966.

1029. V. Z. Vizkanta, "I-Q Loop Receivers," TRW 10C 7323.2-289, September 1968.

1030. A. M. Voyce, "Ultimate Sensitivity of a Detector," *EEE*, January 1963.

1031. J. A. Kivett and G. F. Bowers, "A Wideband Modem for Command and Control of Remote Vehicles," *Proc. Symp. Sp. Spec. Comm.*, March 1973.

1032. R. B. Lowry, "PSK-FSK Spread Spectrum Modulation/Demodulation," *Proc. Symp. Sp. Spec. Comm.*, March 1973.

1033. M. L. Schiff and D. M. Dilley, "A Surface Acoustic Wave Spread Spectrum Modem," *Proc. Symp. Sp. Spec.*, March 1973.

1034. P. Bello, "Demodulation of a phase-modulated noise carrier," *IRE Trans. Info. Th.*, vol. IT-7, pp. 19–27, January 1961.

1035. M. K. Simon, "Two-Channel Costas Loop Tracking Performance for UQPSK Signals with Arbitrary Data Formats," *IEEE Trans. Comm.*, September 1981.

1036. M. K. Simon, G. K. Huth, and A. Polydoros, "Differentially Coherent Detection of QASK for Frequency-Hopping Systems—Parts I and II," *IEEE Trans. Comm.*, January 1982.

1037. W. C. Lindsey and M. K. Simon, "Optimum Design for Performance of Costas Receivers Containing Soft Bandpass Limiters," *IEEE Trans. Comm.*, Aug. 1977.

1038. C. Cahn, D. Leimer, R. Marsh, F. Huntowski, G. LaRue, "Software Implementation of a PN Spread Spectrum Receiver To Accommodate Dynamics," *IEEE Trans. Comm.*, August 1977.

1039. N. F. Krasner, "Optimal Detection of Digitally Modulated Signals," *IEEE Trans. Comm.*, May 1982.

1040. A. Polydoros and C. L. Weber, "Optimal Detection Considerations for Low Probability of Intercept," *IEEE Milcom. Conf. Proc.*, October 1981.

1041. W. Engeler, R. D. Baertsch, and H. G. Parks, "A Binary-Analog Correlator," *IEEE Milcom. Conf. Proc.*, October 1981.

1042. S. G. Glisic, "Noncoherent Demodulation of Spread Spectrum Signals Using Charge Coupled Devices," *IEEE Milcom. Conf. Proc.*, October 1981.

1043. J. G. Proakis, P. H. Anderson, and J. W. Ketchum, "Receiver Processing Techniques for DS Spread Spectrum in Dispersive Channels," *IEEE Milcom. Conf. Proc.*, October 1981.

1044. P. F. McDenzie, D. B. Coomber, and B. H. Hutchinson, "Design and Performance of a SAW-Based MFSK Demodulator," *IEEE Milcom. Conf. Proc.*, October 1981.

1045. M. Simon and W. C. Lindsey, "Optimum Performance of Suppressed Carrier Receivers With Costas Loop Tracking," *IEEE Trans. Comm.*, February 1977.

1046. D. G. Messerschmitt, "Frequency Detectors for PLL Acquisition in Timing and Carrier Recovery," *IEEE Trans. Comm.*, September 1979.

1047. J. K. Holmes and L. Biederman, "Delay-Lock-Loop Mean Time to Lose Lock," *IEEE Trans. Comm.*, November 1978.

1048. F. Amoroso, "Rotating Phase Demodulation," Hughes Aircraft Company 74/1461.20/40.

1049. S. A. Gronemeyer and A. L. McBride, "MSK and Offset QPSK Modulation," *IEEE Trans. Comm.*, August 1976.

1050. F. Amoroso, "Pulse and Spectrum Manipulation in the Minimum (frequency) Shift Keying (MSK) Format," *IEEE Trans. Comm.*, March 1976.

1051. M. B. Pursley, F. D. Garber, and J. S. Lehnert, "Analysis of Generalized Quadriphase Spread-Spectrum Communications," *Proc. IIII Int. Conf. Comm.*, June 1980.

1052. R. LaRosa, T. J. Marynowski, and K. J. Henrich, "A Simple Modulator For Sinusoidal Frequency Shift Keying," *IEEE Trans. Comm.*, May 1982.

1053. F. Amoroso, "Enhancing Direct Sequence Pseudonoise (DSPN) Modulation by Increasing the Bandwidth per Chip Rate," *IEEE Milcom. Conf. Proc.*, October 1981.

1054. P. Stynes, J. Lovell, H. Sunkenberg, and A. Girodano, "Waveform Alternatives for a Frequency-Hopped Simulcast Radio Network," *IEEE Milcom. Conf. Proc.*, October 1981.

1055. R. C. Dixon, "Simplified Approach to Direct Sequence Carrier Modulation and its Selection," *IEEE Milcom Conf. Proc.*, October 1981.

1056. R. Gorur, A. Payzin, and H. S. Oranc, "Multi-Subchannel Spread Spectrum Modulation," *IEEE Milcom. Conf. Proc.*, October 1981.

1057. R. W. Nettleton, "Addressing and Modulation for Frequency-Hopping Multiple Access," *IEEE Milcom. Conf. Proc.*, October 1981.

1058. T. T. N. Bucher, "Spectrum Occupancy of Pulsed FSK," *IEEE Milcom. Conf. Proc.*, October 1981.

1059. W. C. Lindsey, L. Beiderman, and R. P. Sherwin, "Coding and Modulation Tradeoffs for Frequency-Hopped Channels," *Int. Conf. Comm. Proc.*, June 1980.

1060. F. Amoroso, "The Bandwidth of Spread Spectrum Signals," *ITC Conf. Proc.*, October 1981.

1061. F. Amoroso, "Enhancing Direct Sequence Pseudonoise (DSPN) Modulation by Increasing the Bandwidth Per Chip Rate," *IEEE Milcom. Conf. Proc.*, October 1981.

1062. H. Yazdani, K. Feher, and W. Steenaart, "Constant Envelope Bandlimited BPSK Signal," *IEEE Trans. Comm.*, June 1980.

1063. J. W. Bayless and R. D. Pedersen, "Efficient Pulse Shaping Using MSK or PSK Modulation," *IEEE Trans. Comm.*, June 1979.

1064. M. K. Simon, "A Generalization of Minimum-Shift-Keying (MSK)-Type Signaling Based Upon Input Data Symbol Pulse Shaping," *IEEE Trans. Comm.*, August 1976.

1065. M. Rabzel and S. Pasupathy, "Spectral Shaping in Minimum Shift Keying (MSK)-Type Signals," *IEEE Trans. Comm.*, January 1978.

1066. F. Amoroso, "The Use of Quasi-Bandlimited Pulses in MSK Transmission," *IEEE Trans. Comm.*, October 1979.

1067. D. H. Morais and K. Feher, "The Effects of Filtering and Limiting on the Performance of QPSK, Offset QPSK, and MSK Systems," *IEEE Trans. Comm.*, December 1980.

11. MULTIPLE ACCESS

1101. E Bedrosian, N. Feldman, G. Northrop, and W. Sollfrey, "Multiple Access Techniques for Communications Satellites: I. Survey of the Problem," Rand Report RM-4298-NASA, September 1964.

1102. W. H. Harman, "Multiple Access and Antijam Capabilities of Direct-Sequence Quadriphase on a Linear Channel," TRW 10C 7323.2-256, June 1968.

1103. S. Udalov, "Pseudo-Random Pulse Position Multiplexing for Random Access," Magnavox Research Laboratories STN-8, January 1964.

1104. S. Udalov, "Threshold Criterion for Virtual Carrier Multiplexing," Magnavox Research Laboratories STN-10, January 1964.

1105. M. B. Pursley and D. V. Sarwate, "Performance evaluation for phase-coded spread spectrum multiple-access communication—Part II. Code sequence analysis," *IEEE Trans. Comm.*, vol. COM-25, pp. 800–803, August 1977.

1106. J. M. Aein and R. D. Turner, "Effect of co-channel interference on CPSK carriers," *IEEE Trans. Comm.*, vol. COM-21, pp. 783–790, July 1973.

1107. D. Laforgia, A. Luvison, and V. Zingarelli, "Exact bit error probability with application to spread-spectrum multiple access communications," in *Proc. IEEE Int. Conf. Comm.*, June 1981, vol. 4, pp. 76.5.1–76.5.5.

1108. D. Raychaudhuri, "Performance Analysis of Random Access Packet-Switched Code Division Multiple Access Systems," *IEEE Trans. Comm.*, June 1981.

1109. M. B. Pursley, "Performance Evaluation for Phase-Coded Spread Spectrum Multiple-Access Communication—Part I: System Analysis," *IEEE Trans. Comm.*, vol. COM-25, pp. 795–799, August 1977.

1110. H. E. Rowe, "Bounds on the Number of Users in Spread-Spectrum Systems," in *Proc. Natl. Telec. Conf.*, Nov. 30–Dec. 4, 1980, pp. 69.6.1–69.6.6.

1111. D. Borth, M. Pursley, D. Sarwate, and W. Stark, "Bounds on Error Probability for Direct Sequence Spread-Spectrum Multiple-Access Communications," in *Proc. MIDCON Conf.*, November 1979, paper 15/1, pp. 1–14.

1112. D. Laforgia, A. Luvison, and V. Zingarelli, "Exact Bit Error Probability with Application to Spread-Spectrum Multiple Access Communications," in *Proc. IEEE Int. Conf. Comm.*, June 1981, vol. 4, pp. 76.5.1–76.5.5.

1113. E. A. Geraniotis and M. B. Pursley, "Error Probability for Binary PSK Spread-Spectrum Multiple Access Communications," in *Proc. Conf. Info. Sci. Syst.*, The Johns Hopkins University, Baltimore, MD, March 1981, pp. 238–244.

1114. M. B. Pursley, "Spread Spectrum Multiple-Access Communications," in *Multi-User Communication Systems*, G. Longo, Ed., Springer-Verlag, New York, 1981, pp. 139–199.

1115. M. B. Pursley, D. V. Sarwate and W. E. Stark, "On the Average Probability of Error for Direct-Sequence Spread-Spectrum Multiple-Access Systems," in *Proc. 1980 Conf. Info. Sci. Syst.*, Princeton University, Princeton, NJ March 1980, pp. 320–325.

1116. K.-T. Wu, and D. L. Neuhoff, "Average Error Probability for Direct Sequence Spread Spectrum Multiple Access Communication Systems," in *Proc. 18th Annu. Allerton Conf. Comm. Comput.*, October 1980, pp. 359–368.

1117. K. Yao, "Error Probability of Asynchronous Spread Spectrum Multiple Access Communication Systems," *IEEE Trans. Comm.* vol. COM-25, pp. 803–809, August 1977.

1118. K. Yao, "Performance of Offset Quadriphase Spread Spectrum Multiple Access Communication," *IEEE Trans. Comm.*, vol. COM-29, pp. 305–314, March 1981.

1119. K.-T. Wu, "Direct Sequence Spread-Spectrum Communications: Applications to Multiple Access and Jamming Resistance," Ph.D. dissertation, University of Michigan, Ann Arbor, 1981.

1120. M. B. Pursley, D. V. Sarwate, and W. E. Stark, "Spread Spectrum Multiple-Access Communications—Part I: Upper and Lower Bounds," *IEEE Trans. Comm.*, May 1982.

1121. M. B. Pursley, "Spread Spectrum Multiple-Access Communications," in Multi-User Communication Systems, G. Longo, Ed. Springer-Verlag, New York, 1981, pp. 139–199.

1122. D. R. Anderson and P. A. Wintz, "Analysis of a Spread-Spectrum Multiple Access System with a Hard Limiter," *IEEE Trans. Comm. Tech.*, vol. COM-17, pp. 285–290, 1969.

See also 111, 115, 116, 121, 124, 26, 29, 214, 220, 218, 222, 31, 310, 313, 315, 515, 66, 94.

1123. N. C. Mohanty, "Spread Spectrum and Time Division Multiple Access Satellite Communications," *IEEE Trans. Comm.*, August 1977.

1124. H. J. Kochevar, "Spread Spectrum Multiple Access Communications Experiment Through a Satellite," *IEEE Trans. Comm.*, August 1977.

1125. H. E. Rowe, "Bounds on the Number of Signals with Restricted Cross Correlation," *IEEE Trans. Comm.*, May 1982.

1126. E. A. Geraniotis and M. B. Pursley, "Error Probability for Direct-Sequence Spread-Spectrum Multiple Access Communications—Part II: Approximations," *IEEE Trans. Comm.*, May 1982.

1127. E. A. Geraniotis and M. B. Pursley, "Error Probabilities for Slow Frequency Hopped Spread Spectrum Multiple-Access Communications Over Fading Channels," *IEEE Trans. Comm.*, May 1982.

1128. G. Huth, B. Batson, and C. Weber, "Performance of Code Division Multiple Access Systems," *Natl. Telec. Conf. Proc.*, November 1980.

1129. M. B. Pursley and E. A. Geraniotis, "Performance Analysis for Fast Frequency Hopped Spread-Spectrum Multiple—Access Communications," *Natl. Telec. Conf. Proc.*, November 1980.

1130. J. M. Aein and R. L. Pickholtz, "A Simplified Unified Phasor Analysis for PN Multiple Access Limiting Repeaters," *IEEE Trans. Comm.*, May 1982.

1131. T. J. Healy, "The Capacity Region of a Spread Spectrum Multiple-Access Channel," *IEEE Milcom. Conf. Proc.*, October 1981.

1132. I. Mayk, and H. Henderson, "Network Synchronization Initialization Performance of a TDMA System in the Ground Environment," *IEEE Milcom. Conf. Proc.*, October 1981.

1133. C. L. Weber, G. Huth, and B. Batson, "Performance Considerations of Code Division Multiple Access Systems," *IEEE Trans. Veh. Tech.*, February 1981.

1134. F. D. Garber and M. B. Pursley, "Performance of Offset Quadriphase Spread-Spectrum Multiple-Access Communications," *IEEE Trans. Comm.*, March 1981.

1135. S. A. Musa and W. Wasylkiwskyj, "Co-Channel Interference of Spread Spectrum Systems in a Multiple User Environment," *IEEE Trans. Comm.*, October 1978.

1136. D. Raychaudhuri, "Performance Analysis of Random Access Packet-Switched Code Division Multiple Access Systems," *IEEE Trans. Comm.* June 1981.

1137. J. M. Aein, "Multiple access to a hard-limiting communication-satellite repeater," *IEEE Trans. Space Electron. Telem.*, vol. SET-10, Dec. 1964.

1138. J. W. Schwartz, J. M. Aein, and J. Kaiser, "Modulation techniques for multiple access to a hard-limiting satellite repeater," *Proc. IEEE*, vol. 54, May 1966.

1139. H. F. A. Roefs, "Binary sequences for spread spectrum multiple access communication," Ph.D. dissertation, Dept. Elec. Eng., University of Illinois, Urbana, and Coord. Sci. Lab., Rep. R-785, August 1977.

1140. J. M. Aein and R. L. Pickholtz, "A simple unified phasor analysis for PN multiple access to limiting repeaters," this issue, pp. 1018–1026.

1141. G. Einarsson, "Address assignment for a time-frequency-coded, spread-spectrum system," *Bell Syst. Tech. J.*, vol. 59, pp. 1241–1255, September 1980.

12. PHASE LOCK

1201. M. G. Davis, Jr., "On the Measurement of Phase Detector Scale Factor," TRW 10C 7325-17, April 1968.

1202. J. A. Develet, Jr., "Threshold Criterion for Phase-Lock Demodulation," *IEEE Proc.*, February 1963.

1203. J. T. Frankle and J. Klappen, "Principles of Phase Lock and Frequency Feedback," *RCA*, October 1967.

1204. J. P. Frazier and J. Page, "Phase-Lock Loop Frequency Acquisition Study," *IRE Trans. Space Elect. Telem.*, September 1962.

1205. W. J. Gruen, "Theory of AFC Synchronization," *IRE Proc.*, August 1953.

1206. L. A. Hoffman, "Receiver Design and the Phase-Lock Loop," *IEEE Elect. Space Exploration Lecture Series*, August 1963.

1207. R. Jaffe and E. Rechtin, "Design and Performance of Phase-Lock Circuits Capable of Near-Optimum Performance Over a Wide Range of Input Signal and Noise Levels," *IRE Trans. Info. Th.*, March 1955.

1208. J. M Janky, "Nomograms Simplify Phase-Lock-Loop Analysis," *Microwaves,* March 1970.

1209. D. R. Lohrmann, "Designing Sampling Phaselock Loops," *Electronic Design,* November 8, 1970.

1210. H. T. McAleer, "A New Look at the Phase-Locked Oscillator," *IRE Trans.*, June 1959.

1211. G. Nash, "Phase-Locked Loop Design Fundamentals," Motorola AN-535, October 1970.

1212. G. W. Preston and J. C. Tellier, "The Lock-In Performance of an AFC Circuit," *IRE Proc.*, February 1953.

1213. R. W. Sanneman and J. R. Rowbotham, "Unlock Characteristics of the Optimum-Type II Phase-Locked Loop," *IEEE Trans. Aerosp. Nav. Elect.*, March 1964.

1214. R. C. Tausworthe, "Theory and Practical Design of Phase-Locked Receivers," *JPL Tech. Report* 32-819, February 1966.

1215. H. L. Van Trees, "Functional Techniques for the Analysis of the Nonlinear Behavior of Phase-Locked Loops," *IEEE Proc.*, August 1964.

1216. A. J. Viterbi, "Acquisition and Tracking Behavior of Phase-Locked Loops," *JPL Ext. Pub.* 673, July 1959.

1217. A. J. Viterbi, "Phase-Locked Loop Dynamics in the Presence of Noise by Fokker-Planck Techniques," *IEEE Proc.*, December 1963.

1218. W. D. Young, "Receiver Lock Loop Theory and Application," *STL* 10C 9331.2-*258*, January 1962.

1219. "Non-Linear PLL's Flunk Test on Vulnerability to Noise and Interference," *Communications Designer's Digest*, November 1969.

1220. W. C. Lindsey and M. K. Simon, "The Effect of Loop Stress on the Performance of Phase-Coherent Communication Systems," *IEEE Trans. Comm. Tech.*, October 1970.

1221. R. L. Didday and W. C. Lindsey, "Subcarrier Tracking Methods and Communication System Design," *JPL Report* 32-1317, August 1968.

1222. W. C. Lindsey and M. K. Simon, "The Performance of Suppressed Carrier Tracking Loops in the Presence of Frequency Detuning," *IEEE Proc.*, September 1970.

1223. W. C. Lindsey and R. C. Tausworthe, "A Bibliography of the Theory and Application of the Phase-Lock Principle," *JPL Report* 32-1581, April 1, 1973.

13. NAVIGATION/RANGING

1301. R. J. Richardson, "Optimum Transponder Design for Pseudonoise-Coded Ranging Systems, Weak Signals," *IEEE Trans. Aerosp. Elect. Syst.*, January 1972.

1302. A. W. Rihaczek and R. M. Golden, "Resolution Performance of Pulse Trains with Large Time-Bandwidth Products," *IEEE Trans. Aerosp. Elect. Syst.*, July 1971.

1303. B. M. Horton, "Noise Modulated Distance Measuring Systems," *IRE Proc.*, May 1959.

1304. R. L. Frank and S. Zadoff, "Phase-coded hyperbolic navigation system," U.S. Patent 3 099 835, July 30, 1963 (filed May 31, 1956).

1305. G. Guanella, "Direction finding system," U.S. Patent 2 166 991, July 25, 1939 (filed in U.S. November 24, 1937; in Switzerland December 1, 1936).

1306. F. Torino, "NAVSTAR Global Positioning System Development," *Natl. Telec. Conf. Proc.*, November 1980.

1307. R. E. Sansom, "Design and Performance Considerations for Military Aircraft and Orbital Satellite GPS Applications," *Natl. Telec. Conf. Proc.*, November 1980.

1308. A. V. VanLeuwen, "Space Shuttle GPS Implementation," *Natl. Telec. Conf. Proc.*, November 1980.

1309. T. Thompson, "GPS Utilization for Trident Missile Testing," *Natl. Telec. Conf. Proc.*, November 1980.

1310. B. Stonestreet, "GPS Missile Accuracy Evaluator for the MX Program," *Natl. Telec. Conf. Proc.*, November 1980.

1311. P. S. P. Hui, and C. M. Neily, Jr., "Modular Architecture and Adaptive Interface Design for a GPS Orbital Terminal," *Natl. Telec. Conf. Proc.*, November 1980.

1312. P. W. Nilsen, "Threshold Performance of a GPS Receiver for Shuttle Applications," *Natl. Telec. Conf. Proc.*, November 1980.

1314. C. W. Helms, and T. Logsdon, "Comparison Between the Navigation Capabilities of the Navstar GPS and Competing Systems," *IEEE Eascon. Proc.*, November 1981.

1315. L. Feit and P. T. Domanico, "Navigation Signal Pseudo Random Noise Generator Design and Performance Verification," *IEEE Eascon. Proc.*, November 1981.

1316. R. J. Esposito, and E. M. Sawtelle, "FAA Tests on Navstar Z Set," *IEEE Eascon Proc.*, November 1981.

1317. W. Euler, "GPS Provides a Revolutionary New Military Capability," *IEEE Eascon Proc.*, November 1981.

1318. P. Kruh, "Build-up and Replacement of NAVSTAR/GPS and the 18 Satellite Constellation," *ITC Conf. Proc.*, October 1981.

1319. P. Ward, "An Inside View of Pseudo Range and Delta Range Measurements in an NAVSTAR GPS Receiver," *ITT Conf. Proc.*, October 1981.

1320. E. Martin, "Applications of GPS Phase Coherency," *ITC Conf. Proc.*, October 1981.

1321. R. Cnossen and J. D. Cardall, "Civil Air Applications Of Differential GPS," *ITC Conf. Proc.*, October 1981.

1322. J. Painter, P. Noe, and T. Rhyne, "New Comparisons Between C/A and Other Marine Navaids," *ITC Conf. Proc.*, October 1981.

See also 215, 219, 232, 325, 330.

14. RF EFFECTS

1401. J. M. Aein, "On the Output Power Division in a Captured Hard-Limiting Repeater," *IEEE Trans. Comm. Tech.*, June 1966.

1402. A Benoit, "Signal Attenuation Due to Neutral Oxygen and Water Vapour, Rain and Clouds," *The Microwave Journal*, November 1968.

1403. N. M. Blackman, "The Output Signal-to-Noise Ratio of a Bandpass Limiter," *IEEE Trans. Aerosp. Elect. Sys.*, July 1968.

1404. J. J. Bussgang and M. Leiten, "Analysis of Phase-Shift Transmission Through Fading Channels," *NEREM Record*, 1965.

1405. C. R. Cahn, "Effects of Phase and Amplitude Distortion on PSK Signal Demodulation," *Magnavox Research Laboratories* MX-TM-3099-71, March 1971.

1406. W. B. Davenport, Jr., "Signal-to-Noise Ratios in Band-Pass Limiters," *J. Appl. Phys.*, June 1953.

1407. W. Doyle, "Elementary Derivation for Bandpass Limiter S/N," *IEEE Trans. Info. Th.*, March 1962.

1408. J. L. Eckstrom, "Coherent Matched Filter Detection of Quadratically Phase Distorted Carrier-Band Pulses, with Application to Trans-Ionospheric Signalling," *IEEE Trans.* SET-10, No. 1, March 1963.

1409. R. S. Elliott, "Pulse Waveform Degradation Due to Dispersion in Waveguide," *IRE Trans. Micro. Th. Tech.*, October 1957.

1410. I. Jacobs, "The Effects of Video Clipping on the Performance of an Active Satellite PSK Communication System," *IEEE Trans. Comm. Tech.*, June 1965.

1411. J. J. Jones, "Filter Distortion and Intersymbol Interference Effects on QPSK," Proc. Hawaii Conf. Syst. Sci., January 1970.

1412. P. R. Karr, "The Effect of Phase Noise on the Performance of Bi-Phase Communications Systems," TRW 07791-6080-R000, September 1966.

1413. R. L. Kirlin, "Hard-Limiter Intermodulation with Low Input Signal-to-Noise Ratio," *IEEE Trans. Comm. Tech.*, August 1967.

1414. R. K. Kwan, "The Effects of Filtering and Limiting a Double-Binary PSK Signal," *IEEE Trans. Aerosp. Elect. Syst.*, July 1969.

1415. J. L. Sevy, "The Effect of Limiting a Biphase or Quadriphase Signal Plus Interference," *IEEE Trans. Aerosp. Elect. Syst.*, May 1969.

1416. P. D. Shaft, "Limiting of Several Signals and Its Effect on Communication System Performance," *IEEE Trans. Comm. Tech.*, December 1965.

1417. H. Staras, "The Propagation of Wideband Signals Through the Atmosphere," *IRE Proc.* **49,** No. 7, July, 1961.

1418. E. D. Sunde, "Pulse Transmission by AM, FM, and PM in the Presence of Phase Distortion," *BSTJ*, March 1961.

1419. M. Y. Weidner, "Analysis of Spread Spectrum Multiple Access Systems Experiencing Linear Distortion," TRW 10C 7323.4-117, May 1968.

1420. "Digital Transmission Study Rates Filtering Effects on Spread-Spectrum PSK," *Communications Designer's Digest*, August 1969.

1421. J. M. Speiser and H. J. Whitehouse, "Signal Design for Synchronization and Multipath Resolution in Delay-Spread Channels," *Proc. Symp. Sp. Spec. Comm.*, March 1973.

1422. C. R. Cahn and C. R. Moore, "Bandwidth Efficiency for Digital Comm Via a Hard Limiting Channel," *ITC Proc.*, **VIII,** 1972.

1423. S. Y. Kwon and R. S. Simpson, "Effect of Hard Limiting on a Quadrature PSK Signal," *IEEE Trans. Comm.*, July 1973.

1424. R. A. Le Fande, "Effects of Phase Non Linearities on a Phase Shift Keyed Pseudonoise/Spread Spectrum System," *IEEE Trans. Comm. Tech.*, October 1970.

1425. O. C. Yue, "Hard-limited Versus Linear Combining for Frequency Multiple-Access Systems in a Rayleigh Fading Environment," *IEEE Trans. Veh. Tech.*, vol. VT-30, pp. 10–14, February 1981.

1426. R. W. Nettleton and G. R. Cooper, "Performance of a Frequency-Hopped Differentially Modulated Spread-Spectrum Receiver in a Rayleigh Fading Channel," *IEEE Trans. Veh. Tech.*, vol. VT-30, pp. 14–29, February 1981.

1427. D. E. Borth and M. B. Pursley, "Analysis of Direct-Sequence Spread-Spectrum Multiple-Access Communication over Rician fading Channels," *IEEE Trans. Comm.*, vol. COM-27, pp. 1566–1577, October 1979.

1428. E. A. Geraniotis, and M. B. Pursley, "Error Probability Bounds for Slow Frequency-Hopped Spread Spectrum Multiple Access Communications over Fading Channels," in *Proc. 1981 Int. Conf. Comm.*

1429. L. V. Milstein and D. L. Schilling, "Performance of a Spread Spectrum Communication System Operating over a Frequency-selective Fading Channel in the Presence of Tone Interference," *IEEE Trans. Comm.*, vol. COM-30, pp. 240–247, January 1982.

1430. L. B. Milstein and D. L. Schilling, "The Effect of Frequency Selective Fading on a Noncoherent FH-FSK System Operating with Partial-Band Interference," *IEEE Trans. Comm.*, May 1982.

1431. D. S. Arnstein, "Power Division in Spread Spectrum Systems with Limiting" *IEEE Trans. Comm.*, vol. COM-27, pp. 574–582, 1979.

1432. P. Monsen, "Fading Channel Communications," *IEEE Comm. Mag.*, vol. 18, pp. 16–25, January 1980.

1433. S. S. George, "Antenna Multiplexers for VHF Frequency Hopping Radio," *IEEE Milcom. Conf. Proc.*, October 1981.

1434. P. F. Sass, "Propagation Measurements for Spread-Spectrum Communications," *IEEE Milcom. Conf. Proc.*, October 1981.

1435. H. Ochsner, "Analysis of Pseudo-Noise Spread-Spectrum Communication over Randomly Time-Variant Channels without Assuming Perfect Synchronization," *IEEE Milcom. Conf. Proc.*, October 1981.

1436. H. C. Kashian, D. C. Rogers, and J. R. Walker, "Propagation Measurements over Geographically Diverse Paths," *IEEE Milcom. Conf. Proc.*, October 1981.

1437. F. D. Garber and M. B. Pursley, "Effects of Frequency-Selective Fading on Slow-Frequency Hopped DPSK Spread Spectrum Communications," *IEEE Milcom. Conf. Proc.*, October 1981.

1438. E. T. Tsui and R. Ibaraki, "An Adaptive Binary DPSK Spread Spectrum Receiver for Multipath/Scatter Channels," *IEEE Milcom. Conf. Proc.*, October 1981.

1439. A. C. Lythe, "Signal Processing Distortion Loss in Spread-Spectrum Communication, Command, Control and Navigation Systems," *Natl. Telec. Conf. Proc.*, November 1980.

1440. J. D. Oetting, "An Analysis of Anti-Jam Communications Requirements in Fading Media," *Natl. Telec. Conf. Proc.*, November 1980.

1441. N. E. Kanayake, "The Effects of Transponder Non-Linearity on the Performance of MSK and Offset QPSK Signal Transmission," *Natl. Telec. Conf. Proc.*, November 1980.

1442. L. Alcuri, G. Mamola, and E. Randazzo, "Signal-to-Noise Ratio in Bandpass Direct-Sequence Spread-Spectrum Modulation Systems," *IEEE Trans. Comm.*, March 1982.

1443. A. Weinberg, "The Effects of Repeater Hard-Limiting, Filter Distortion, and Noise on a Pseudo-Noise, Time-of-Arrival Estimation System," *IEEE Trans. Comm.*, September 1979.

1444. R. Price, "Notes on ideal receivers for scatter multipath," M.I.T. Lincoln Lab., Group Rep. 34-39, May 12, 1955 (AD 224559).

1445. G. L. Turin, "Communication through noisy, random-multipath channels," in *IRE Conv. Rec.*, New York, NY, Mar. 19–22, 1956, part 4, pp. 154–166.

1446. P. Monsen, "Fading channel communications," *IEEE Comm. Mag.*, vol. 18, pp. 16–25, January 1980.

1447. R. Price and P. E. Green, Jr., "An anti-multipath communication system," M.I.T. Lincoln Lab., Tech. Memo. 65, Nov. 9, 1956 (not available from M.I.T.).

1448. R. Price and P. E. Green, Jr., "A communication technique for multipath channels," *Proc. IRE*, vol. 46, pp. 555–570, March 1958.

1449. R. Price and P. E. Green, Jr., "Anti-multipath receiving system," U.S. Patent 2 982 853, May 2, 1961 (filed July 2, 1956).

1450. D. E. Sunstein and B. Steinberg, "Communication technique for multipath distortion," U.S. Patent 3 168 699, February 2, 1965 (filed June 10, 1959).

1451. W. G. Ehrich, "Common channel multipath receiver," U.S. Patent 3 293 551, December 20, 1966 (filed December 24, 1963).

1452. R. M. Fano, "Anti-multipath communication system," U.S. Patent 2 982 852, May 2, 1961 (filed November 21, 1956).

1453. P. A. Bello and B. P. Nelin, "The effect of frequency selective fading on the binary error probabilities of incoherent and differentially coherent matched filter receivers," *IEEE Trans. Comm. Syst.*, vol. CS-11, pp. 170–180, June 1963.

1454. J-F. Chang, "The Effect of Multipath Interference on the Performance of a Digital Matched Filter," *IEEE Trans. Comm.*, April 1979.

1455. T. C. Huang and W. C. Lindsey, "PN Spread Spectrum Signaling Through a

Nonlinear Satellite Channel Disturbed by Interference and Noise," *IEEE Trans. Comm.*, May 1982.

1456. E. A. Gerionotis and M. B. Pursley, "Effects of Specular Multipath Fading on Noncoherent Direct Sequence Spread-Spectrum Communications," *I.C.C. Proc.*, 1982.

15. SYNCHRONIZATION

1501. C. R. Cahn, "Synchronization Scheme with Sequential Detection," Magnavox Research Laboratories, MX-TM-3084-70, October 1970.

1502. R. M. Gagliardi, "A Geometrical Study of Transmitted Reference Communications Systems," *IEEE Trans. Comm. Tech.*, December 1964.

1503. S. W. Golomb, et al., "Synchronization," *IEEE Trans. Comm. Syst.*, December 1963.

1504. H. Kaneko, "A Statistical Analysis of the Synchronization of a Binary Receiver," *IEEE Trans. Comm. Syst.*, December 1963.

1505. W. C. Lindsey, *Synchronization Systems in Communication and Control*, Prentice-Hall, Englewood Cliffs, New Jersey, 1972.

1506. R. W. Mifflin and J. P. Wheeler, "Transmitted Reference Synchronization System," U.S. Patent 3,641,433, February 8, 1972.

1507. "Coherent System Speeds Sync Acquisition in Military TDMA Systems," *Communications Designer's Digest*, January 1970.

1508. C. R. Carter and S. S. Haykin, "A New Synchronization Technique for TDMA Satellite Systems," *Proc. Symp. Sp. Spec. Comm.*, March 1973.

1509. R. M. Gagliardi, "Rapid Acquisition Signal Design in a Multiple-Access Environment," *Proc. Symp. Sp. Spec. Comm.*, March 1973.

1510. R. LaRosa, "Switchable and Fixed-Code Surface Wave Matched Filters," *Proc. Symp. Sp. Spec. Comm.*, March 1973.

1511. J. L. Ramsey, "Effective Acquisition of FH TDMA Signals in Jamming," *Proc. Symp. Sp. Spec. Comm.*, March 1973.

1512. R. Cole, Jr., "Synchronization of a Frequency-Hopped Spread Spectrum Signal," *Proc. Symp. Sp. Spec. Comm.*, March 1973.

1513. G.A. De Couvreur, "Effect of Random Synchronization Errors in PN and PSK Systems," *IEEE Trans. Aerosp. Elect. Syst.*, January 1970.

1514. J. R. Sergo and J. F. Hayes, "Analysis and Simulation of a PN Synchronization System," *IEEE Trans. Comm. Tech.*, October 1970.

16. TRACKING

1601. W. J. Gill, "A Comparison of Binary Delay-Lock Tracking-Loop Implementations," *IEEE Trans. Aerosp. Elect. Syst.*, July 1966.

1602. S. S. Haykim and C. Thorsteinson, "A Quantized Delay-Lock Discriminator," *IEEE Proc.*, June 1968.

1603. R. J. Huff and K. L. Reinhard, "A Sampled-Data Delay-Lock Loop for Synchronizing TDMA Space Communications Systems," *EASCON* 1968 Record.

1604. J. J. Spilker, Jr., "Delay-Lock Tracking of Binary Signals," *IEEE Trans. Sp. Elect. Telem.*, March 1963.

1605. R. B. Ward, "Application of Delay-Lock Radar Techniques to Deep-Space Tasks," *IEEE Trans. Sp. Elect. Telem.*, June 1964.

1606. F. Bruno, "Tracking Performance and Loss of a Carrier Loop Due to the Presence of a Spoofed Spread Spectrum Signal," *Proc. Symp. Sp. Spec. Comm.*, March 1973.

1607. J. J. Freeman, "The Action of Dither in a Polarity Coincidence Correlator," *IEEE Trans. Comm.*, June 1974.

1608. H. P. Hartman, "Analysis of a Dithering Loop for PN Code Tracking," *Proc. Symp. Sp. Spec. Comm.*, March 1973.

1609. M. K. Simon, "Non-Linear Analysis of an Absolute Value Type of an Early-Late Gate Bit Synchronizer," *IEEE Trans. Comm. Tech.*, October 1970.

1610. D. M. Grieco, "The Application of Charge-Coupled Devices to Spread Spectrum Systems," *IEEE Trans. Comm.*, September 1980.

1611. D. G. O'Clock, Jr. and M. T. Duffy, "Acoustic Surface Wave Properties of Epitaxially Grown Aluminum Nitride and Gallium Nitride on Sapphire," *Appl. Phys. Lett.*, 15 July 1973.

1612. G. D. O'Clock, Jr., "Acoustic-Wave Filters Improve Radar," *Electronic Design*, 5 July 1973.

1613. G. S. Kino and H. Matthews, "Signal Processing in Acoustic Surface-Wave Devices," *IEEE Spectrum*, August 1971.

1614. R. D. Colvin, "Spread-Spectrum Devices Cater to New Systems," *Microwaves*, February 1981.

1615. S. Reible, "Acoustoelectric Convolver Technology for Spread Spectrum Communications," *IEEE Trans. MTT*, May 1981.

1616. K. Dostert and M. Pandit, "Performance of a SAW Tapped Delay Line in an Improved Synchronization Circuit," *IEEE Trans. Comm.*, January 1982.

1617. V. C. M. Leung and R. W. Donaldson, "Confidence Estimates for Acquisition Times and Hold-In Times for PN-SSMA Synchronizer Employing Envelope Correlation," *IEEE Trans. Comm.*, January 1982.

1618. R. B. Ward and K. P. Yiu, "Acquisition of Pseudonoise Signals by Recursion-Aided Sequential Estimation," *IEEE Trans. Comm.*, vol. COM-25, pp. 784–794, August 1977.

1619. P. M. Hopkins, "A Unified Analysis of Pseudonoise Synchronization by Envelope Correlation," *IEEE Trans. Comm.*, August 1977.

1620. H. K. Holmes and C. C. Chen, "Acquisition Time Performance of PN Spread Spectrum Systems," *IEEE Trans. Comm.*, August 1977.

1621. M. K. Simon, "Noncoherent Pseudonoise Code Tracking Performance of Spread Spectrum Receivers," *IEEE Trans. Comm.*, March 1977.

1622. R. A. Yost and R. W. Boyd, "A Modified PN Code Tracking Loop: Its Performance and Implementation Sensitivities," *Proc. Natl. Telec. Conf.*, December 1980.

1623. W. P. Baier, H. Grammuller, and M. Pandit, "Combined Acquisition and Fine Synchronization System for Spread Spectrum Receivers Using a Tapped Delay Line Correlator," *Proc AGARD Conf.*, 1977.

1624. M. Pandit, "The Mean Acquisition Time of Active and Passive Correlation Acquisition Systems for Spread Spectrum Communication Systems," *Proc. Inst. Elec. Eng.*, 1981.

1625. P. W. Baier, K. Dostert, and M. Pandit, "A Novel Spread-Spectrum Receiver Synchronization Scheme Using a SAW-Tapped Delay Line," *IEEE Trans. Comm.*, May 1982.

1626. P. M. Hopkins, "A Unified Analysis of Pseudonoise Synchronization by Envelope Correlation," *IEEE Trans. Comm.*, August 1977.

1627. J. K. Holmes and C. C. Chen, "Acquisition Time Performance of PN Spread Spectrum Systems," *IEEE Trans. Comm.*, August 1977.

1628. R. B. Ward and K. P. Yiu, "Acquisition of Pseudonoise Signals by Recursion-Aided Sequential Estimation," *IEEE Trans. Comm.*, August 1977.

1629. L. B. Milstein and P. K. Das, "Spread Spectrum Receiver Using Surface Acoustic Wave Technology," *IEEE Trans. Comm.*, August 1977.

1630. S. L. Maskara, "An Active RC Delay Line for Matched Filter Correlators," *IEEE Trans. Comm.*, August 1977.

1631. S. C. Munroe, "The Use of CCD Correlators in the SEEK COMM Spread Spectrum Receivers," *IEEE Milcom. Conf. Proc.*, October 1981.

1632. K. H. Annecke, "Adaptive Regeneration of Combination Spreading Functions by Nonlinear Control Loops," *IEEE Milcom. Conf. Proc.*, October 1981.

1633. R. Peterson and R. Ziemer, "Multiple Dwell Serial Acquisition for Direct Sequence Spread Spectrum Systems," *IEEE Milcom Conf. Proc.,* October 1981.

1634. K. T. Woo, "Acquisition of Fast Frequency Hopping Spread Spectrum Systems," *IEEE Milcom Conf. Proc.*, October 1981.

1635. S. Davidovici, L. B. Milstein and D. L. Schilling, "A Fast PN Synchronization Scheme," *IEEE Milcom Conf. Proc.*, October 1981.

1636. K. Dostert and M. Pandit, "Synchronization of Spread Spectrum Binary Orthogonal Keyed (BOK) Burst Transmission Systems," *IEEE Milcom Conf. Proc.*, October 1981.

1637. R Singh and S. Kochar, "Acquisition and Network Response Times for Centrally Controlled SS Networks," *IEEE Milcom Conf. Proc.*, October 1981.

1638. G. D. O'Clock, "Comparison of Microwave and Optical Signal Processors for Spread Spectrum Correlation," *IEEE Trans. Comm.*, August 1977.

1639. R. LaRosa, S. Kerbel, T. Fowler and P. Corcoran, "Bandwidth Partitioned Programmable Matched Filter for 54,000 Time-Bandwidth Product Direct Sequence," *IEEE Milcom Conf. Proc.*, October 1981.

1640. P. W. Baier, J. Meyer, and H. Waibel, "Power Level Adaptive Synchronization Circuit for Spread-Spectrum BOK Receivers in Burst Transmission Systems," *IEEE Milcom Conf. Proc.*, October 1981.

1641. A. Polydoros and C. L. Weber, "Analysis and Optimization of Correlative Code Tracking Loop for Spread Spectrum Systems," *IEEE Milcom Conf. Proc.*, October 1981.

1642. J. R. Luecke and R. A. Yost, "Decision Directed AFC for Noncoherent Detectors Experiencing Large Signal Dynamics," *IEEE Milcom Conf. Proc.*, October 1981.

1643. C. A. Putnam, S. S. Rappaport, and D. L. Schilling, "Tracking of Frequency Hopped Spread Spectrum Signals in Adverse Environments," *IEEE Milcom Conf. Proc.*, October 1981.

1644. N. K. Broome, "An All Digital Maximum A Posteriori Based Synchronizer for MSK," *IEEE Milcom Conf. Proc.*, October 1981.

1645. G. W. Judd and R. B. Nelson, "Surface Acoustic Wave Device Applications to Communication Equipment," *Natl. Telec. Conf. Proc.*, November 1980.

1646. R. R. Rhodes, W. K. Hutchinson, and B. H. Hutchinson, "Correlators and Convolvers Used in Spread Spectrum Systems," *Natl. Telec. Conf. Proc.*, November 1980.

1647. J. H. Goll, "A SAW Convolver-Based Spread Spectrum Communications Subsystem," *Natl. Telec. Conf. Proc.*, November 1980.

1648. W. M. Bowles, "GPS Code Tracking and Acquisition Using Extended-Range Detection," *Natl. Telec. Conf. Proc.*, November 1980.

1649. J.-F. Chang, "The Performance of a Spread-Spectrum, Multiple-Access Communication System Using Digital Matched Filters," *Natl. Telec. Conf. Proc.*, November 1980.

1650. D. M. Gieco, "The Application of Charge-Coupled Devices to Spread-Spectrum Systems," *Natl. Telec. Conf. Proc.*, November 1980.

1651. R. A. Yost, "A Modified PN Code Tracking Loop: Its Performance and Implementation Sensitivities," *Natl. Telec. Conf. Proc.*, November 1980.

1652. J. A. Ponnusamy and M. D. Srinath, "Acquisition of FH PN Codes Using Partial Correlation Coefficients," *Natl. Telec. Conf. Proc.*, November 1980.

1653. W. R. Shreve, "Signal Processing Using Surface Acoustic Waves," *Hewlett-Packard Journal*, January 1982.

1654. M. K. Simon, "Noncoherent Pseudonoise Code Tracking Performance of Spread Spectrum Receivers," *IEEE Trans. Comm.*, March 1977.

1655. W. K. Alem, G. K. Huth, J. K. Holmes, and S. Udalov, "Spread Spectrum Acquisition and Tracking Performance for Shuttle Communication Links," *IEEE Trans. Comm.*, November 1978.

1656. S. S. Rappaport and D. L. Schilling, "A Two-Level Coarse Code Acquisition Scheme for Spread Spectrum Radio," *IEEE Trans. Comm.*, September 1980.

1657. A. K. Elhakeem, G. S. Takhar, and S. C. Gupta, "New Code Acquisition Techniques in Spread-Spectrum Communication," *IEEE Trans. Comm.*, February 1980.

1658. G. D. O'Clock, Jr., C. L. Grasse, and D. A. Gandolfo, "Switchable Acoustic Surface Wave Sequence Generator," *IEEE Proc.*, October 1971.

1659. G. D. O'Clock, Jr. and M. T. Duffy, "Acoustic Surface Wave Properties of Epitaxially Grown Aluminum Nitride and Gallium Nitride On Sapphire," *Appl. Phys. Lett.*, July 15, 1973.

1660. G. D. O'Clock, Jr., "Acoustic-Wave Filters Improve Radar," *Electronic Design*, July 5, 1973.

1661. G. S. Kino, and H. Matthews, "Signal Processing in Acoustic Surface-Wave Devices," *IEEE Spectrum*, August 1971.

1662. R. D. Colvin, "Spread-Spectrum Devices Cater to New Systems," *Microwaves*, February 1981.

1663. S. A. Reible, "Acoustoelectric Convolver Technology for Spread Spectrum Communications," *IEEE Trans. MTT*, May 1981.

1664. D. M. Grieco, "The Application of Charge-Coupled Devices to Spread Spectrum Systems," *IEEE Trans. Comm.*, September 1980.

1665. D. Casasent, "A Review of Optical Signal Processing," *IEEE Comm. Mag.*, September 1981.

1666. W. R. Shreve, "Signal Processing Using Surface Acoustic Waves," *Hewlett-Packard Journal*, January 1982.

1667. K. Dostert and M. Pandit, "Performance of a SAW Tapped Delay Line in an Improved Synchronization Circuit," *IEEE Trans. Comm.*, January 1982.

1668. V. C. M. Leung and R. W. Donaldson, "Confidence Estimates for Acquisition Times and Hold-In Times for PN-SSMA Synchronizer Employing Envelope Correlation," *IEEE Trans. Comm.*, January 1982.

1669. D. Casasent, "Optical Signal Processing," *Electro-Optical Systems Design*, September 1981.

1670. M. K. Simon, "Two-Channel Costas Loop Tracking Performance for UQPSK Signals With Arbitrary Data Formats," *IEEE Trans. on Comm.*, September 1981.

1671. A. Kohlenberg, S. M. Sussman, and D. Van Meter, "Matched filter communication systems," U.S. Patent 3 876 941, April 8, 1975 (filed June 23, 1961).

1672. E. Guillemin, "Matched filter communication systems," U.S. Patent 3 936 749, February 3, 1976 (filed June 23, 1961). S. M. Sussman, "A matched filter communication system for multipath channels," *IRE Trans. Info. Th.*, vol. IT-6, pp. 367–373, June 1960.

1673. J. J. Spilker, Jr. and D. T. Magill, "The delay-lock discriminator—An optimum tracking device," *Proc. IRE*, vol. 49, pp. 1403–1416, September 1961.

1674. R. A. Yost and R. W. Boyd, "A modified PN code tracking loop: Its performance and implementation sensitivities," in *Proc. Nat. Telecommun. Conf.*, Houston, TX, December 1980, pp. 61.5.1–61.5.5.

1675. B. J. Hunsinger, "Surface acoustic wave devices and applications—3. Spread spectrum processors, "*Ultrasonics*, pp. 254–262, November 1973.

1676. W. P. Baier, H. Grammuller, and M. Pandit, "Combined acquisition and fine synchronization system for spread spectrum receivers using a tapped delay line correlator," in *Proc. AGARD Conf.*, 1977, no. 230.

1677. K. Dostert and M. Pandit, "Performance of a SAW tapped delay line in a synchronizing circuit," *IEEE Trans. Comm.*, vol. COM-30, pp. 219–222, January 1982.

1678. M. Pandit, "The mean acquisition time of active and passive correlation acquisition systems for spread spectrum communication systems," *Proc. Inst. Elec. Eng.*, Part F, 1981.

1679. P. M. Hopkins, "A unified analysis of pseudonoise synchronization by envelope correlation," *IEEE Trans. Comm.*, vol. COM-25, pp. 770–778, August 1977.

1680. J. K. Holmes and C. C. Chen, "Acquisition time performance of PN spread-spectrum systems," *IEEE Trans. Comm.*, vol. COM-25, pp. 778–783, August 1977.

1681. C. A. Putman, S. S. Rappaport, and D. L. Schilling, "A Comparison of Schemes for Coarse Acquisition of Frequency-Hopped Spread-Spectrum Signals," *IEEE Trans. Comm.*, Feb. 1983.

1682. D. M. DiCarlo and C. L. Weber, "Multiple Dwell Serial Search: Performance and Application to Direct Sequence Code Acquisition," *IEEE Trans. Comm.*, May 1983.

1683. H. Meyr, "Delay-Lock Tracking of Stochastic Signals," *IEEE Trans. Comm.*, March 1976.

1684. W. A. Crofut, "SAWs Critical to Avionics," *MSN Magazine*, November 1980.

1685. W. R. Shreve, "Signal Processing Using Surface Acoustic Waves," *Hewlett-Packard Journal*, January 1982.

1686. J. A. Rajan, "Adaptive Acquisition of Multiple Access Codes," *I.C.C. Proc.*, 1982.

17. MISCELLANEOUS

1701. R. W. Burnell and L. N. Ma, "Fourier Analysis of an Imperfect PSK Signal," TRW 10C 7322.14-7, October 1967.

1702. C. R. Cahn and C. R. Moore, "Bandwidth Efficiency for Digital Comm. Via a Hard Limiting Channel," *ITC Proc.* **VIII,** 1972.

1703. J. C. Hancock, *An Introduction to the Prinicples of Communication Theory,* McGraw-Hill, New York, 1961.

1704. J. Klapper, "The Effect of the Integrator-Dump Circuit on PCM/FM Error Rates," *IEEE Trans. Comm. Tech.,* June 1966.

1705. R. K. Kwan, "The Effects of Filtering and Limiting a Double-Binary PSK Signal," *IEEE Trans. Aerosp. Elect. Sys.,* July 1969.

1706. Mario Petrich, "On the Number of Orthogonal Signals which Can Be Placed in a WT-Product," *J. Soc. Indust. Appl. Math.,* December 1963.

1707. R. J. Richardson, "The Signal-Suppression Threshold of a Frequency Doubler," *IEEE Trans. Comm. Tech.,* December 1964.

1708. H. M. Sierra, "The Matched Filter Concept," *Electro-Technology,* August 1964.

1709. N. Solat, "Comparison of Power Spectra," Space General Corp. Interoffice Memo, October 1963.

1710. J. J. Stiffler, *Theory of Synchronous Communications,* Prentice-Hall, Englewood Cliffs, New Jersey, 1971.

1711. G. L. Turin, "An Introduction to Matched Filters," *IRE Trans. Info. Th.*, June 1960.

1712. C. S. Weaver, "An Adaptive Communications Filter," *IRE Proc.,* October 1961.

1713. W. K. Victor and M. H. Brockman, "The Application of Linear Servo Theory to the Design of AGC Loops," *JPL Ext. Pub. 586,* December 1958.

1714. "Spectrum Analysis," *Hewlett-Packard AN63*, July 1964.

1715. "Digital Filter Studies Anticipate Potential of Spread Spectrum Designs," *Communications Designer's Digest*, July/August 1971.

1716. "Reference Data for Radio Engineers," H. W. Sams, 1968.

1717. D. D. Buss, W. H. Bailey, and L. R. Hite, "Spread-Spectrum Communications Using Charge Transfer Devices," *Proc. Symp. Sp. Spec. Comm.*, March 1973.

1718. B. G. Glazer, "Spread Spectrum Concepts—A Tutorial," *Proc. Symp. Sp. Spec. Comm.,* March 1973.

1719. D. J. Gooding, "Increasing the Utility of the Digital Matched Filter," *Proc. Symp. Sp. Spec. Comm.,* March 1973.

1720. G. L. Matthaei, "Acoustic Surface-Wave Transversal Filters," *IEEE Trans. Ckt. Th.,* September 1973.

1721. N. C. Mohanty, "Signal Design for Small Correlation," University of Southern California Report USCEE 438, February 1973.

1722. M. P. Ristenbatt, "Estimating Effectiveness of Covert Communications," *Proc. Symp. Sp. Spec. Comm.,* March 1973.

1723. S. M. Sussman and E. J. Ferrari, "The Effects of Notch Filters on the Correlation Properties of a PN Signal," *IEEE Trans. Aerosp. Elect. Syst.*, May 1974.

1724. F. M. Torre, "A Unifed Description and Design Format for Spread Spectrum Waveforms," *Proc. Symp. Sp. Spec. Comm.,* March 1973.

1725. N. P. Muraka, "Spread Spectrum Systems Using Noise Band Shift Keying," *IEEE Trans. Comm.,* July 1973.

1726. C. R. Cahn, "Performance of Digital Matched Filter Correlators with Unknown Interference," *IEEE Trans. Comm. Tech.*, December 1971.

1727. C. P. Hatsell and L. W. Nolte, "Detectability of Burst-Like Signals," *IEEE Trans. Aerosp. Elect. Syst.,* March 1971.

1728. R. E. Millett, "A Matched-Filter Pulse-Compression System Using a Nonlinear FM Waveform," *IEEE Trans. Aerosp. Elect. Syst.,* January 1970.

1729. W. C. Lindsey and J. L. Lewis, "Modeling, Characterization and Measurement of Oscillator Frequency Instability," *NRL Report* NRL-1.1351, June 14, 1974.

1730. J. Schiffer, "EW Systems Synthesizers Find a New Design Approach," *MSN,* December 1981.

1731. D. Casasent, "Optical Signal Processing," *Electro-Optical Systems Design,* September, 1981.

1732. S. L. Johnston, "Radar ECCM History," in *Proc. NAECON,* May 1980.

1733. A. J. Viterbi and I. M. Jacobs, "Advances in Coding and Modulation for Noncoherent Channels Affected by Fading, Partial Band, and Multiple Access Interference," *Advances in Communication Systems,* 1975.

1734. P. S. Henry, "Spectrum Efficiency of a Frequency-Hopped-DPSK Spread Spectrum Mobile Radio System," *IEEE Trans. Veh. Tech.,* November 1979.

1735. J. K. Holmes, *Coherent Spread Spectrum Systems,* Wiley, New York, 1982.

1736. R. T. Compton, Jr., "An Adaptive Array in a Spread Spectrum Communication System," *Proc. IEEE,* March 1978.

1737. G. R. Cooper and R. W. Nettleton, "A Spread Spectrum Technique for High-Capacity Mobile Communications," *IEEE Trans. Veh. Tech.,* November 1978.

1738. P. S. Henry, "Spectrum Efficiency of a Frequency-Hopped-DPSK Spread Spectrum Mobile Radio System," *IEEE Trans. Veh. Tech.*, November 1979.

1739. J. D. Edell, "Wideband, Noncoherent, Frequency-Hopped Waveforms and their Hybrids in Low Probability-of-Intercept Communications," *Naval Res. Lab.*, Washington, D.C., November 8, 1976.

1740. G. Einarsson, "Address Assignment for a Time-Frequency-Coded Spread Spectrum System," *Bell Syst. Tech. J.*, September 1980.

1741. C. R. Cahn, "Spread Spectrum Applications and State-of-the Art-Equipments," *Magnavox Comm. Navigation*, November 22, 1972.

1742. A. J. Viterbi, "Spread Spectrum Communications—Myths and Realities," *IEEE Comm. Soc. Mag.*, May 1979.

1743. R. A. Scholtz, "The Origins of Spread-Spectrum Communications," *IEEE Trans. Comm.*, May 1982.

1744. D. L. Schilling, L. B. Milstein, R. L. Pickholtz, and R. Brown, "Optimization of the Processing Gain of an M-ary Direct Sequence Spread Spectrum Communication System," *IEEE Trans. Comm.*, August 1980.

1745. G. K. Huth, "Optimization of Coded Spread Spectrum Systems Performance," *IEEE Trans. Comm.*, August 1977.

1746. F. Ananasso, "Matched Filters Up Power, Reduce Noise in Radar," *Microwaves*, April 1982.

1747. L. B. Milstein, R. L. Pickholtz, and D. L. Schilling, "Optimization of the Processing Gain of an FSK-FH System," *IEEE Trans. Comm.*, July 1980.

1748. R. A. Scholtz, "The Origins of Spread Spectrum Communications," *IEEE Trans. Comm.*, May 1982.

1749. R. L. Pickholtz, D. L. Schilling, and L. B. Milstein, "Theory of Spread-Spectrum Communications—A Tutorial," *IEEE Trans. Comm.*, May 1982.

1750. J. H. Winters, "Spread Spectrum in a Four-Phase Communication System Employing Adaptive Antennas," *IEEE Trans. Comm.*, May 1982.

1751. J. R. Smith, "Tradeoff Between Processing Gain and Interference Immunity in Co-Site Multichannel Spread Spectrum Communications," *IEEE Trans. Comm.*, May 1982.

1752. R. A. Scholtz, "The Spread Spectrum Concept," *IEEE Trans. Comm.*, August 1977.

1753. M. P. Ristenbatt and J. L. Daws, Jr., "Performance Criteria for Spread Spectrum Communications," *IEEE Trans. Comm.*, August 1977.

1754. W. J. DeVore, "Computer-Aided Spread Spectrum Signal Design," *IEEE Trans. Comm.*, August 1977.

1755. R. P. Eckert and P. M. Kelly, "Implementing Spread Spectrum Technology in Land Mobile Radio Services," *IEEE Trans. Comm.*, August 1977.

1756. A. V. Lukosevicius, "VHSIC Impact on Spread Spectrum Communications," *IEEE Milcom. Conf. Proc.*, October 1981.

1757. P. A. Major, "Spread Spectrum EMC: Measurement and Analysis," *IEEE Milcom. Conf. Proc.*, October 1981.

1758. A. El-Osmany and M. Abdel Kader, "Interference Effects of Direct Sequence

Spread Spectrum Signals on Voice AM and FM Communications Systems," *IEEE Milcom. Conf. Proc.*, October 1981.

1759. A. J. Budreau, "Compact Fast-Hopping Frequency Synthesizer for Spread-Spectrum Systems," *IEEE Milcom. Conf. Proc.*, October 1981.

1760. A. J. Budreau, "Frequency-Hopping Filters and Programmable Matched Filters for Spread-Spectrum Systems," *IEEE Milcom. Conf. Proc.*, October 1981.

1761. R. Bernardo, "SAW Delay Line Discriminators and Their Applications to Low Noise Agile UHF Frequency Sources," *IEEE Milcom. Conf. Proc.*, October 1981.

1762. L. B. Milstein, P. K. Das, and J. Gevargiz, "Processing Gain Advantage of Transform Domain Filtering DS-Spread Spectrum Systems," *IEEE Milcom. Conf. Proc.*, October 1981.

1763. R. S. Orr, "FH/MFSK Analyses," *IEEE Milcom. Conf. Proc.*, October 1981.

1764. R. Agustic and G. Junyent, "Performance of a MH Multilevel FSK for Mobile Radio," *IEEE Milcom. Conf. Proc.*, October 1981.

1765. M. Kennedy, "Regulatory Aspects of Spread Spectrum," *IEEE Milcom. Conf. Proc.*, October 1981.

1766. L. F. Chesto, "Spread Spectrum and Radio Frequency Management," *IEEE Milcom. Conf. Proc.*, October 1981.

1767. H. Fienstein, "Spread Spectrum Experiments for Fixed and Land Mobile Applications," *IEEE Milcom. Conf. Proc.*, October 1981.

1768. D. Dodson, M. Huang, and R. Kagiwada, "Fast Hopping SAW Frequency Synthesizer," *IEEE Milcom. Conf. Proc.*, October 1981.

1769. R. R. Rhodes, W. K. Hutchinson, and B. H. Hutchinson, "Frequency Agile Phase-Locked Loop Synthesizer for a Communication Satellite," *Natl. Telec. Conf. Proc.*, November 1980.

1770. H. DePedro and J. K. Wolf, "On the Design of Spread Spectrum Communication Networks," *Natl. Telec. Conf. Proc.*, November 1980.

1771. G. D. O'Clock and L. P. Erickson, "Impact of Molecular Beam Epitaxy Technology on Spread Spectrum Systems," *ITC Conf. Proc.*, October 1981.

1772. W. K. Masenton, "Adaptive Interference Rejection Filters for Spread Spectrum Communication Systems," *ITC Conf. Proc.*, October 1981

1773. R. A. Reilly and C. R. Ward, "The Influence of Technology Advances on Integrated CNI Avionics," *Natl. Telec. Conf. Proc.*, November 1980.

1774. R. Moore, "Spread Spectrum Techniques," *Natl. Telec. Conf. Proc.*, November 1980.

1775. C. H. Haber and E. J. Nossen, "Analog Versus Digital Antijam Video Transmission," *IEEE Trans. Comm.*, March 1977.

1776. D. M. Grieco, "Inherent Signal-to-Noise Ratio Limitations of Charge Coupled Device Pseudonoise Matched Filters," *IEEE Trans. Comm.*, May 1980.

1777. F. Ananasso, "Matched Filters Up Power, Reduce Noise In Radar," *Microwaves*, April 1982.

1778. H. D. Lewis, "Array Radars Solve Communication Jams," *Microwaves*, April 1982.

1779. C. L. Cuccia, "Spread Spectrum Systems Serve Nearly All C^3 Aspects," *MSN,* April 1982.

1780. P. M. Goggans, D. A. Jelinek, and W. F. Nielsen, "SECURICOM: An Antijam, Secure-Message, Security Force Radio Communication System. Technical Summary," Sandia Laboratories SAND80-0347, March 1981.

1781. R. C. Dixon, "SECURICOM: An Anti-Jam, Secure-Message, Security Force Radio Communication System. Design Feasibility Study," Sandia Laboratories SAND80-7049, March 1981.

1782. G. B. Jordan, et al., "Hybrid Spread Spectrum Systems: Theory, Advantages, and Limitations," Philco-Ford Report IRDP G692.FBLB.

1783. M. Spellman, "Spread-Spectrum Radios Thwart Hostile Jammers," *Microwaves,* September 1981.

1784. C. F. N. Cowan and P. M. Grant, "Adaptive Filters Await Breakthroughs," *MSN,* August 1981.

1785. M. Lewis, "PLLs Upconvert Chirp Radar Signals," *Microwaves,* June 1981.

1786. K. Custance, "Hop, Skip, and Jump. Who Wins?" *Telecommunications,* August 1981.

1787. N. E. Bolen, "JTIDS Coordinates Tactical Elements," *Defense Electronics,* January 1982.

1788. J. Schiffer, "EW Systems Synthesizers Find a New Design Approach," *MSN,* December 1981.

1789. W. C. Melton, "Time-of-Arrival Measurements Possible With Use of Global Positioning System," *Defense Electronics,* December 1981.

1790. E. Ribchester, "Frequency-Hopping Radios Outwit 'Smart' Jammers" *Microwaves,* November 1979.

1791. W. Hirt and S. Pasupathy, "Continuous Phase Chirp (CPC) Signals For Binary Data Communication—Parts I and II," *IEEE Trans. Comm.,* June 1981.

1792. B. L. Basore, "Noise-like signals and their detection by correlation," M.I.T. Res. Lab. Electron. and Lincoln Lab., Tech. Rep. 7, May 26, 1952 (AD 004641).

1793. B. J. Pankowski, "Multiplexing a radio teletype system using a random carrier and correlation detection," M.I.T. Res. Lab. Electron. and Lincoln Lab., Tech. Rep. 5, May 16, 1952 (ATI 168857; not available from M.I.T.).

1794. "Engineering study and experimental investigation of secure directive radio communication systems," Sylvania Elec. Products, Buffalo, NY, Interim Eng. Rep., Contr. AF-33(616)-167, Aug. 5–Nov. 5, 1952 (AD 005243).

1795. R. C. Dixon, Ed., *Spread Spectrum Techniques,* IEEE Press, New York, 1976.

1796. J. P. Costas, "Poisson, Shannon, and the radio amateur," *Proc. IRE,* vol. 47, pp. 2058–2068, December 1959.

1797. G. K. Huth, "Optimization of coded spread spectrum systems performance," *IEEE Trans. Comm.,* vol. COM-25, pp. 763–770, August 1977.

1798. B. Goldberg, "Applications of statistical communications theory," presented at the Army Sci. Conf., West Point, NY, June 20–22, 1962 (AD 332048); republished in *IEEE Comm. Mag.,* vol. 19, pp. 26–33, July 1981.

1799. H. G. Lindner, "Communication security method and system," U.S. Patent 4 184 117, January 25, 1980 (filed April 16, 1956). J. W. Craig and R. Price, "A secure voice communication system," *Trans. Electron. Warfare Symp.,* 1959.

17100. L. S. Schwartz, "Wide-bandwidth communications," *Space Aeronautics,* pp. 84–89, December 1963.

17101. J. J. Spilker, Jr., "Nonperiodic energy communication system capable of operating at low signal-to-noise ratios, U.S. Patent 3 638 121, January 25, 1972 (filed December 20, 1960).

17102. R. Lowrie, "A secure digital command link," *IRE Trans. Space Electron. Telem.,* vol. SET-6, pp. 103–114, Sept.-Dec. 1960.

17103. E. Rechtin, "An annotated history of CODORAC: 1953–1958," Jet Propulsion Lab., Pasadena, CA, Rep. 20–120, Contr. DA-04-495-Ord 18, August 4, 1958 (AD 301248).

17104. "Lincoln F9C-A radio teletype system," Sylvania Electron. Defense Lab., Mountian View, CA, Instruction Manual EDL-B8, December 21, 1956.

17105. M. G. Nicholson, Jr., "Time delay apparatus," U.S. Patent 2 401 094, May 28, 1946 (filed June 23, 1944).

17106. E. M. Deloraine, H. G. Busignies, and L. A. deRosa, "Facsimile system," U.S. Patent 2 406 811, September 3, 1946 (filed December 15, 1942).

17107. E. M. Deloraine, H. G. Busignies, and L. A. deRosa, "Facsimile system and method," U.S. Patent 2 406 812, September 3, 1946 (filed January 30, 1943). L. A. deRosa, "Random impulse system," U.S. Patent 2 671 896, Mar. 9, 1954 (filed Dec. 13, 1942).

17108. E. M. Deloraine, "Protected communication system," Fed. Radio Tel. Lab., New York, NY, Rep. 937-2, April 28, 1944 (from the National Archives, Record Group 227; this report was written to Division 15 of the National Defense Research Committee, Office of Scientific Research and Develpment, on Project RP-124).

17109. H. Busignies, S. H. Dodington, J. A. Herbst, and G. R. Clark, "Radio communication system protected against interference," Fed. Tel. Radio Corp., New York, NY, Final Rep. 937-3, July 12, 1945 (same source as [54]).

17110. C. H. Hoeppner, "Pulse communication system," U.S. Patent 2 999 128, September 5, 1961 (filed November 4, 1945).

17111. E. H. Krause and C. E. Cleeton, "Pulse signalling system," U.S. Patent 4 005 818. February 1, 1977 (filed May 11, 1945).

17112. J. R. Pierce, "The early days of information theory," *IEEE Trans. Info. Th.,* vol. IT-19, pp. 3–8, January 1973.

17113. "The WHYN guidance system," Phys. Lab., Sylvania Elec. Products, Bayside, NY, Final Eng. Rep., Modulation Wave Form Study & F-M Exciter Develop., Contr. W28-099ac465, June 1949 (AD895816).

17114. "The WHYN guidance system," Phys. Lab., Sylvania Elec. Products, Flushing, NY. Interim Eng. Rep. 5, Contr. W28-099ac465, October 1948 (ATI 44524).

17115. "The WHYN guidance system," Phys. Lab., Sylvania Elec. Products, Bayside, NY, Final Eng. Rep., Equipment Develop. & East Coast Field Test, Contr. W28-099ac465, June 1950 (AD 895815).

17116. "Radio countermeasures," Natl. Defense Res. Committee, Office Sci. Res. Develop., Washington, DC, Summary Tech. Rep. Div. 15, vol. I, 1946 (AD 221601).

17117. J. H. Green, Jr. and J. Gordon, "Selective calling system," U.S. Patent 3 069 657. December 18, 1962 (filed June 11, 1958).

17118. V. C. Oxley and W. E. De Lisle, "Communications and data processing equipment," U.S. Patent 3 235 661. February 15, 1966 (filed July 11, 1962).

17119. L. A. deRosa, M. J. DiToro, and L. G. Fischer, "Signal correlation radio receiver," U.S. Patent 2 718 638, September 20, 1955 (filed January 20, 1950).

17120. R. L. Frank, "Phase-coded communication system," U.S. Patient 3 099 795, July 30, 1963 (filed April 3, 1957).

17121. C. E. Shannon, "Communication in the presence of noise," *Proc. IRE,* vol. 37, pp. 10–21, January 1949.

17122. ——, "A mathematical theory of communication," *Bell Syst. Tech. J.,* vol. 27, pp. 379–423, July, and 623–656, October 1948.

17123. Y. W. Lee, J. B. Wiesner, and T. P. Cheatham, Jr., "Apparatus for computing correlation functions," U.S. Patent 2 643 819, June 30, 1953 (filed August 11, 1949).

17124. Y. W. Lee, T. P. Cheatham, Jr., and J. B. Weisner, "The application of correlation functions in the detection of small signals in noise," M.I.T. Res. Lab. Electron., Tech. Rep. 141, October 13, 1949 (ATI 066538, PB 102361).

17125. H. E. Singleton, "A digital electronic correlator," M.I.T. Res. Lab. Electron., Tech. Rep. 152, February 21, 1950.

17126. H. C. Harris, Jr., M. Leifer, and D. W. Cawood, "Modified cross-correlation radio system and method," U.S. Patent 2 941 202, June 14, 1960 (filed August 4, 1951).

17127. "The WHYN guidance system," Phys. Lab., Sylvania Elec. Products, Bayside, NY, Final Eng. Rep., Equipment. Syst. Lab. Tests & Anal., Contr. W28-099ac-465, June 1953 (AD 024044).

17128. M. Leifer and W. Serniuk, "Long range high accuracy guidance system," presented at the RDB Symp. Inform. Theory Appl. Guided Missile Problems, California Inst. Technol., Pasadena, February 2–3, 1953.

17129. A N. Goldsmith, "Radio signalling system," U.S. Patent 1 761 118, June 3, 1930 (filed November 6, 1924).

17130. C. B. H. Feldman, "Wobbled radio carrier communication system," U.S. Patent 2 422 664, June 24, 1947 (filed July 12, 1944).

17131. N. Marchand and M. Leifer, "Cross-correlation in periodic radio systems," presented at the IRE Conf. Airborne Electron., Dayton, OH, May 23–25, 1951.

17132. W. E. Budd, "Analysis of correlation distortion," M.E.E. thesis, Polytech. Inst. Brooklyn, Brooklyn, NY, May 1955.

17133. P. Kotowski and K. Dannehl, "Method of transmitting secret messages," U.S. Patent 2 211 132, August 13, 1940 (filed in U.S. May 6, 1936; in Germany May 9, 1935).

17134. M. B. Pursley, F. D. Garber, and J. S. Lehnert, "Analysis of generalized quadriphase spread-spectrum communications," in *Proc. IEEE Int. Conf. Comm.,* June 1980, vol. 1, pp. 15.3.1–15.3.6.

17135. R. T. Compton, Jr., "An adaptive array in a spread-spectrum communication system," *Proc. IEEE,* vol. 66 p. 289, March 1978.

17136. G. R. Cooper and R. W. Nettleton, "A spread-spectrum technique for high-capacity mobile communications," *IEEE Trans. Veh. Tech.,* vol. VT-27, pp. 264–274, November 1978.

17137. G. R. Cooper and R. W. Nettleton, "Spectral efficiency in cellular land-mobile communications: A spread-spectrum approach," TR-EE 78-44, Final Rep., NSF Grant ENG 76-80536, October 31, 1978.

17138. J. K. Holmes, *Coherent Spread Spectrum Systems,* Wiley, New York, 1982.

17139. D. J. Torrieri, *Principles of Military Communication Systems,* Artech House, Dedham, MA, 1981.

17140. A. J. Viterbi, "Spread spectrum communications—Myths and realities," *IEEE Comm. Soc. Mag.,* vol. 17, pp. 11–18, May 1979.

17141. C. R. Cahn, "Spread spectrum applications and state-of-the-art equipments," *Magnavox Comm. Navigation,* November 22, 1972.

17142. C. R. Cahn *et al., AGARD Lecture Series No. 58—Spread Spectrum Communications,* 1973.

17143. R. C. Dixon and L. A. Gerhardt (Eds.), Special Issue, *IEEE Trans. Comm.,* vol. COM-25, August 1977.

17144. C. E. Cook and H. S. Marsh, "An Introduction to Spread Spectrum," *IEEE Comm. Mag.,* March 1983.

17145. M. R. Campbell, L. E. Hoff, and R. E. Ziemer, "A Large Time-Bandwidth Product Signaling Technique for Nonwhite Noise Channels," *IEEE Trans. Comm.,* October 1976.

17146. I. M. Gottlieb, "From Smoke Signals to Spread Spectrum, Military Communcations Decides Who Wins the War," *Military Electronics/Countermeasures,* December 1982.

17147. R. Moser, "Generation and Reception of Spread Spectrum Signals," *Microwave Journal,* May 1983.

17148. R. Moser and G. J. Gross, "Spread Spectrum Techniques," *Microwave Journal,* October 1982.

17149. J. Oetting, "Spread Spectrum Communications," *Sea Technology,* May 1981.

17150. C. L. Cuccia, "Spread Spectrum Systems Solve Nearly All C^3 Aspects," *MSN Magazine,* April 1982.

17151. T. J. Moore, R. L. McKinley, V. D. Mortimer, and C. W. Nixon, "Evaluation of Word Intelligibility of Two Modulator/Demodulator Systems of a Spread Spectrum Communication System in Presence of Simulated Cockpit Noise," USAF Aerospace Medical Report AMRL-TR-78-54.

17152. D. L. Nielsen, "Microwave propagation and Noise Measurements for Mobile Digital Radio Applications," Stanford Research Institute Report for contract DAHC15-73-C-0187.

17153. J. J. Patti and A. W. Roeder, "An Adaptive Spread Spectrum Correlation Receiver" Presented at Natl. Conf. on R.P.V's, June 1975.

17154. C. E. Cook, F. W. Ellersick, L. B. Milstein, and D. L. Schilling, (Eds.) Special Issue, *IEEE Trans. Comm.*, May 1982.

17155. P. Newhouse, "Procedures for Analyzing Interference Cause by Spread Spectrum Signals," IIT Research Inst. ESD-TR-77-003A, August 1978.

17156. R. Singh, A. K. Elkaheem, and S. C. Gupta, "Hybrid Diversities in a Spread Spectrum Mobile Communication System," *IEEE I.C.C. Proc.*, 1982.

APPENDIX 3

EQUIVALENCE OF SIMPLE AND MODULAR SEQUENCES

Assume a linear sequence $|n, p|_s$ and its alleged equivalent $|n, n-p|_m$. Writing equations and drawing diagrams for them, we get

SSRG

$$\lambda_s^0 = \lambda^n \oplus \lambda^p$$
$$= \lambda^0 D_n \oplus \lambda^0 D_p$$

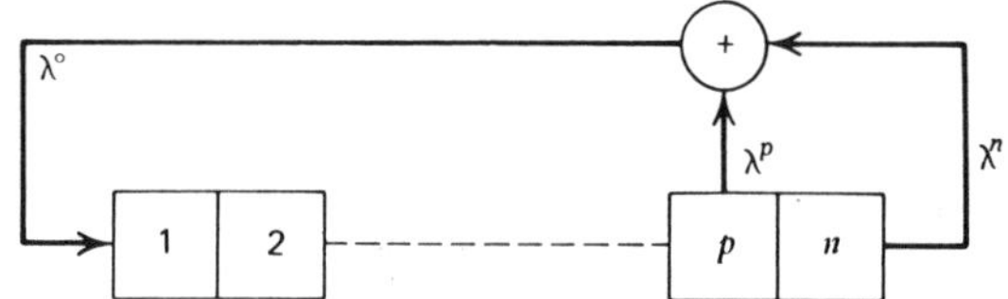

MSRG

$$\lambda_m^0 = (\lambda^0 D_{n-p} \oplus \lambda^0) L_p$$
$$= \lambda^0 D_{n-p} D_p \oplus \lambda^0 D_p$$
$$= \lambda^0 D_n \oplus \lambda^0 D_p$$

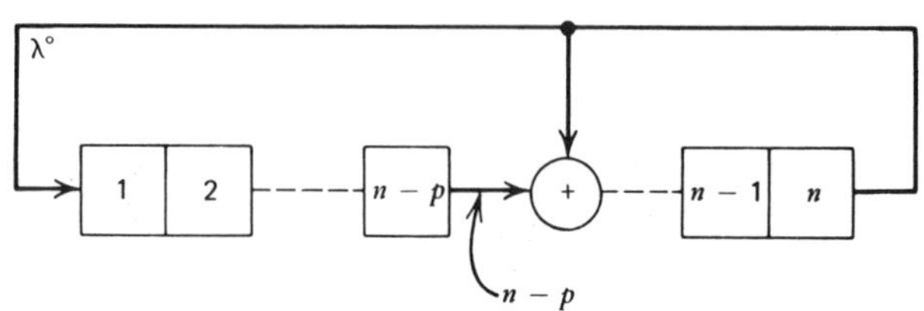

where $\lambda^{n-p} = \lambda^0 D_{n-p}$, and $D_{n-p} D_p = D_n$, because $(n-p)+p=n$. Therefore, because $\lambda_s^0 = \lambda_m^0$, the two are equivalent. Now, because this proof is so austere, let us try a more concrete example to help us to remember the equivalence of certain simple and modular sequence generator connections, at least for one specific case.

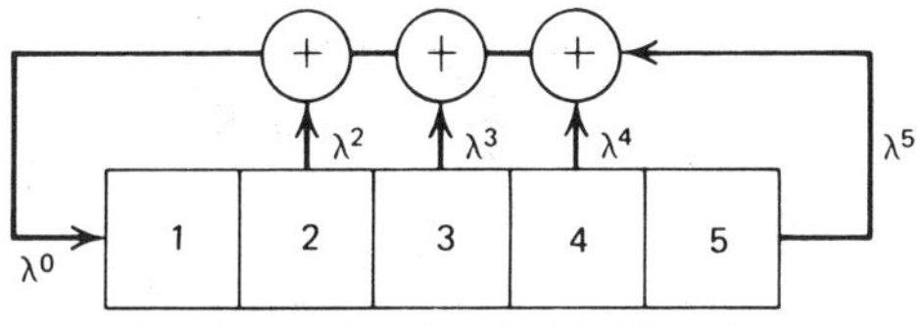

Simple generator for [5, 4, 3, 2]s code

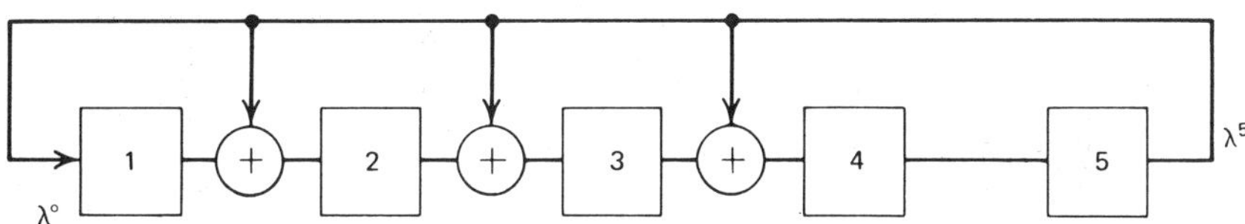

Modular SRG for [5, 3, 2, 1]m code

Given $[5, 4, 3, 2]_s$ and its equivalent $[5, 3, 2, 1]_m$, we find that the equation for the simple generator is just $\lambda^0 = \lambda^5 \oplus \lambda^4 \oplus \lambda^3 \oplus \lambda^2$.

Alternatively, we can write $\lambda^0 = ((\lambda^5 \oplus \lambda^4) \oplus \lambda^3) \oplus \lambda^2$. This is just the previous λ^0, once the brackets have been removed. We can write the equation for the modular sequence generator, noting that $\lambda^5 = \lambda^0$, as

$$\begin{aligned}
\lambda^0 &= (((\lambda^0\, D \oplus \lambda^0)\, D \oplus \lambda^0)\, D \oplus \lambda^0)\, D_2 \\
&= ((\lambda^0\, D \oplus \lambda^0)\, D \oplus \lambda^0)\, D_3 \oplus \lambda^2 \\
&= (\lambda^0\, D \oplus \lambda^0)\, D_4 \oplus \lambda^3 \oplus \lambda^2 \\
&= \lambda^5 \oplus \lambda^4 \oplus \lambda^3 \oplus \lambda^2,
\end{aligned}$$

and the modular generator is equivalent to the simple generator, as previously stated.

APPENDIX 4

MULTIPLICATION OF DIRECT SEQUENCE SIGNALS

Multiplication of a direct sequence signal is of interest from at least two aspects. One is in the context of Chapter 7, in which the possibility of multiplying a signal up to a desired transmitting frequency is discussed. The second is from the standpoint of a receiver who wishes to know if a DS signal is present. The example that follows simply shows that a biphase DS signal, after squaring, becomes a CW carrier (as does any symmetrically modulated signal). Though examples are not given here for all possible multiples, it should be obvious from the result that multiplication can be employed to detect a signal or, in the other extreme, should be avoided when the signal must be preserved. Table 7.1 shows the result of various multiplications of phase-modulated signals. The example given here is for squaring a biphase modulated signal. The same result may be expected of squaring a quadriphase modulated signal twice.

Given a direct sequence biphase signal, $A \cos \omega_c t \pm 90$ (where the 90° term is a function of the code modulation). By squaring this signal (neglecting the amplitude coefficient A), we have

$$(\cos \omega_c t \pm 90)^2 = (\cos \omega_c t \cos 90 \pm \sin \omega_c t \sin 90)^2$$

$$= (\cos \omega_c t \cos 90)^2 \pm \sin \omega_c t \cos \omega_c t \sin 90 \cos 90$$

$$+ (\sin \omega_c t \sin 90)^2,$$

and, considering term by term,

$$(\cos \omega_c t \cos 90)^2 = \cos^2 \omega_c t = \tfrac{1}{2}(1 + \cos 2\omega_c t)$$

$$(\sin \omega_c t \sin 90)^2 = 0$$

$$\sin \omega_c t \cos \omega_c t \sin 90 \cos 90 = 0,$$

which leaves only a $\frac{1}{2}(1 + \cos 2\omega_c t)$ term, a CW signal at twice the original modulated carrier frequency. There is no modulated carrier at twice the input frequency. Therefore we see that squaring or doubling is not practical for use with a direct sequence signal.

APPENDIX 5
LINEAR CODE SEQUENCE VULNERABILITY

Linear* code sequences are often employed in communications systems for various uses, such as error correction, addressing, or for spectrum spreading. Such codes have great utility, and in the case of the linear maximal codes (or *m*-sequences) can offer characteristics that are difficult to achieve with any other type of sequence.

It must be realized, however, that linear codes are limited by their vulnerability to simple analysis. That is, a system which uses a linear code for spectrum spreading can offer no advantage against the would-be interferor who can receive his or her transmitted signal, analyze it, and retransmit a replica of it to "spoof" the receiver. We will not consider the interferor's difficulty in receiving and detecting the code used. Suffice it to say that he or she must be able to procure any $2n + 1$ consecutive chips from the code sequence, and that given those $2n + 1$ chips, he or she can build a code generator that generates an identical code in its entirety. For example, if a $2^{31} - 1$ chip code is being employed, then any $(2 \times 31) + 1 = 63$ consecutive chips from the two billion-chip-long code sequence can furnish all of the information needed to reproduce the entire code.

To be sure that the import of this note is clear—no linear code sequence can offer protection to information beyond the level of the casual observer. Any determined person who can procure $2n + 1$ consecutive chips has ready access to any information enbedded in the signal, as well as knowledge of all future code states. Furthermore, the process for determining this information is straightforward and simple, though perhaps somewhat cumbersome for those who must do the required operations by hand.

*Any code sequence is considered to be linear if it is generated through the use of only those operations that are linear for the arithmetic field of interest (i.e. binary codes may be generated through modulo-2 addition, ternary codes through modulo-3 addition, etc.) Otherwise, the codes are nonlinear, and such codes are not considered here.

A tabular technique that produces code tap connections for an equivalent simple shift register code sequence generator is given in the following pages, with an example to show its use. One assumption is made: that the analyst knows the length of the code sequence generator that his or her sample comes from. This is a simplifying assumption, necessary only to make the example given more readily applicable to the question at issue—that of showing the vulnerability of linear codes.

For our first example, let us assume that the code generator in the transmitter is five stages long ($n = 5$). Typical code sequence generators are much longer than this, but here we choose five stages because of the need for a simple example. The same principles apply, no matter what the register length. Now, let us suppose that we have just detected a subsequence 01110011011, 11 chips long. ($2n + 1$ chips = 11. Code length is $2^5 - 1 = 31$ chips, overall.)

First, we arrange the received subsequence in a column, with the last detected chip at the top. Then, shifting the same subsequence down by one row per column, generate more columns until the number of columns equals one more than the number of stages in the assumed generating register (five plus one).

The result should be as follows:

Q_5	Q_4	Q_3	Q_2	Q_1	λ
1					
1	1				
0	1	1			
1	0	1	1		
1	1	0	1	1	
0	1	1	0	1	1
0	0	1	1	0	1
1	0	0	1	1	0
1	1	0	0	1	1
1	1	1	0	0	1
0	1	1	1	0	0

This table is a listing of the state of shift register stages one through five, and the feedback (λ) necessary to feed into stage one, to produce the 11 chips that were received, using a five-stage shift register generator. This equivalent simple shift register generator is shown in Figure A5.1, which follows.

Now, noting two properties of shift register generators:

1. Code length is a function of the number of stages in the register, and the last stage, which determines the number, is always a member of the set in the feedback.

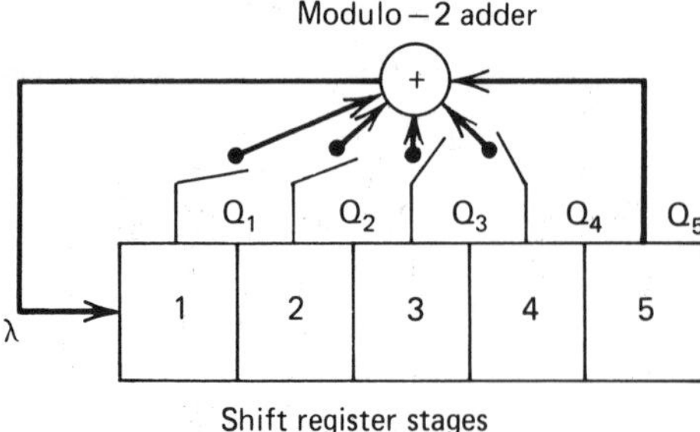

Figure A5.1 Five-stage simple shift register generator.

2. Stable, recurrent sequences do not exist for odd numbers of feedback taps (i.e., one, three, five, etc.).

Observe the feedback column, λ. A listing of shift register outputs, Q_1, Q_2, . . . , can be generated that shows, for each register state, which stages may be added modulo-2 to the last stage output to give the desired λ for insertion to stage 1 in the next state. This produces the following result:

Q_5	Q_4	Q_3	Q_2	Q_1	λ	Possible feedback
1						
1	1					
0	1	1				
1	0	1	1			
1	1	0	1	1		
0	1	1	0	1	1	4, 3, 1
0	0	1	1	0	1	3, 2
1	0	0	1	1	0	2, 1
1	1	0	0	1	1	3, 2
1	1	1	0	0	1	2, 1
0	1	1	1	0	0	1

To be more specific: if the Q_5 output is modulo-2 added to Q_4, Q_3, or Q_1 outputs, then the result is the desired λ.

Now observe the possible feedback column. From this column we may determine the feedback taps that can be used to generate an *m*-sequence that contains the received 11-bit subsequence. The first entry in the λ column, a one, would result from modulo-2 addition of Q_5 (a zero) with Q_4, Q_3, or Q_1. Similarly, the second λ would result from modulo-2 addition of Q_5 with either Q_3 or Q_2, and all other λ's would be generated by combination of Q_5 with the appropriate Q_n.

The last entry in the possible feedback column is a set with only one member, which means that for this state, only Q_1 could be added to Q_5 to produce the desired λ, a zero. Therefore, one of the feedback taps used to

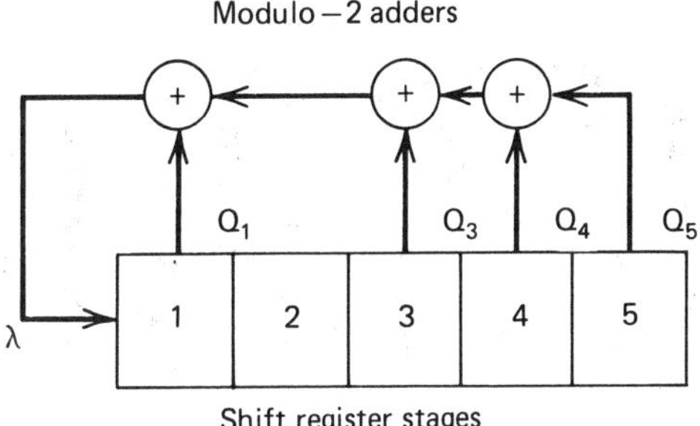

Figure A5.2 Five-stage shift register sequence generator derived from subsequence analysis.

generate the code of which our 11-bit subsequence is a sample, is at the output of shift register stage 1.

Remembering that no code is possible with an odd number of feedback taps, and that stage five is always a member of the tap set, we see that there are four rows in the possible feedback column that are invalid: (3,2) in rows two and four, and (2,1) in rows three and five. Since taps at stage 2 and stage 1 cannot both be valid, and we know already that tap 1 is valid, the possibility of a tap at stage 2 can be discarded. We know, then, that feedback taps exist in the set we are seeking, at register stages 3 and 1, because 2 is also invalid in row two and four.

Again, we know that one more tap must exist because odd numbers of feedback taps are prohibited. Therefore, the remaining tap we seek must be at register stage 4, since it is the only one remaining, and it is contained in the top row of possible feedback taps. The code sequence generator we are seeking is seen to be a five-stage register with feedback taps at stages 4, 3, and 1, as pictured in Figure A5.2.

Does this configuration generate a sequence containing the subset we started with? Indeed it does. The $2^5 - 1$ chip sequence produced by this generator is . . .1111101000100101011000011100110 . . . , in which our 11 chips start at chip 25 and run through chip 4. (The overall 31 chip sequence is cyclic.)

At this point, it is well that the rules used for evaluating possible feedback sets be enumerated.

The procedure is:

1. Examine the possible feedback column for single member sets. Any single member is a feedback tap.
2. Determine if any one possible tap is found in all sets in the possible feedback column. If so, that tap is the only one used (i.e., the configuration is a single-tap generator).
3. If step 2 does not show that a single-tap generator is correct, then examine the possible feedback column for sets which contain two members, one of which is the tap found in step 1. Any two member set

that contains the tap already identified contains a second member which is not a valid tap, and that tap can be eliminated from further consideration.

4. Proceed to higher-order tap sets (3 to 4 to 5, etc.), eliminating invalid tap possibilities in higher-order sets through the same process as was used in step 3.
5. Assemble the results. All possible taps not eliminated by the process described are valid, and necessary to generating the desired code sequence.

The two properties of shift register sequence generators that were previously listed are all that were required to derive the taps used, and we have shown that a sequence generator can readily be constructed to simulate an unknown generator, if only its length is known.

Had it happened that the 11 chip input sequence was not a valid subsequence, then unresolvable ambiguities would have occurred in searching the possible feedback sets. It is obvious that for longer sequences, finding feedback taps by this handcrank method becomes more and more tedious and subject to error. The methods described are easily adaptable for computer solution, however, for any foreseeable code sequence length.

It has been assumed that the analyst knows the length of the register being used to generate the code. This is not too far fetched a notion, since the population of one/zero runs in linear codes is fixed, and bounded in run length at n ones and $n - 1$ zeros. Analysis of the run length population in a received, coded signal could produce an estimate of the most likely code length, which in turn would support a set of trial solutions.

It is worthy of note that the analyst does not really care what the true length of the transmitted code is, or what the generator's feedback taps really are, as long as he or she can synthesize the code segment and project future states of the code.

With this thought in mind, we point out that a nonmaximal code produced by an n-stage register could just as well be a maximal length code from a shorter register. For instance, two of the nonmaximal codes produced by a (5,4) connected generator are identical to (2,1) and (3,1) connected m-sequences.

The analyst then would not care if a particular sequence was from a long register with nonmaximal connections or a shorter one with maximal connections. This view, in which the analyst does not really care how the sequence was generated, brings up a new point: Given any binary subsequence, some linear generator can be built to reproduce it. Therefore, why bother to build nonlinear sequence generators?

The question is answered by recalling that the analyst's real interest is in finding out what the code sequence will do *in the future*. Even though a linear sequence or subsequence may reproduce a particular nonlinear subsequence perfectly, the future state of the nonlinear sequence cannot be linearly derived from the presently known subsequence. We have, on the other hand, shown that linear sequences can be reproduced by a simple process of analysis and generation by an equivalent simple shift register sequence generator.

APPENDIX 6
THE RELATION BETWEEN ω_N AND $\omega_{3\text{ dB}}$

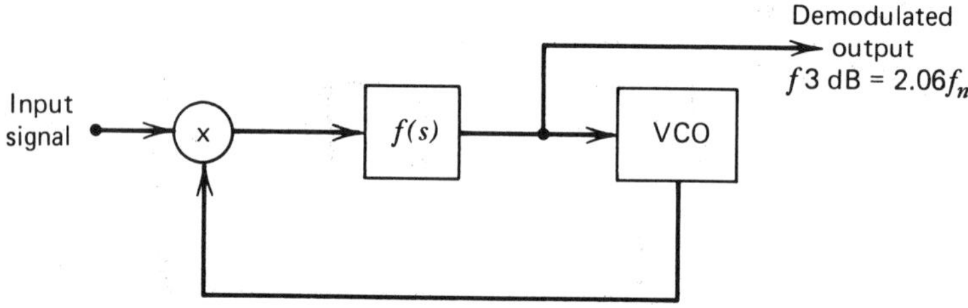

Figure A6.1 Phase lock loop model.

(A) Given that it is desired that an FM-demodulating phase-lock loop be designed that has a bandpass f_{3db}. Hoffman[1206] gives a relation between B_N and ω_n:

$$B_N = \frac{4\zeta^2 + 1}{4\zeta}\,\omega_n \text{ cps}; \qquad K \gg \omega_n/\zeta$$

which for $\zeta = \sqrt{2}/2 = 0.7$ gives $B_N = 1.06\omega_n$ cps.

In this same paper, Hoffman states that

$$f\,(3\text{ db}) = 0.3088\ B_N \text{ cps} \quad (\zeta = 0.7)$$

Expressing B_N in Hertz, we get B_N in terms of f_n:

$$B_N = 2\pi \times 1.06 = 6.658\ f_n \text{ cps}$$

From the f_0 $(-3\text{ db}) = 0.3088\ B_N$ equation, we get

$$B_N = \frac{f\,(3\text{ db})}{0.3088}$$

Substituting, then, we have

$$\frac{f\,(3\text{ db})}{0.3088} = 6.658\ f_n$$

and $f\,(3\text{ db}) = 2.055\ f_n \approx 2.06\ f_n$. A derivation confirming this relation follows.

(B) Derivation of the relationship between ω_n and $\omega_{3\text{db}}$.

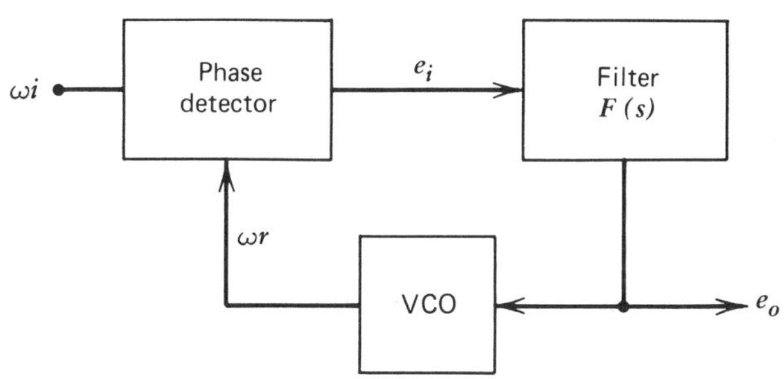

Definitions:

$$\omega_i = \omega_c + \omega_m$$

$$K_1 = \text{multiplier gain in V/rad} = \frac{AB}{2}\sin\theta_m$$

$$\begin{aligned} e_i &= K_1\,(\phi_i - \phi_r) \\ &= K_1 \textstyle\int (\omega_1 - \omega_r)\,dt \\ e_{o(s)} &= e_{i(s)}\ F_{(s)} \\ K_2 &= d\,\omega_{\text{VCO}}/d_{eo} = \text{VCO gain in rad/V} \\ \omega_r &= \omega_c + K_2 e_o \\ e_i &= K_1\,(\omega_c + \omega_m - \omega_c - K_2 e)\,dt \\ \omega_m &= \dot\omega_m \sin\omega_m t = \text{modulating signal} \\ e_i &= K_1 \textstyle\int (\dot\omega_m \sin\omega_m t - K_2 e_o)\,dt \\ e_{i(s)} &= \frac{K_1}{s}\left[\dot\omega_m\left(\frac{s}{s^2+\omega_m^2}\right) - K_2 e_{o(s)}\right] \\ e_{o(s)} &= e_{i(s)}\ F_{(s)}\ (F_{(s)} = \text{loop filter response}) \\ e_{i(s)} &= \frac{e_{o(s)}}{F_{(s)}} \end{aligned}$$

thus

$$\frac{e_{o(s)}}{F_{(s)}} = \frac{K_1}{s}\left[\frac{\dot\omega_m s}{s^2+\omega_m^2} - K_2 e_{o(s)}\right]$$

and

$$e_{o(s)} = \frac{K_1\ F_{(s)}}{s}\left[\frac{\dot{\omega}_m s}{s^2 + \omega_m^2} - K_2\ e_{o(s)}\right]$$

$$= \frac{K_1\ F_{(s)}\ \dot{\omega}_m}{s^2 + \omega_m^2} - \frac{K_1 K_2\ e_{o(s)}\ F_{(s)}}{s}$$

gathering terms

$$e_{o(s)} + \frac{K_1 K_2\ e_{o(s)}\ F_{(s)}}{s} = \frac{K_1\ F_{(s)}\ \dot{\omega}_m}{s^2 + \omega_m^2}$$

$$e_{o(s)} = \frac{s\ K_1\ F_{(s)}\ \dot{\omega}_m}{(s_2 + \omega_m^2)\ (s + K_1\ K_2\ F_{(s)})}$$

Now, since the loop filter is a lag-*RC* type, as is called for by most tracking phase lock loops,

$$F_{(s)} = \frac{1 + T_2\ s}{1 + T_1\ s}$$

substituting this value for $F_{(s)}$ into the above equation for $e_{o(s)}$

$$e_{o(s)} = \frac{s\ K_1\ \dot{\omega}_m(1 + T_2\ s)}{(s^2 + \omega_m^2)\ s\ (1 + T_1\ s) + K_1\ K_2\ (1 + T_2\ s)\ \dot{\omega}_m}$$

rearranging terms

$$e_{o(s)} = \frac{K_1\ s\ T_2\ (1/T_2 + s)\ \dot{\omega}_m}{(s^2 + \omega_m^2)\ [(s\ T_1\ (1/T_1 + s) + K_1\ K_2\ T_2\ (1/T_2 + s)\ \dot{\omega}_m)}$$

and converting to the time domain

$$e_o(\mathrm{t}) = \frac{K_1\ T_2\ \dot{\Delta}_\omega}{T_1}\left[\frac{\omega_m^2 + 1/T_1^2}{(K_1\ K_2/T_1 - \omega_m)^2 + 4[(1 + K_1\ K_2\ T_2)/2\ T_1]^2\ \omega_m^2}\right]^{1/2}$$

$$\sin\ (\omega_m^t + \phi)$$

at $\omega_m = 0$

$$e_o(t) = \frac{\dot{\omega}_m}{K_2}$$

at the point where $e_o(t)$ is down 3 db

$$e_o(t) = \frac{\sqrt{2}\,\dot{\omega}_m}{2\,K_2}$$

Now, setting

$$e_o\,(t - 3\text{ dB}) = \frac{\sqrt{2}\,\dot{\omega}_m}{2\,K_2}$$

$$= \frac{K_1\,T_2\,\dot{\omega}_m}{T_1}\left[\frac{(\omega_{3\,\text{dB}})^2 + (1/T_2)^2}{[K_1\,K_2/T_1 - (\omega_{3\,\text{dB}})^2]^2 + 4[(1 + K_1\,K_2\,T_2)/2\,T_1]^2\,(\omega_{3\,\text{dB}})^2}\right]^{1/2}$$

$$\sin\,(\omega_s\,t + \phi)$$

defining

$$\omega_n = \sqrt{\frac{K_1\,K_4}{R_1\,C}}$$

$$\zeta = \frac{1 + K_1\,K_2\,T_2}{2\,T_1\,\omega_n}$$

substituting the above and squaring

$$1 = 2\,(\omega_n)^4\,\frac{(\omega_{3\,\text{dB}})^2\,T_2^2 + 1}{(\omega_3)^4 + (\omega_3)^2\,(4\,\omega_n^2\,\zeta^2 - 2\,\omega_n^2)\,\omega_n^4}$$

$$= \frac{2[(\omega_{3\,\text{dB}})^2\,T_2^2 + 1]}{1 + (\omega_{3\,\text{dB}}^2/\omega_n^2)\,(4\,\zeta^2 - 2) + \omega_{3\,\text{dB}}/\omega_n^4}$$

and substituting

$$\zeta = 0.7, \quad T_2 = \sqrt{2}/\omega_n$$

$$\frac{\omega_{3\,\text{dB}}^4}{\omega_n^4} - \frac{4\,\omega_{3\,\text{dB}}^2}{\omega_n^2} - 1 = 0$$

$$\omega_{3\,dB}^2 = \frac{4/\omega_n^2 \pm \sqrt{16/\omega_n^4 + 4/\omega_n^4}}{2/\omega_n^4}$$

$$= \frac{4 \pm 4.48}{2}\,\omega_n^2$$

$$\omega_{3\,dB}^2 = 4.24\,\omega_n^2$$

$$\omega_{3\,dB} = 2.06\,\omega_n$$

This is the same result previously derived from Hoffman's values given for B_N as related to ω_n and $f_{(3\,dB)}$.

APPENDIX 7
GOLD CODE SELECTION

Any code can be represented by a polynomial, where the binary codes are represented by a polynomial of the form

$$1 + AX + BX^2 + CX^3 + \cdots + Z\,x^n$$

Here, each coefficient (A, B, . . . , Z) is either 0 or 1, each term of the polynomial (except for the first, 1) corresponds to a stage of a binary shift register, and there are n stages in the register. That is, each term in the polynomial containing an X corresponds one-to-one with a stage in a binary shift register.

The feedback connections in the code generator are defined by the terms in the polynomial whose coefficient is 1. For example, a code generator whose characteristic polynomial is $1 + 1X + 1X^2 + 1X^3 + 1X^7$ would have seven stages with feedback taken from its first, second, third, and seventh stages. Comparing this notation with that we have used previously, we see that the same code could be expressed as

$$[7, 3, 2, 1]_s = [7, 6, 5, 4]_m = 1 + X + X^2 + X^3 + X^7$$

Linear maximal codes, in which we have major interest, have characteristic polynomials that are primitive. That is, the primitive or nonfactorable polynomials each define a different linear maximal code. Fortunately, the tables of polynomials mentioned in section 3.6 both define primitive polynomials and provide information that allows proper selection of pairs of codes for use in generating Gold codes.

Before going further, let us make it clear that generation of proper Gold codes requires that the codes need be properly chosen. Arbitrary selection of code pairs from the tables can result in very poor correlation

Table A7.1 Performance of Preferred Pairs Compared with Worst Case Pairs

		Undesired Correlation		
Degree	Period	Worst Case	Preferred Pair	Difference (DB)
5	31	11	9	1.7
6	63	23	15	3.7
7	127	41	17	7.6
8	255	95	31	9.7
9	511	113	33	10.7
10	1023	383	63	15.7
11	2047	287	65	12.9
12	4095	1407	127	20.9
13	8191	703	127	14.9

performance, as is demonstrated by the results shown in Table A7.1. The "preferred" pairs of codes, as selected by the Gold-derived algorithm, always give undesired correlation that is bounded at $2^{(n+1)/2} + 1$ (n odd) and $2^{(n+2)/2} - 1$ (n even), however.

The Gold-derived algorithm for selection of preferred pairs requires the use of code tables that list the polynomial roots (as do Peterson's tables[326]). The algorithm is used as follows:

1. Select a polynomial of the proper degree from the table (an n-stage shift register requires an nth degree polynomial).
2. Read the number (k) in the polynomial roots column associated with the polynomial selected.
3. If the code generator has an odd number of stages, then calculate $2^k + 1$. If the number of stages is even, calculate $2^{(k+2)/2} + 1$.
4. The number calculated in step 3 is the polynomial root of a second code that completes a preferred pair.

Use of any polynomial (code) with the polynomial root calculated in step 4 will produce Gold codes when combined with the original code that has properly bounded correlation with all members of the set.

Example: For 19-stage codes, suppose we select a code whose polynomial is 2000047, which converts to 11100100000000000010 or $1 + X + X^2 + X^5 + X^{19}$. The polynomial has 1 as its polynomial root.

With the algorithm from the foregoing definition (step 4), we calculate the second polynomial root required as $2^k + 1 = 2^1 + 1 = 3$. A second polynomial having polynomial root = 3 is 2020471 or $1 + X^3 + X^4 + X^5 + X^8 + X^{13} + X^{19}$,

Table A7.2 Description of Available Tables of Binary Polynomials

	Irreducible Polynomials	Number of Maximals	Degree	Polynomial Roots	Correlation Function of Sequences
Marsh (1957)	Yes	All	19	No	No
Peterson (1961)	Yes	All	16	Yes	No
		Partial	17–34		
Watson (1961)	No	1	100	No	No
Gold (1964)	No	All	13	Yes	Yes
Bradford (1965)	No	Partial	58	No	No

This pair of codes would, when combined, produce Gold codes, every one of which would have cross-correlation bounded at $2^{(n+1)/2} + 1 = 1025$, which is $20 \log 2^{19} / 1025 = 54$ dB below the peak of autocorrelation. A listing of some readily available tables of primitive polynomials is given in Table A7.2. Note that of those given, however, only two list the polynomial roots.

ANSWERS TO PROBLEMS

CHAPTER ONE

1. 10 log 250 = **24 dB.**
2. $M_J + S/N + L_{\text{sys}}$ = **34 dB** + system losses.
3. $\text{BW}_{\text{RF}} \geqslant 6 \times 2.5 \times 10^6$ = **15 MHz.** (Remember to allow for the 10 dB output S/N.)
4. $\dfrac{1.5 \times 10^7}{2} = 7.5 \times 10^6$ bps.
5. **⩾2500** ($\log^{-1} 3.4 = 2500$).
6. TW = 20 dB or 100; $T = (10{,}000)^{-1} = 1 \times 10^{-4}$; $W = (1 \times 10^2)/(1 \times 10^{-4}) = 1 \times 10^6$ = 1 MHz.
7. $G_p = 10 \log (2 \times 10^6)/(1 \times 10^5) = 13$ dB. If output S/N must be 6 to 10 dB and losses are 1 to 2 dB, then M_J is a maximum of 6 dB. Probably not worth the trouble. Decide for yourself.
8. With 100 khop/sec rate minimum channel spacing should be 100 kHz. In 5 MHz this would allow for 50 channels. This number of channels would permit only a 17-dB process gain and high vulnerability to partial band interference.
9. Chip rate is a function of information rate but sweep bandwidth is usually the maximum of which the generator is capable.
10. Transmission bandwidth is a function of the code sequence used in a direct sequence system.
11. Frequency hop transmission bandwidth is a function of the number of channels employed. Channel spacing may or may not be a function of the hop rate.
12. The simple time-hopping system transmits at a constant center frequency and thus does not avoid a single-frequency interferer. Reducing duty cycle does increase potential process gain but often

forces an increase in transmitted power to maintain E_b/N_0 for the demodulator.

CHAPTER TWO

1. Primarily to suppress the RF carrier while providing a constant signal envelope.
2. (a) Let $0.5 = (\sin^2 x)/x^2$; $0.707\ x = \sin x/x$; 0.707 times $7\pi/16 \sim \sin 7\pi/16$; bandwidth $\approx 0.44 \times$ clock rate (one-sided).
3. Approximately 90%.
4. (a) $\dfrac{P_T}{2 \times 10^7}$ W/Hz.

   (b) $\dfrac{P_T}{2 \times 10^7} \dfrac{(\sin \pi/4)^2}{(\pi/4)^2}$ W/Hz.
5. $\log^{-1} 30/10 = 1000$; the number of jammed channels $= 1000$: $1000/0.01 =$ **100,000 channels.**
6. 3 kHz $\times$ 100,000 = **300 MHz.**
7. S/N @ 1×10^{-2} error rate $\approx$ 7 dB (allow 2 dB for implementation loss): $30 + 7 + 2 = 39$ dB; $\log^{-1} 39/10 = 8000$; $1000 \times 8000 = 8$ MHz.
8. $10 \log 100 = 20$ dB. $10 \log (4 \times 10^7)/(3 \times 10^3) = 41$ dB. Process gain $= 20 + 41 =$ **61 dB.**
9. A coherent frequency hopping receiver maps the received signal to a single phase coherent frequency, pulsed at the data rate. The noncoherent hopper performs the same frequency mapping but the carrier pulses corresponding to data received are at random phases with respect to one another; that is, the noncoherent receiver receives a carrier signal that is both pulse- and random-phase modulated. The demodulator must be approximately twice as wide to accept the same signal when the carrier pulses are not coherent, so that a 3-dB penalty is paid for noncoherent reception.
10. The 3-chip-per-bit demodulator sees an effective chip rate three times that of the 1-chip-per-bit system and its demodulator bandwidth must be three times as great. The resulting 4.7-dB loss due to increased noise bandwidth must be considered when calculating interference rejection.
11. In a chip-matched filter a particular delay is associated with each increment of frequency input. If each frequency in a particular sweep is offset by R* Hz, then each increment at the filter output is delayed a different amount. Thus the compressed output pulse is offset by an

*R is the mean frequency offset due to Doppler for a particular frequency sweep.

amount $R\Delta t/\Delta f$(seconds). If the slope of Δf is positive, then positive Doppler causes increased delay. For Δf negative positive Doppler causes decreased delay. Upchirp–downchirp delays can therefore be combined to develop a signal unaffected by Doppler shift.

12. Multipath signals consist of components arriving from two or more paths—the direct path and one or more indirect paths. Signals arriving with delay greater than one code chip time are uncorrelated in either the direct sequence or frequency hopping receivers and are treated the same as any other interference. If delay is less than a code chip time (a hop interval for frequency hoppers), the signals do interfere directly with one another and the receiver will track either one or a combination of the signals that appear. If the receiver is a noncoherent frequency hopper, it accepts the vector sum of its inputs and makes its decisions on that basis.
13. Time-frequency hopping implies that from time slot to time slot different frequencies are used, and an alternating set of transmitters takes turns making use of a common channel. Because only one is on the air at any one time, there is no possibility that more powerful transmitters will overpower the weaker ones.
14. Direct path = 100 miles. Indirect path = 103.2 miles. For propagation delay of 6.1 μsec per nautical mile differential path delay = $3.2 \times 6.1 \times 10^{-6} = 19.52$ μsec. The hopper should change frequencies before the repeater can recognize the signal and repeat it to the receiver. Maximum chip time is 19.52 μsec. Minimum hop rate is therefore $(19.52\ \mu\text{sec})^{-1} = 51{,}229$ khps.
15. The power spectrum of a chirp signal is an equal-level set of lines separated by an amount corresponding to the frequency related to the sweep period (i.e., frequency separation = $1/\Delta t$). Therefore the power level in one line is $P/\Delta f\,\Delta t$ and average power density is $P/\Delta f$ for a linear chirp signal. FH signals place all their transmitted power in each channel, as it is chosen. Power density is $P/\text{BW}_{\text{channel}}$. DS signals have power density that varies according to $(\sin x)/x)^2$. Power density varies as a function of distance from center frequency, with respect to code rate. On the basis of approximating the $(\sin x/x)^2$ main lobe with a triangular spectrum, the power density at any point is approximately $(PR_c - f_d)/R_c$, where R_c = code clock rate and f_d = distance from center frequency (Hz).

CHAPTER THREE

1. $2^{23} - 1 =$ **8,388,607.**
2. $\phi(2^n - 1) = 131070$; $131070/17 =$ **7,710.**

3.

T	D_1	D_2	D_3	D_4	$D_4 \oplus D_2$
1	1	1	0	1	0
2	0	1	1	0	1
3	1	0	1	1	1
4	1	1	0	1	0

The sequence is **101.**

4. $2^{11} - 1 =$ **2047.**
5. The sequence is

. . . 1111100110100100001010111011000. . .

Therefore the sequence is maximal $((2^5 - 1) = 31)$ and $\psi_{max} =$ **31;** $\psi_{min} =$ **−1.**

6. For initial conditions = 1111111:
. . . 111111100110010110100011101011000100101000010000001010 101110111001001111010011011110001100000 . . . (length = 93 chips).
For initial conditions = 1100110:
. . . 0110011100001101010010001011111 . . . (length = 31 chips).
For initial conditions = 1101101:
. . . 101 . . . (length = 3 chips).
7. $[9, 6, 4, 3]_s$:

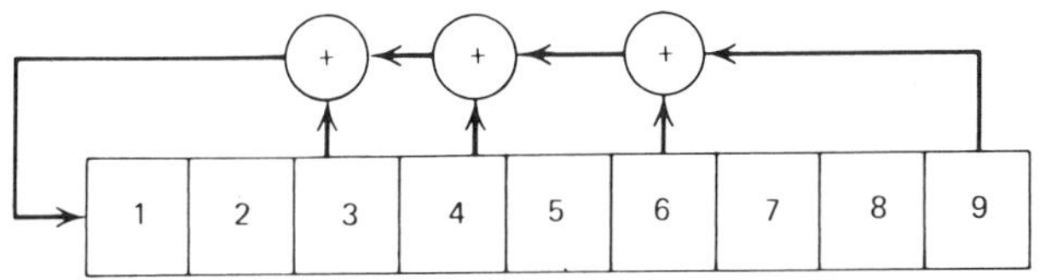

$[9, 8, 4, 1]_s$:

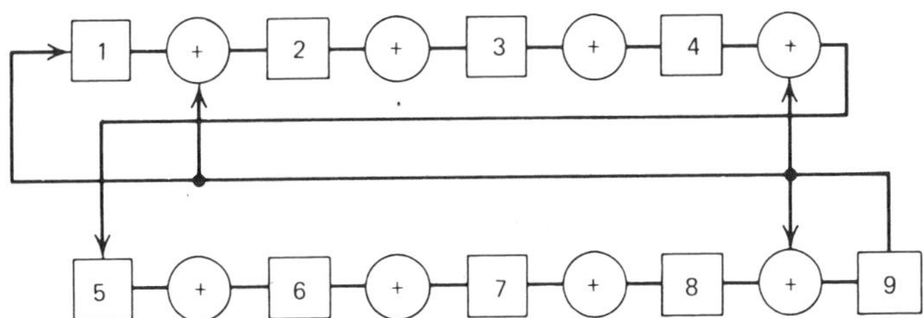

8. $[5, 3]_s$:
. . . 1111100110100100001010111011000 . . .
$[5, 4, 3, 2]_s$:
. . . 1111100100110000101101010001110 . . .

Shift	A–D	Shift	A–D
0	7	16	7
1	−1	17	−3
2	−1	18	−11
3	7	19	−9
4	7	20	7
5	−1	21	−1
6	7	22	5
7	−1	23	7
8	−1	24	−1
9	−9	25	−1
10	−5	26	−1
11	−1	27	−9
12	−1	28	7
13	−3	29	−9
14	−9	30	7
15	−1		

$A\text{–}D_{max} = \mathbf{7}$ $\quad A\text{–}D_{min} = \mathbf{-11.}$

9. **0.0131072 sec; 76.293 Hz.**
10. $2^{n-(p+2)}\Big|_{\substack{n=13\\p=3}} = 2^{13-(3+2)} = 2^8 = \mathbf{256.}$
11. **256.**
12. Any modulo-2 combination of two or more 12 stage, m-sequence generators with selected sequences.
13. Let $[\varphi(2^n - 1)]/n \geqslant 1000$. It is known that there are only 630 sequences generated by a 13-stage register in the m-sequence configuration (8191 chips). For $n = 15$, $2^n - 1 = 32{,}767$; 32,767 has prime factors 7, 31, and 151. $[(7-1) \times (31-1) \times (151-1)]/15 = 1800$. A 15-stage register has enough m-sequences available.
14. Using 5, 3_s, the sequence produced is

    . . . 1111100011011101010000100101100 . . .

    Selecting a pair of arbitrary phases and adding them to the above,

```
     1111100011011101010000100101100
(a)  0010110011111000110111010100001
     -------------------------------
     1101010000100101100111110001101

     1111100011011101010000100101100
(b)  1001011001111100011011101010000
     -------------------------------
     0110111010100001001011001111100
```

Both (a) and (b) are just new phase shifts of the same sequence. The same case holds for all possible phase shifts. Therefore for the operation of addition the set of phase shifts of an *m*-sequence connected generator is closed.

15. Given a code sequence $\lambda^0 = \lambda^n \oplus \lambda^{n-m}$, where $\lambda^n \oplus \lambda^{n-m}$ is a maximal register connection, let there be n zeros in the register: $\lambda^n(0) \oplus \lambda^{n-m}(0) = 0 = \lambda^0(0)$. At t_1, with λ^n and λ^{n-m} both containing zeros, the feedback input to the first register stage (λ^0) is zero. For all $t_n (t_n = t_1, t_2, \ldots, t_n)\lambda^0 = 0$. Because register input is always zero, the register never leaves the all-zeros state and no *m*-sequence is generated.
16. Assume a code sequence $\lambda^0 = \lambda^n \oplus \lambda^{n-m} \oplus \lambda^{n-p}$. If the sequence is maximal, it must contain the all-ones state at some point in its progression. When all-ones are inserted in the n-stage register, λ^0 is always a one. Therefore the generator can never leave the all-ones state and an *m*-sequence cannot be generated.
17. Because each state occurs once and only once in each sequence, sensing any state and triggering an oscilloscope when that state occurs allows the oscilloscope to be synchronized so that the code can be observed.

CHAPTER FOUR

1. $BW_{BPSK} = 2BW_{QPSK}$.
2. Let $16^x \geqslant 8000$, $16^3 = 4096$, $16^4 = 65{,}536$. **Four.**
3.

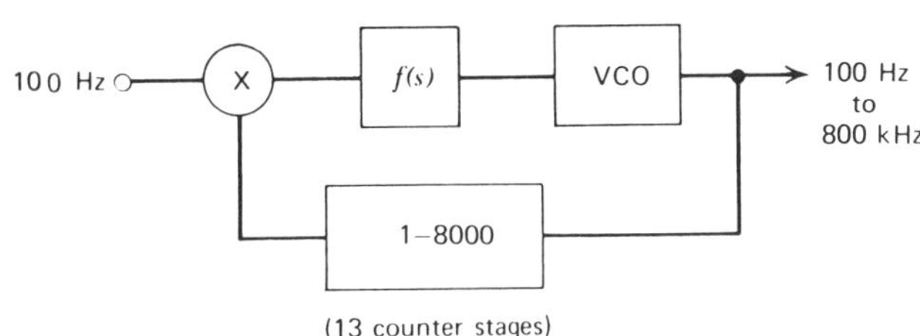

4. Assume that minimum sampling rate is two samples per Hertz: Data rate $= 2 \times 3 \times 3 \times 10^3 =$ **18 Kbps.** Forty decibels correspond to 10,000 times RF-to-baseband ratio. Therefore the bandwidth is $18 \times 10 \times 10^6 = 180$ MHz.
5.

	Direct	Indirect
Hop rate	Determined by inter-stage filters	Loop must relock at each hop; loop BW determines rate

	Direct	Indirect
Frequency stability		
Long term	Set by reference	Same
Short term	Set by reference	Depends on loop jitter
Spurious	Many spurious signals generated by mixing processes; interstage filters determine output	Usually close-in noise caused by loop jitter and internal noise

6. For 64 frequencies use four add-and-divide modules: $K^A = 4^4 = 64$. Four base frequencies are needed. Let $M = 4$; $f_0 = 70$ MHz (given); $f_m = f_0 - f_0/M = 70 - 70/4 = 52.5$ MHz. Base separation $= K^{A-1} \times 10$ kHz $= 4^3 \times 10^4 = 640$ kHz.

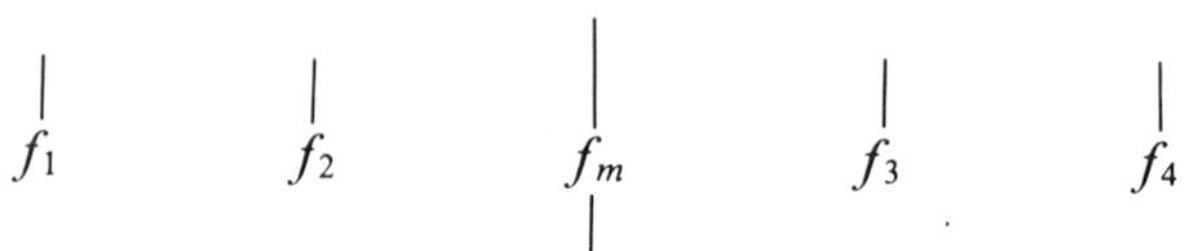

$$f_2 \text{ and } f_3 = f_m \mp \frac{\text{base separation}}{2}$$

$$= 52.5 \times 10^6 \mp \frac{6.4 \times 10^5}{2} = 52.5 \text{ MHz} \mp 320 \text{ kHz}$$

$f_2 = 52.18$ MHz,
$f_3 = 52.82$ MHz,
$f_1 = 52.18 - 0.64 = 51.54$ MHz,
$f_4 = 52.82 + 0.64 = 53.46$ MHz.

7. Because squaring the direct sequence signal produces a twice-frequency carrier (see Appendix 4), the modulated carrier could be detected by a single squaring process.

8. $\log_2 2000 =$ **10.966.** Eleven binary bits are required.

9. $K^A \geqslant$ **256** is desired. How many modules (A) with K base frequencies should be used to make the simplest synthesizer?
Solving $\log_K X$ for A as the next larger integer:

K	A
2	8
3	6
4	4
5	3
⋮	⋮
15	3
16	2

$K = 4$ or 6 appears to be optimum, because an even number of frequencies balances the spectrum around f_m and f_0 and little is gained by reducing K once 5 or more base frequencies are used.

10. Flip-flop output is identical to the suppressed-clock PDM signal (i.e., the carrier clock may be subtracted by simple division by two).

CHAPTER FIVE

1. $30 - 20 =$ **10 dB.**
2.

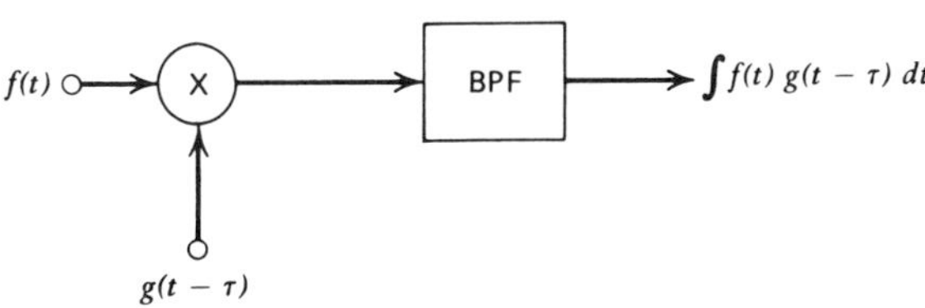

3. **0.3.**
4.

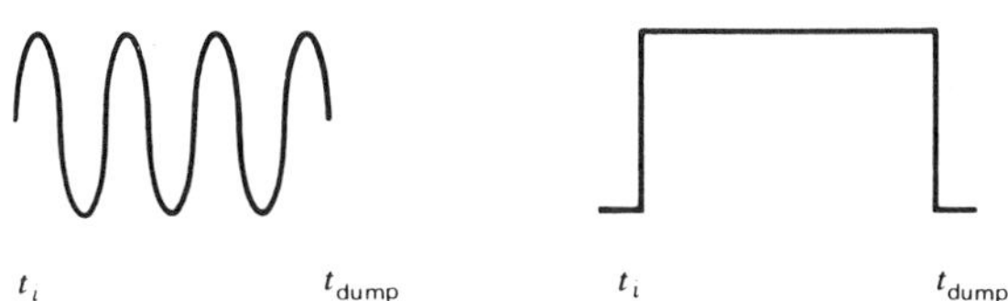

5. m-ary FH system has n/m frequency cells available, compared with n cells for a binary FH system. Binary system must hop faster to convey information at the same rate.
6. $\log_2 m$ bit PCM or m-level analog (for the base band information).
7. If signal 1 has a structure $A \cos(\omega_{c1} \pm \omega_{m1})t$ and this is multiplied with $B \cos(\omega_{c2} \pm \omega_{m2})t$, the output bandwidth produced corresponds to the maxima of ω_{m1} and ω_{m2}. (BW $= g(\omega_{m1}, \omega_{m2})$). The output signal is AB/2 $\cos(\omega_{c1} \pm \omega_{c2} \pm \omega_{m1} \pm \omega_{m2})t$. Therefore at either $\omega_{c1} + \omega_{c2}$ or $\omega_{c1} - \omega_{c2}$ the modulation bandwidth is $2(\omega_{m1} + \omega_{m2})$ maximum. The multiplier output due to a similar cochannel signal would be twice as wide as the local reference bandwidth.
8. $\left(\frac{S+I}{I}\right)_{\text{out}} = (SG_p + I)/I = (1000 + 10)/10 = 1010/10 = 101$; $10 \log 101 =$ **20.04 dB.**
9. Code repetition would be at a $1 \times 10^6/4095 = 244.2$ repetitions per second rate. Line spacing would be 244.2 Hz.
10. $\sin = \omega_{\text{offset}}/K = (2 \times 10^4)/(1 \times 10^5) = 0.628$; $\delta =$ **38.9°**.

11. Subcarrier FM or FSK modulation for direct sequence signals is a DSB suppressed-carrier technique. Therefore the output of a correlator when such a signal is being sent is a pair of mirror-image modulated signals, separated by twice the subcarrier frequency, around the output IF center frequency. Costas or squaring loops are designed to combine coherently such DSB signals and regenerate the missing carrier. A phase-lock loop can track only one of the pair of sidebands and is therefore at a 3-dB disadvantage. Two independent phase-lock loops, each tuned to one of the two sidebands (the upper and lower subcarrier-produced frequencies), can be employed, with summing of their demodulated outputs. (This alternate does provide for recovering the lost 3 dB.)
12. Assuming that the frequency shift of the carrier is the same as that applied to the subcarrier, we can generate the following matrix:

Shift (f_c, f_s)	$++$	$+-$	$-+$	$--$
f_c	δ	δ	$-\delta$	$-\delta$
f_{us}	2δ	0	0	-2δ
f_{1s}	0	2δ	-2δ	0

where f_c = carrier,
f_s = subcarrier,
$+$ signifies a positive shift (increase in frequency),
$-$ signifies a negative shift (decrease in frequency),
δ is the amount shifted.

If the demodulator is such that it can accommodate the unsymmetrical signal, the two can be separately demodulated. A Costas demodulator, for instance, can track carrier and DSB subcarrier shifts separately. Two separate baseband channels can be accommodated with minimum mutual interference.

CHAPTER SIX

1. Drift per day (from Figure 6.1) = 1.1×10^2 chips; $1.1 \times 10^2 \times 2.5 = 2.75 \times 10^2$ chips. Total uncertainty is **±275 chips.**
2. Drift per day = 1.1 chips. Total uncertainty = $1.1 \times 2.5 =$ **±2.75 chips.**
3. Allowing 2-dB implementation loss, BW_{RF}/BW_{info} corresponding to 33 dB is 2×10^3. Information bandwidth then is $(2 \times 10^7)/(2 \times 10^3) = 10$ kHz. This is also assumed to be the synch recognition bandwidth. Recognition rise time is approximately $0.35/(1 \times 10^4) = 0.35 \times 10^4 = 35$ μsec. To search 4095 chips, at $\frac{1}{2}$ chip per search, increment would require at least $8190 \times 35 \times 10^{-6} =$ **0.286 sec.**

4. 2.5×10^{-2} = **0.025** (from Figure 6.15).
5. Search offset is ±0.25 chip. Therefore maximum loss corresponds to a 0.25 chip offset. Correlation loss is also 0.25 or 1.2 dB. Average loss should be much less.
6. $10 \times 1.48 \times 10^{-3} \times 9.8 \times 10^{2}$ = **14.5.** Travel is away from us and Doppler shift is negative. Apparent clock rate is 9,999,985.5 Hz.
7. $10 \times 1.48 \times 10^{-3} \times 5.8 \times 10^{2}$ = **8.58.** He is approaching. Therefore Doppler shift is +8.58 Hz. Apparent clock rate is now 10,000,008.5 Hz.
8. Case 1: $3.75 \times 10^{2} \times 1.48 \times 10^{-3} \times 9.8 \times 10^{2}$ = **543.9.** Shift is **−543.9 Hz.**
   Case 2: $3.75 \times 10^{2} \times 1.48 \times 10^{-3} \times 5.8 \times 10^{2}$ = **321.9.** Shift is **+321.9 Hz.**
9. The period of each hop is $\frac{1}{10}^{5} = 10^{-5}$ sec. Propagation time from transmitter is $6.1 \times 10^{-6} \times 1.12 \times 10^{2}$ = 683.2 μsec. The receiver must search $(683.2 \times 10^{-6})/(1 \times 10^{5})$ = 68.32 chips (assuming that $R_{hop} = R_{clock}$).
10. Twice as far (two times round trip delay) or 136.64 chips, without reset. With reset, 68.32 chips.
11. Code period = $(5 \times 10^{6})^{-1} = 2 \times 10^{-7}$ sec. $1023 \times 2 \times 10^{-7} = 2.046 \times 10^{-4}$. Line length is 0.2046 msec. Taps are 200 nsec apart.
12. **11.5 dB** or greater (see Figure 6.15).
13. Each chirp signal, or pulse, is usually independent of all others and synchronized on the basis over one chirp interval.

CHAPTER SEVEN

1. Noise power in 10 MHz = 3.5×10^{-14} W (from Figure 7.3). Noise power in 50 kHz = 1.8×10^{-16} W: 10 log $(3.6 \times 10^{-14})/(1.8 \times 10^{-16})$ = 10 log 2×10^{2} = 20 dB. Assuming 10-dB output S/N ratios, 30 percent average modulation, 6-dB noise figures, and 2-dB DS system implementation loss, AM sensitivity = 32 dB above 1.8×10^{-16} W, or 2.88×10^{-13} W = 2.05 μV. DS sensitivity = 8 dB below 3.6×10^{-14} W, or 5.76×10^{-15} W = 0.411 μV.
2. 20 dB corresponds to $J/S = 100$; 10 dB corresponds to $J/S = 10$. Total worst case $J/S = 110$. Difference in overall interference due to the second interference source is 10 percent, or approximately 0.4 dB.
3. For −20-dB signal input, a 20-dB jamming signal reduces the output S/N ratio by 3 dB (see Figure 7.2).
4. Pulsed effects are enhanced.
5. (a) 400 MHz center frequency and 40 kHz spacing. (b) 400 MHz CW carrier only.

6. Biphase signal power loss = 3.4 dB. Quadriphase signal power loss = 0 dB (if two 2.5 Mbps data streams are used for modulation). Same as biphase if two 5.0 Mbps data streams are used.
7. If there are 100 link users, then $S/N_{\mathrm{RF}} = -20$ dB. An additional interfering signal 15 dB larger than the desired signal appearing at a receiver would be less effective than the 20-dB friendly interference level already existing there. 20 dB = 100 times the desired signal power; 15 dB = 31 times the desired signal power. Total interference = 131 times the desired signal; 10 log 131 = **21.17 dB.** The signal would be degraded by only 1.7 dB by the 15-dB jammer, beyond the loss already existing due to code division multiplex.
8. (See Figure 7.3) Noise power = 3.6×10^{-14} W. NF = 10 dB. Minimum signal is 20 dB below 3.6×10^{-13} W, or 3.6×10^{-15} W.
9. Noise power increases to $\mathbf{3.6 \times 10^{-13}}$ **W.** M_J increases to **30 dB.** Minimum signal is 30 dB below 3.6×10^{-12} W, or $\mathbf{3.6 \times 10^{-15}}$ **W.** (Note that sensitivity is not improved).
10. Modulation is $\sin (\omega_c \pm \omega_m)t$, where $\omega_m t$ = 0, 90, 180, and 270°.

$\omega_m t =$	0	90	180	270
$64 \times \sin (\omega_c \pm \omega_m)t =$	0	0	0	0

There would be only a ×64 carrier remaining.

CHAPTER EIGHT

1. 10 Mcps period is 100 nsec. Worst case resolution then is ±100 nsec.
2. 20 Mcps corresponds to a 50 nsec period. Range resolution is approximately 1 nsec/ft. 350 nautical miles 2.1×10^6 ft; $(2.1 \times 10^6/50 = 42{,}000$.
   Therefore code length must be at least **42,000 chips.**
3. $25 \times 0.161875 \times 10^6 =$ **4.046875 Mcps.**
4. $\cos \theta = (c_1 - c_2)/(a_1 - a_2) = (192.05 - 192.03)/01 = 0.2$; $\theta =$ **78.5°**.
5. Assuming $\pm\frac{1}{10}$ chip resolution accuracy; 0.5 mile corresponds to approximately 150 chips at 50 Mcps. Directional increments = 150/01 = 1500 quadrant. Directional resolution = 90/1500 = **±0.06.**
6. With simple ranging by code position, resolution would be to within $\frac{1}{10}^5 = 10^{-5}$ sec, or $10^{-5}/(6.1 \times 10^{-6}) =$ **1.64 miles.**
7. Carrier frequency offset does not directly effect range accuracy. Clock rate offset appears as a direct range inaccuracy to the extent that transmitting and receiving code sequences drift apart. Doppler and oscillator inaccuracies have the same effect.

8. Given that the estimate

$$\cos\theta = \frac{\text{antenna separation in chips}}{\text{distance difference in chips}}$$

is employed, errors of 1 and 0.1% produce the following angle measurements:

True $\theta°$	$\cos\theta$	θ est*	θ (0.01)	θ (0.001)
5	0.996	5.126	0	4.442
10	0.984	10.26	5.926	9.94
15	0.966	14.98	12.66	14.76
30	0.866	30.002	28.99	29.9
45	0.707	45.008	44.43	44.95
60	0.5	60.0	59.67	59.96
75	0.259	74.98	74.84	74.97
89	0.017	89.02	88.98	89.02

* θ est is given only to show the amount of error generated by rounding the cosine to three places, as compared with the errors generated by perturbing an angle measurement by a 1 or 0.1% error in the distance measurements used to estimate it.

In practice, both distance measurements used for an angle estimate would be made with the same clocks, and little error would be seen except for that caused by a simple chip counting error.

INDEX